Treatise on Materials Science and Technology

VOLUME 11

Properties and Microstructure

TREATISE ON MATERIALS SCIENCE AND TECHNOLOGY

VOLUME 11

PROPERTIES AND MICROSTRUCTURE

EDITED BY

R. K. MacCrone

Materials Engineering Department
Rensselaer Polytechnic Institute
Troy, New York

 1977

ACADEMIC PRESS   New York   San Francisco   London

A Subsidiary of Harcourt Brace Jovanovich, Publishers

ACADEMIC PRESS, INC.
111 Fifth Avenue, New York, New York 10003

United Kingdom Edition published by
ACADEMIC PRESS, INC. (LONDON) LTD.
24/28 Oval Road, London NW1

LIBRARY OF CONGRESS CATALOG CARD NUMBER:

Library of Congress Cataloging in Publication Data

Main entry under title:

Properties and microstructure.

(Treatise on materials science and technology ;
v. 11)
Includes bibliographies.
1.   Materials—Defects—Addresses, essays, lec-
tures.   2.   Microstructure—Addresses, essays, lec-
tures.   I.   MacCrone, R. K.   II.   Series.
TA403.T74   vol. 11   [TA418.5]   620.1'1'08s
ISBN 0–12–341811–9   [620.1'1299]   77-3882

Contents

Direct Observation of Defects

R. Sinclair

Crystal Defects in Integrated Circuits

C. M. Melliar-Smith

Microstructure of Glass

L. D. Pye

Microstructure Dependence of Mechanical Behavior

Roy W. Rice

Microstructure and Ferrites

G. P. Rodrigue

List of Contributors

Numbers in parentheses indicate the pages on which the authors' contributions begin.

C. M. MELLIAR-SMITH (47), Bell Laboratories, Murray Hill, New Jersey

L. D. PYE (151), New York State College of Ceramics, Alfred University, Alfred, New York

ROY W. RICE (199), Naval Research Laboratory, Washington, D.C.

G. P. RODRIGUE (383), School of Electrical Engineering, Georgia Institute of Technology, Atlanta, Georgia

R. SINCLAIR* (1), Department of Materials Science and Engineering, Stanford University, Stanford, California

* *Formerly at:* Department of Materials Science and Engineering, University of California, Berkeley, California.

Preface

Some time ago an idle remark was made (while sitting on a sailboat, in fact) to the effect that, in contrast to the well-studied and well-documented relation between mechanical properties and microstructures in metals and alloys, very little attention had been paid to other physical properties in this respect. In recent years we have seen the blossoming of conventional concepts of metallurgy into materials science. There can be little doubt that no matter how "materials science" is defined, microstructural effects lie close to the kernel of the subject. This volume is intended to be a developmental step in this direction.

From a pedagogical stance an enormous difficulty becomes apparent: In conventional metallurgy, mechanical behavior is so important that the theory is adequately covered in most curricula. Consequently a basis for discussing microstructure dependence is firmly established. Where other properties are concerned, it cannot be assumed that these are adequately covered and thus are familiar in sufficient detail to those interested in microstructure effects in general.

To overcome this difficulty, each author has prefaced his chapter with a very comprehensive, but concise, development of the physical theory of the property of concern. There are several advantages. First, those interested in microstructural effects will be able to enter several different fields rather painlessly. The parameters important to understanding microstructural effects will be emphasized. Also, a consistent discussion relating the microstructure and its effects can be given in relation to the physical principles involved.

The first chapter in the volume is devoted to microstructure determination in general to emphasize the theme of the volume and to establish that microstructural characterization is in itself a continuously developing field. With each advance of microstructural characterization, an advance in understanding its effects on some physical property is to be expected.

This is followed by chapters on microstructural defects in the important semiconductors silicon and germanium, microstructural effects in glasses, microstructural effects on the mechanical properties of ceramics, and finally, microstructures in ferrites.

The subject matter of this volume should be of interest to materials scientists in general and to students from their senior year on. The physical properties and materials covered in this volume are not only interesting, but useful. There is no sin in practicality.

The individual contributors have worked hard. The volume is theirs, and to them belongs the sole credit and our gratitude.

R. K. MacCrone

Contents of Previous Volumes

Treatise on Materials Science and Technology

VOLUME 11

Properties and Microstructure

Direct Observation and Characterization
of Defects in Materials

R. SINCLAIR

Department of Materials Science and Engineering
Stanford University
Stanford, California

I. Introduction

The purpose of this chapter is to examine various experimental techniques whereby structural information may be obtained *directly* about defects in materials. Thi is an important subject as the properties of perfect materials are modified by the faults that are present. In order to understand their resultant effects the defects must be characterized to as great a degree as possible, the procedures for which are rapidly expanding as more detailed information is required. Obviously in a short chapter it is impossible to cover in any detail these various techniques. The approach taken here is to emphasize two important factors: the *contrast* mechanism by which faults are detected and the *resolution* which is necessary to study them.

1

To indicate the optimum use of the many methods currently available the characteristics of various faults are first discussed (Section II) together with the spatial scale over which they extend. Conventional methods of defect observation are then reviewed in Section III and their appropriate applications outlined. Finally, in Section IV a relatively recent development of transmission electron microscopy is described, the lattice or structure imaging technique. As this method possesses the capability of imaging the structure of most materials at the atomic level it may yield more valuable information than that gathered by existing methods.

II. Lattice Defects

Most materials are crystalline and it is convenient to describe defects in terms of the lattice structure of crystals. They may be categorized according to the number of dimensions over which they extend (zero-dimensional, one-dimensional, etc.). The distortion of the crystal lattice generally occurs over a volume considerably greater than the actual center of the individual fault and thus information should be obtained not only at the core but also in areas far removed from it. The latter is more readily attainable except when ultra-high-resolution techniques are applied. Figure 1 indicates the approximate sizes of the various faults.

There are several types of point or zero-dimensional defects: vacant lattice sites, self-interstitial atoms, substitutional solute atoms, and interstitial solute

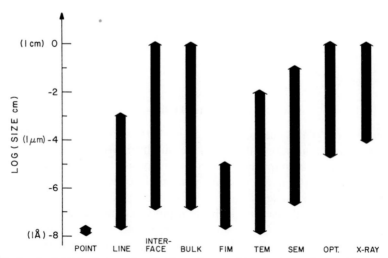

Fig. 1. A schematic diagram showing the scale over which the various lattice defects extend and the useful range of the investigative techniques.

atoms. Self-point defects, being equilibrium defects, must always exist in a concentration dependent on the ambient temperature, unless a nonequilibrium concentration is introduced by quenching from a higher temperature or by knock-on displacement damage in a radiation environment. They are the smallest faults (of the order of atomic dimensions) and consequently are the most difficult to study directly. Their presence, however, can markedly affect atomic diffusion rates and, hence, time-dependent processes in the solid state (e.g., precipitation, coarsening, and creep), properties associated with regular crystalline periodicity can be modified (e.g., electrical and magnetic properties), and their lattice strain may alter mechanical properties and their interaction with other defects.

Dislocations in the lattice are one-dimensional defects. Their behavior under stress provides the basis for the plasticity and toughness of metals, and their presence may cause breakdown in electrical devices. Along the line of the fault they may extend considerable distances (e.g., the width of the crystal in which they exist), but perpendicular their dimensions are at the atomic level (see Fig. 1). Thus information is more readily available about the dislocation lines rather than the distortions very close to the dislocation core.

Interfaces and stacking faults represent two-dimensional lattice defects. These may be free surfaces, grain or crystal boundaries, and interphase interfaces, with dimensions varying from small (~ 20 Å $\times$ 20 Å for small precipitate interfaces), medium ($\sim 100 \times 100$ μm for grain boundaries), to very large ($\sim$ mm^2 for surfaces and cracks). They are relatively straightforward to detect although it is often those features which occur in the third dimension which most affect the properties of the material (and are the most difficult about which to obtain data). The three-dimensional defects are voids, second-phase particles, and inclusions, and these may also exist over a wide range of scale (with small precipitates having a size of ~ 20 Å, inclusions often being ~ 1 μm and voids ranging from ~ 100 Å to as large as ~ 1 mm). The effect of two- and three-dimensional defects on properties is generally better understood as it is easier to correlate the associated modifications with these larger scale faults.

Information may be obtained on two levels; either about individual defects and their effect on the crystal lattice or about the mutual interaction and distribution of the faults. The latter is known as the *microstructure* of the material and it is often possible to identify the effects of various microstructures on the properties of components. To observe the microstructure it is only necessary to detect, image, and identify the appropriate faults which exist; this is an approach generally well-known to materials scientists. Details about their fine-scale structure are then generally not required. However, for a more sophisticated understanding of the effect of a particular fault, a detailed characterization is necessary, often at the highest resolution possible. Thus there are two principal philosophies behind defect characterization of materials, both of which are covered in the present chapter.

The range of use of the various investigative techniques is also shown in Fig. 1. It can be seen that optical microscopy is sufficient for detail down to $\sim 0.5 \, \mu m$, that scanning electron microscopy is appropriate for $\sim 100 \, \text{Å}$ detail, but for extremely high resolution in which the atomic lattice itself may be resolved, transmission electron microscopy or field-ion microscopy are required.

III. Review of Investigative Techniques[†]

A. Optical Microscopy

In materials science the optical method is most often applied by reflection of light from the surface of the specimen. This surface is suitably prepared to reveal the microstructure of the material (e.g., by polishing and subsequent chemical etching). Contrast arises from the varying intensity of reflected light across the surface, which may be due to natural changes in reflectance caused by composition or structural changes, or to local changes of surface angle caused by preferential etching. A schematic example is shown in Fig. 2. Various techniques such as polarized light microscopy, dark-field microscopy, etc., may be used to enhance contrast or to produce additional effects (Phillips, 1967, 1971). The ultimate resolution is limited by the wavelength of visible light and is normally given by the Rayleigh criterion:

$$\text{Resolution} = 0.61 \lambda / \text{N.A.}$$

where λ is the wavelength and N.A. the numerical aperture.

As $\lambda \sim 0.5 \, \mu m$ and the maximum feasible aperture has N.A. ~ 1.4, it is not possible to reduce the resolution below about $0.2 \, \mu m$. The method is not applicable to gaining information about distortions in the atomic lattice, for detecting the fine-scale defects such as vacancies or dislocations [unless they can be revealed by a large-scale effect such as an etch pit (Thomas, 1973)], nor for gaining crystallographic information. However, it is extremely useful for obtaining a low-resolution view of microstructure and should precede more high-resolution specialist techniques to ensure that the appropriate level of examination is to be chosen. It has found a wide range of use in materials science in the past although the more detailed information currently required generally needs application of more sophisticated techniques.

A micrograph illustrating the application and limitation of optical microscopy is shown in Fig. 3, taken of a Cu–Ni–Fe alloy aged at 840°C (Gronsky,

[†] For a more detailed account of the various experimental methods, see for instance Phillips (1971).

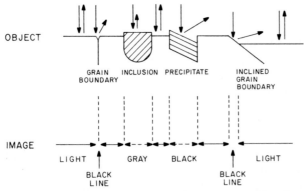

OBJECT

GRAIN   INCLUSION   PRECIPITATE   INCLINED
BOUNDARY   GRAIN
                                                BOUNDARY

IMAGE

LIGHT   GRAY   BLACK   LIGHT

BLACK   BLACK
LINE   LINE

Fig. 2. A schematic representation of the origin of some contrast effects in optical microscopy of polished and etched specimens.

Fig. 3. Optical micrograph of 69.3% Cu, 19.4% Ni, 11.3% Fe aged for 150 hr at 840°C showing the effect on the microstructure of enhanced precipitate coarsening at grain boundaries. (Courtesy of Gronsky, 1974.)

1974). The region where the interparticle spacing is larger than that in the surrounding material has been created by an enhanced precipitate coarsening at the grain boundary. This effect causes intergranular embrittlement of the alloy, which otherwise could have a homogeneous and very desirable microstructure. The scale of this reaction is clearly shown by the micrograph but no information is given about whether a grain with a new orientation has been nucleated or whether grain boundary migration has occurred. This may be obtained by transmission electron microscopy, and careful experimentation has shown the latter to be the case (Gronsky and Thomas, 1975). Thus information may be obtained about grain sizes, shape and the distribution of phases in a material, but not about their relative orientation nor about details on a scale finer than ~ 0.2–0.5 μm.

B. Scanning Electron Microscopy (SEM)

Electrons possess several advantages over visible light in forming an image of an object. They are light, negatively charged particles a beam of which can be readily focused by electromagnetic lenses. Their de Broglie wavelength, dependent on their energy, is considerably lower, being 0.122 Å at 10 keV and 0.037 Å at 100 keV. This imposes no limitation on the resolution of atoms (i.e., 2 Å resolution), the minimum resolution being set by lens aberrations, microscope stability, and the mode of image formation. Electron microscopy is thus preferable to optical microscopy for high-resolution purposes.

In the scanning electron microscope an image of the specimen surface is formed by the use of a fine electron beam (Thornton, 1968; Oatley, 1972; Hearle et al., 1972; Holt et al., 1974). Besides higher resolution, several further advantages accrue from this technique. First, the depth of focus is increased by about three orders of magnitude owing to a narrow angle of illumination. Second, the instrument may be used in a variety of operational modes by which different signals from the specimen can be processed. Third, x-ray or electron signals induced from the specimen by the incident electron beam may be characterized to yield data on the local composition. Fourth, crystallographic information may be obtained about the orientation from electron channeling patterns (Booker, 1970; Joy, 1974), or about accurate local lattice parameters by use of Kossel patterns from emitted x rays (Dingley and Steeds, 1974; Yakowitz, 1974).

The region near the specimen is shown schematically in Fig. 4. The fine electron beam penetrates the sample to a depth of about a micron depending on the applied voltage and specimen material. Three emitted electron signals are produced. These are the high-energy incident electrons which are backscattered with no energy loss, low-energy secondary electrons which can only escape from a region very close to the surface (within about 100 Å), and the

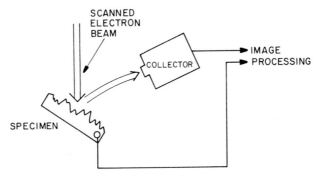

Fig. 4. A schematic representation of the area near the specimen in the scanning electron microscope.

Auger electrons (MacDonald, 1975) which have low energy and are characteristic of atoms in the area irradiated. These signals may be collected and processed separately. The incident beam is scanned across the specimen and the collected signal will vary in intensity in accordance with the particular features of the examined area. The signal is displayed in synchronization on a cathode ray tube (CRT) so that an image is obtained of the total area scanned. The magnification achieved is simply the ratio of CRT size to the scanned distance on the sample.

As an alternative to examining the emitted electron signal, the current passing through the sample to earth may be utilized in image formation. Since this is the difference between the incident and emitted currents, the contrast is readily interpretable. In studying semiconductor devices, a biassing voltage may also be applied to the specimen and the contrast from the specimen current utilized in plotting the current flow in the device. This application is an important diagnostic tool in the semiconductor industry. In addition, emitted electromagnetic radiation [x rays, or visible light in the case of cathodoluminescent materials (Holt, 1974)] can be used with suitable detectors to form the image. SEM is obviously an extremely versatile technique, especially when one also considers instrumental variables such as applied voltage (5–50 kV), electron probe size, scanning speed, specimen tilt, and collector angle. For *in situ* studies, hot, cold, deformation, and environmental stages are available.

Contrast arises from the variation of the examined signal with the topography and chemical composition of the specimen surface. The most common operational mode in materials science is the emissive mode, whereby the secondary emitted electrons are collected. The intensity of this signal is roughly proportional to sec θ (Oatley, 1972), where θ is the angle of beam incidence, and so the brightness of the image varies according to the local surface inclination. As only a small depth is examined, a picture is obtained of the specimen surface. The eye is sensitive to relatively small differences in brightness (and can detect the appropriate contrast), but is less sensitive to larger scale differences. Although the image may represent very large intensity differences from one region to another, the eye reduces this to roughly what would be perceived from an optical micrograph of the surface under diffuse lighting conditions (Oatley, 1972). Thus we can interpret the image intuitively as we would a high-magnification optical picture, with the contrast often being more marked (as in Fig. 5). The considerably increased depth of field allows detailed examination of highly complex materials. This is also exemplified by Fig. 5 in which it may be appreciated that even parts of the relatively flat epoxy matrix would be out of focus in an optical picture whereas the carbon fibers, standing out from the surface, are quite dramatically shown in the scanning micrograph.

The intensity of the backscattered electron beam varies with both inclination and the average atomic number of the illuminated region (the secondary

Fig. 5. Scanning electron micrograph (secondary emission mode) of a fractured carbon fiber–epoxy composite. The protrusion of 8-μm diameter fibers from the epoxy matrix is vividly revealed owing to the large depth of field of the SEM. Small holes which weaken the fibers can be seen in several cases. [Courtesy of D. R. Clarke (1975) private communication.]

current is relatively insensitive to atomic number) so that by collecting these higher energy electrons information is obtained about both topography and chemical composition. As the electrons come from a larger volume of material the clarity of the picture is reduced compared to the emissive mode. A polished specimen surface may be used for qualitative information in the image about composition alone. Conversely, the specimen current image (the difference between incident and total emitted currents) also yields information about both aspects of the sample.

It is evident that the resolution limit is set roughly by the probe size since the signal is processed from a region instantaneously irradiated by the electron beam. In commercial instruments possessing a tungsten, thermionic emission source this is about 100 Å, reducing to about 50 Å in the best circumstances (Broers, 1969). Although further reduction in probe size is possible electron optically, the commensurate decrease in image intensity makes recording times prohibitively long. A source with higher brightness can overcome this difficulty. Practical resolutions of ~25 Å are expected for a lanthanum hexaboride emitter and a resolution of about 3 Å may be achieved with a field-emission source (Muir, 1974; Crewe, 1974; Veneklasen, 1975). However, despite these possibilities it is doubtful whether this resolution capability is desirable for scanning electron microscopes, especially when one considers the increased expense of the more refined systems. The primary purpose of the instrument is to view the surface of the specimen, from which information may be obtained about the effect of the larger scale faults. Images may be formed in appropriate circumstances of the underlying finer scale defects such as dislocations and stacking faults (Clarke, 1971; Stern et al., 1972; Joy et al., 1974), but are inferior to those obtained by transmission electron microscopy. It would seem more worthwhile to utilize the technique in its optimum range (100 Å–100 μm) rather than attempt to compete with transmission electron systems where the latter are superior.

In materials science SEM finds widespread application, for instance in the study of fracture surfaces (e.g., Fig. 6), eutectic and composite materials, electrodeposition and oxidation processes, sintering, and complex three-dimensional structures such as textile materials. The effect of magnetic fields on the emitted electrons allows contrast from magnetic domains and the effect of electrical fields permits examination of semiconducting devices in operation. [For examples, see references in this section and also Lifshin et al., (1969), the proceedings of the IIT conferences (Johari, 1968–1975), and for a bibliography Johnson, (1975).] It is an invaluable investigative tool, providing superior information to the optical microscope and complementary information to the transmission electron microscope.[†]

[†] A comparison of the three techniques has been given by Hearle (1972).

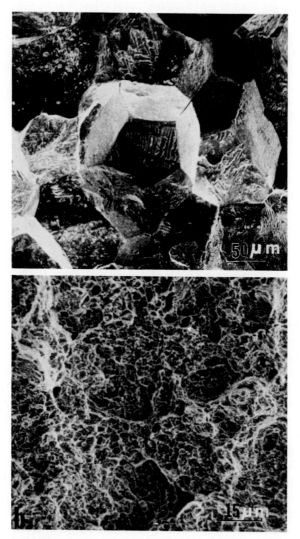

Fig. 6. SEM fractographs (secondary emission mode) of Fe–4% Cr–0.4% C martensitic steels. (a) Conventionally heat treated material with large prior austenitic grain size (∼300 μm) suffers quench cracking and fails by an intergranular mechanism, as clearly shown by the fractograph. (b) The fracture toughness may be increased by nonconventional heat treatments which give rise to a material which undergoes ductile transgranular failure with a dimple-like fracture surface [Courtesy of B. V. Narasimha Rao, (1975), Narasimha Rao and Thomas (1975).]

C.   *Transmission Electron Microscopy (TEM)*

The principle of image formation in the transmission electron microscope involves magnifying the image produced in an electron beam which has passed through a thin foil specimen of the material (Hirsch *et al.*, 1965; Amelinckx *et al.*, 1970; Thomas *et al.*, 1972; Glauert, 1972, 1974; Siegel and Beaman, 1975). The resolution depends on the stability of the lens and source voltage supply, mechanical stability of the microscope column, and chromatic and spherical aberration (particularly of the objective lens which forms the first image of the sample). Microscopes currently available possess a guaranteed line resolution of $\lesssim 2$ Å with which it is possible to image atomic planes in alloys and individual atoms in ceramic materials. Micrographs can be taken in a magnification range of 10^2–10^6 times, overlapping with the effective scales of optical and scanning microscopy but also achieving atomic level resolution. Thus on the same specimen it is possible to characterize microstructure and to obtain detailed information about individual defects. Furthermore, the electron diffraction pattern of areas 1–100 μm in diameter can be obtained in a straightforward manner to provide standard crystallographic data and to aid in the interpretation of the electron image. The technique is probably the most important one for the direct characterization of defects in materials.

There are three main modes of image formation (Fig. 7):

i. Bright-field imaging, whereby an aperture is positioned in the back focal plane of the objective lens to allow the electron beam transmitted through the sample to form the image.

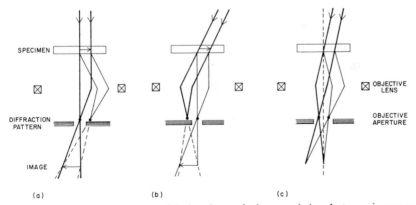

Fig. 7.   Schematic representation of the imaging modes in transmission electron microscopy: (a) bright-field imaging, (b) dark-field imaging with tilted illumination, and (c) lattice imaging with tilted illumination.

ii. Dark-field imaging, in which a beam diffracted by the crystal may be employed. This beam may be chosen by suitable positioning of the objective aperture but for highest resolution the illumination is tilted with respect to the specimen so that the diffracted beam is parallel to the optical axis of the microscope (Fig. 7b).

iii. Lattice or structure imaging, which may be produced by allowing transmitted and diffracted beams to form the image using a suitably large objective aperture.

The diffraction pattern may be observed by focusing the microscope's intermediate lens on the back focal plane of the objective lens rather than on the image plane.

Contrast arises from several sources. "Mass thickness" or "structure factor" contrast occurs owing to differing electron absorption in adjacent parts of the specimen. This is particularly useful in two-phase materials where chemical composition variations may cause corresponding variations of electron scattering power or may give rise to foil thickness variations by differential electropolishing behavior during sample preparation.

"Diffraction" contrast arises because the intensity of the transmitted and diffracted electron beams is locally altered near a defect. For instance if the specimen is oriented for high electron transmission (i.e., light area in bright field, dark in dark field) the lattice distortion caused by a dislocation may bring the nearby atomic planes into an orientation where strong diffraction can take place, with the Bragg condition locally satisfied. This region will appear dark against the bright background in the bright-field image and vice versa in the dark-field image. A dislocation line thus appears as a dark line in the TEM bright-field image. A similar contrast mechanism operates for other lattice defects which give rise to lattice distortions (e.g., interstitial and vacancy loops, coherent precipitates). A grain boundary is readily recognized by changes of contrast arising from the different orientations of the adjacent crystals. Stacking faults produce contrast as the crystal periodicity is disturbed relative to surrounding perfect crystal and they appear as a series of fringes or dark lines parallel to the intersection of the fault with the specimen surface. Second-phase particles are revealed by the mass thickness or structure factor contrast outlined above, but preferably they are studied in dark field by placing the aperture around a reflection originating from the second phase itself. Only the particles then appear bright in the dark-field image. It is clear that one-, two-, and three-dimensional lattice defects readily produce characteristic contrast effects in TEM which can be studied at extremely high resolution (down to ~ 5 Å). This is the principal reason why the technique is so powerful for studying the microstructure of materials.

Interpretation of the precise contrast from lattice defects in TEM is a complex subject requiring detailed understanding of the dynamical theory of electron diffraction (see Hirsch *et al.*, 1965; Brown, 1971). However, by systematic experimentation it is possible to identify uniquely the character of the fault. The habit of the fault (i.e., the line of a dislocation, the plane of a stacking fault or grain boundary, or the shape of a precipitate) may be found by trace analysis, with the specimen orientation established from the diffraction pattern of the area of interest. A three-dimensional view of the microstructure may be obtained by a stereo pair of micrographs of the same area. The nature of the fault (e.g., Burger's vector of a dislocation, displacement vector of a stacking fault, misfit of a coherent particle) is found by a series of contrast experiments performed on the same area of the foil. The specimen is generally oriented close to "two-beam diffracting conditions" (i.e., only one diffracted beam **g** has high intensity and is close to the Bragg condition) and the behavior of the defect image contrast in this and in a series of such diffracting conditions is carefully recorded. By suitable choice of **g** the character of the fault may be found. For example the Burger's vector (**b**) of a dislocation may be determined by finding conditions under which the dislocation image disappears (or possesses weak contrast). This occurs when $\mathbf{g} \cdot \mathbf{b} = 0$, i.e., the Burgers vector is perpendicular to the operating reflection. Figure 8 shows the variation of contrast of interfacial dislocations in partially coherent precipitates in a Cu–Mn–Al alloy in different diffracting conditions. Figure 9 shows how such experiments can be used to identify a fault in a more complex material, in this case lithium iron spinel.

In addition to defect characterization, more detailed information about faults may be derived with the aid of computer simulation. Application of dynamical electron diffraction theory for interpretation of contrast is well established (Head *et al.*, 1973). It is then possible to adjust parameters describing the defect (e.g., lattice distortions near a dislocation) until exact fit is found between the simulated and experimental image. A detailed model may then be found for the configuration of the lattice defect itself. Further refinement is possible due to recent advances in image formation. The image width of a fault is often much greater than the true width (e.g., dislocation images are ∼100 Å wide, whereas we envisage a dislocation core as being only ∼3 atomic diameters wide). It can be considerably reduced, not by imaging conventionally in strong two-beam conditions, but by forming either a dark-field image using a weakly diffracted beam [the weak beam method (Cockayne *et al.*, 1969, Cockayne, 1974)] or by taking a bright-field micrograph with a high-order reflection close to the Bragg condition (Goringe *et al.*, 1972). Dislocation images can thus be reduced to about 15 Å and a much a clearer picture of the defect is revealed (e.g., Figs. 9f and 10). Accordingly, a more satisfactory model of the fault can be used for the simulated

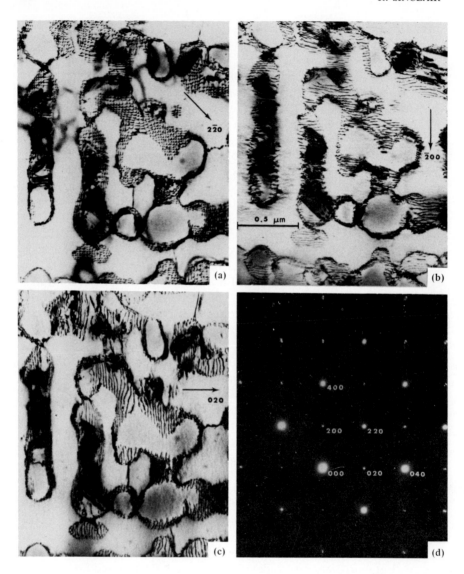

Fig. 8. Bright-field electron micrographs (a–c) and corresponding selected area electron diffraction pattern (d) of $Cu_{2.5}Mn_{0.5}Al$ aged for 10^4 min at 300°C. The network of lines represents dislocations at the precipitate–matrix interface in the alloy. With the operating reflection $\mathbf{g} = (\bar{2}00)$, (020) (b, c) one set of dislocations is revealed, whereas both sets appear in the image when $\mathbf{g} = (2\bar{2}0)$. These and similar contrast experiments show that the Burger's vectors of the dislocations are of the type $a/2 \langle 100 \rangle$ (Courtesy of Bouchard et al., 1972.)

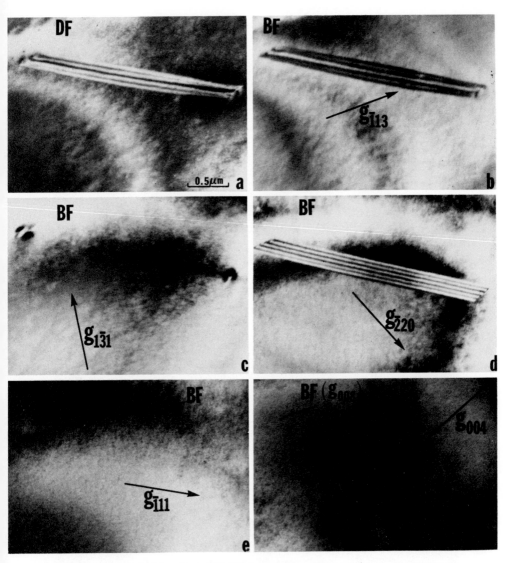

Fig. 9. A series of high-voltage (650 kV) transmission electron microscopy contrast experiments is shown to establish the nature of a stacking fault in lithium iron spinel. The fault (described by a displacement vector **R**) gives rise to a phase change of $\pm\pi$ in the electron waves as evidenced by the symmetry of the stacking fault fringes. It goes out of contrast (c, e, f) when $\mathbf{g} \cdot \mathbf{R} = $ zero or integer, whereas the bounding partial dislocations disappear when $\mathbf{g} \cdot \mathbf{b} = 0$ (e). The fault vector is thus found to be $\mathbf{R} = \pm\frac{1}{4}[0\bar{1}1]$, the Burger's vectors of the partial dislocations have $\mathbf{b} = \frac{1}{4}[0\bar{1}1]$ and the fault plane is $(0\bar{1}1)$. The images, however, cannot distinguish between extrinsic and intrinsic π-faults in this material (i.e., whether material has locally been "inserted" or "removed" from the structure) and direct lattice imaging would be necessary to establish this aspect of the fault. In addition, by comparing (c) and (f) it can be seen that the dislocation image widths are decreased by imaging in high order bright-field conditions (f). [Courtesy of Mishra (1976); for method, see van der Biest and Thomas (1974).]

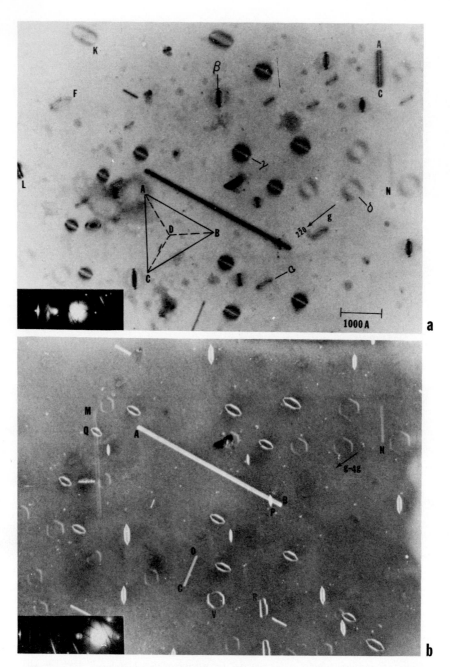

Fig. 10. A comparison of the conventional bright-field image (a) and the weak-beam dark-field image (b) of ion-implantation damage in silicon. The morphology of the radiation damage is much clearer in the weak beam image. (Courtesy of Seshan, 1975; Seshan and Washburn, 1972.)

image (e.g., Cockayne and Vitek, 1974). It may be noted that this approach for finding suitable defect models is subject to the limitations imposed by approximations in the appropriate theories. A more satisfactory method is direct imaging of the atomic lattice, as described in Section IV.

Most materials can now be thinned sufficiently for transmission of 100 kV electrons [either electrolytically (metals), chemically, by cleavage, or by ion bombardment (ceramics)]. A range of specimen stages is available for *in situ* studies, as for SEM, and the increasing availability of high-voltage instruments (up to 1.5 MeV, commercially) allows more sophisticated experimentation (Swann *et al.*, 1974). Not only can thicker samples be examined, which is useful for materials difficult to thin and for experiments on foils which are more representative of bulk materials (e.g., Butler and Swann, 1975), but an environmental stage can also be incorporated due to the greater space between the objective lens pole pieces. Thus it is now possible to study solid–gas reactions in the microscope in a controlled atmosphere at high temperature (e.g., Flower and Swann, 1974) and phase transformations in materials which would degrade in the normal vacuum of the microscope (e.g., van der Biest *et al.*, 1975). Furthermore, the high-energy electrons are capable of displacing atoms from their lattice sites so that *in situ* radiation damage experiments are also possible (Makin, 1971; Laidler and Mastel, 1975).

In summary, the current applications of TEM include complete characterization of one-, two-, and three-dimensional lattice defects, study of phase transformation processes either in selected samples or *in situ* in appropriate high-temperature stages, phase analysis of complex materials from electron diffraction patterns, crystallographic analysis of defects and transformation processes, environmental reactions, and radiation damage. Much of the desired information concerning the relationship of structure and properties of materials may be gained from TEM.

D. Scanning Transmission Electron Microscopy (STEM)

The scanning transmission electron microscope combines various advantages of the transmission and scanning microscopes (Crewe *et al.*, 1968; Crewe, 1974; Cowley, 1975; Thompson, 1975; Crewe *et al.*, 1975). A thin sample is used (as for TEM), a fine beam is scanned across the specimen (as for SEM), and the transmitted electrons are collected and processed for display on a CRT in synchronization with the scanned incident beam (hence STEM). Under normal circumstances the contrast in the image is identical to that in TEM, if a small enough collector angle is used, according to the principle of reciprocity (Cowley, 1969). Thus, it is possible to image defects at high resolution directly in the STEM. Standard electron diffraction patterns may also be obtained by rocking the beam on the specimen and, because of

the small probe size, extremely small areas can be selected for analysis
[$\sim$30-Å diameter compared to $\sim$1-μm diameter for TEM (Geiss, 1975a)].

Several commericial TEMs have attachments to allow conversion to
STEM operation but their resolution is limited to about 30 Å by the probe
size and thermionic source brightness. By use of a field-emission electron
source the probe size can be reduced to $\sim$3 Å with high intensity. Consequently, the resolution achieved can be similar to that of present TEM and
with higher accelerating voltages it is predicted that resolutions lower than
1 Å will be attained (Crewe, 1974). One of the first commercial STEM with
field-emission source has recently been announced (Banbury *et al.*, 1975)
with 3.1 Å lattice fringes of silicon readily resolved (Fig. 11).

Apart from the microdiffraction capability of STEM there are several other
advantages over TEM. First, an image can be obtained on the CRT at
magnifications up to 20×10^6 times, and relatively rapid exposures taken
at this level, compared to typical maximum TEM magnifications of $0.5-1 \times 10^6$ times on the final viewing screen. Second, chromatic aberration does not
limit resolution from thick specimen areas as it does in TEM at similar
applied voltages (80–120 kV) because there are no lenses after the specimen.
The inelastically scattered electrons can be separated from the unscattered
beams prior to image processing to improve the image contrast. This yields
a considerable increase in the specimen thickness that can be examined

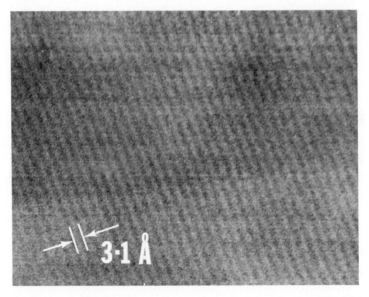

Fig. 11. Lattice image of the (111) planes in silicon as resolved by a commercial scanning
transmission electron microscope (Courtesy of I. L. F. Ray, 1975, private communication; see
also Banbury *et al.*, 1975.)

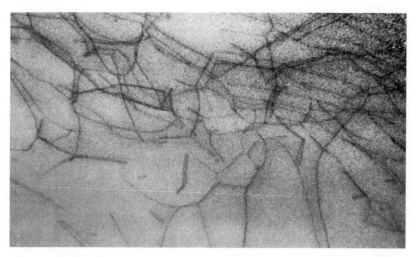

Fig. 12. Scanning transmission electron micrograph of dislocations imaged in a thick area of a silicon specimen. (Courtesy of I. L. F. Ray, 1975, private communication.)

at high resolution which is thought comparative to the gain achieved by high-voltage TEM (e.g., Fig. 12). Third, the energy loss spectrum from the specimen is readily available in the STEM and may be obtained with a commercial instrument, whereas energy loss analysis in TEM is confined to a few prototype instruments. Fourth, energy dispersion analysis of characteristic x rays produced by the electron irradiation (for composition determination) is also readily available on the STEM, as it is for SEM. Fifth, the gun chamber of a field-emission STEM must operate at an extremely good vacuum (10^{-11} Torr) so that the specimen may also be maintained in good vacuum with a corresponding very low contamination rate. Contamination does present a problem, though, with small, stationary probes.

The advantages of a STEM instrument over a conventional TEM appear to be significant. However, the cost is correspondingly greater and so it seems that defect characterization will continue to be performed on TEM, with STEM remaining a specialist instrument for the next few years. Ultimately, it is predicted that work currently carried out on TEM will be taken over by STEM.

E. X-Ray Microscopy[†] (*Topography*)

As the wavelength of x rays is considerably lower ($\lambda \sim 1$ Å) than that for optical light it might be expected that x-ray microscopy (often known as x-ray topography) would also improve the resolution possible for defect studies.

[†] The information that may be *indirectly* derived from analysis of diffracted x-ray beams is not considered in this article.

However, it is extremely difficult practically to focus x rays in the same way as light or electrons for the following reasons. First, their refractive index is close to unity for all materials and they are rapidly absorbed. Hence, the design of optical-type lenses with low aberrations is not possible. Second, being electromagnetic radiation their beam path is unaffected by electrostatic or magnetic lenses. Furthermore, the detection and informational display of x rays is less efficient than for light or electrons so that instantaneous information about sample areas would not be possible with current technology. X-ray microscopy is, therefore, confined either to direct one-to-one images which may be enlarged photographically from high-resolution photographic film, or to magnifications of up to 100 times with various geometrical arrangements. The best resolution possible is about 1 μm which is on the same scale as optical microscopy.

The reason that x-ray micrscopy is used is that direct micrographs can be taken, at low magnification, of lattice defects in large areas of bulk materials. Optical microscopy relies on indirect methods such as etch pitting for information about finer scale defects such as dislocations. However, the x-ray image is comparable to the electron image, with similar contrast for dislocations, stacking faults, vacancy loops, low angle boundaries, etc. (Young

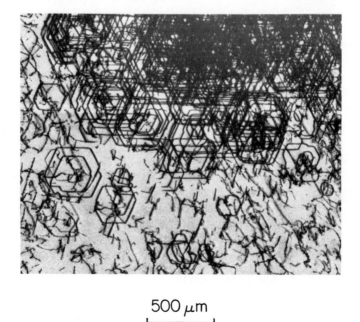

500 μm

Fig. 13. X-ray micrograph of dislocations in a silicon single crystal. (Courtesy of E. Meieran, 1975, private communication.)

et al., 1967; Authier, 1970; Lang, 1970; Wu and Armstrong, 1975). Dislocations, for instance, appear as a black line in incident beam images but lack of resolution limits the line width to a minimum of about 2 μm (dislocation image widths as narrow as 15 Å may be produced by TEM, see Section III,C). It is important to realize that x-ray microscopy does not, and can not, compete with transmission electron microscopy for high resolution detail. Its advantage comes from the ability to produce a picture of the defect distribution in bulk sized specimens, confined to those cases where low defect densities are expected, for example, silicon single crystals for semiconductor devices. (An estimate shows that for a dislocation image width of 10 μm, individual dislocations will not be resolved if their population density is greater than 10^6 cm^{-2}). Micrographs may be taken in the back-reflection (Berg–Barrett) mode or the transmission (Lang) mode for thin samples (see Lang, 1970; Armstrong and Wu, 1973) and the picture may be recorded photographically or by diffractometer methods. An example showing dislocations in silicon is illustrated in Fig. 13. Thus for specialist applications, most notably examination of crystals for electrical devices, the x-ray microscopy method may be preferred for general examination over optical or electron microscopy, although for detail on a high-resolution level the latter would be utilized.

F. *Field-Ion Microscopy (FIM)*

A technique which can resolve individual atoms deserves consideration as a tool for studying lattice defects. The field-ion microscope (Muller and Tsong, 1971) forms an image of a highly curved, approximately hemispherical surface of metals and where this surface is parallel to high index planes in the lattice (e.g., {420}, {531}) the individual more widely spaced atoms ($\gtrsim$ 3-Å separation) can be discerned. Image formation occurs by the ionization of a suitable gas (H, He, Ne, Ar) at positions on the positively charged surface where atoms protrude, generally the outermost atoms on a plane layer (e.g., Fig. 14). The newly formed positive gas ions are then attracted away from the specimen to a fluorescent screen at ground potential or more commonly to an image intensification system. The resultant image is approximately a stereographic projection of the surface magnified by the ratio of tip–screen distance to tip radius, typically 10^6–10^7 times. A series of sections can be studied by removal of a controlled amount of material, atom by atom or plane by plane, on raising the tip voltage to the level where the specimen atoms themselves are attracted away (field evaporation) thus giving a three-dimensional view of the sample. In the high-index pole regions a column of material ($\sim$ 20–40 Å in diameter) may be examined atom by atom.

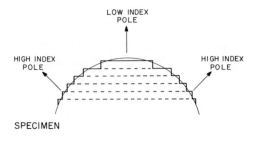

SPECIMEN

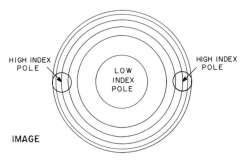

IMAGE

Fig. 14.   Schematic diagram showing the origin of the characteristic ring pattern in field-ion micrographs of metals. The curved section is an isopotential contour and atoms close to this section are imaged.

Contrast can occur on two levels: at the atomic level either by preferential gas ionization over one atomic species in an alloy [i.e., a brighter image spot (Fig. 15)] or by preferential removal of atom species; at a lower level by higher or lower ionization [hence image brightness (Fig. 16)] at the defect or by modification of the crystallographic nature of the image.

The technique is a highly effective one for studying atomic distributions but still remains a specialist research tool which does not appear to be readily available. There is no profitable industry behind FIM and hence instrumental development is slow and confined to prototype instruments and the number of competent users is also correspondingly low. Only a very small statistical sample can be analyzed and it is often not possible to gain an adequate perspective of microstructure. The correlation of structure and properties is more readily obtained by TEM, SEM, and optical microscopy, and these latter techniques find more widescale use as they can be employed to solve immediate problems which may arise in the production of components as well as providing a basis for understanding the fundamentals of materials science.

Several further difficulties arise in applying FIM. Although the specimen may be studied atom by atom, the image of the atom positions is distorted

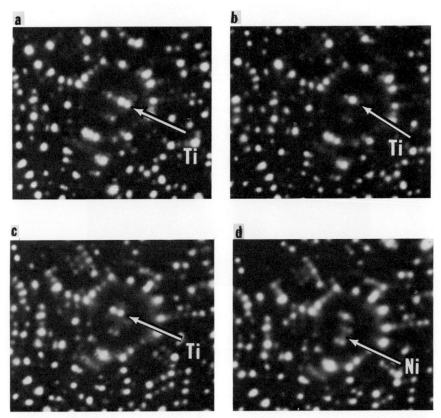

Fig. 15. A field-evaporation sequence showing the stripping, a few atoms at a time, of a (420) plane in a partially ordered γ' precipitate in Ni–14% Ti, aged for 4 hr at 600°C. Alternate rows of atoms are Ni and Ti, the latter appearing brighter and being more susceptible to field evaporation. A nickel atom misplaced in a titanium atom site is indicated in (d) (see Sinclair *et al.*, 1973).

from the true lattice positions by the extremely high electrical field. Thus the distortion of the crystal lattice near defects may not be studied. Second, most of the area of the image is taken up by the characteristic low-index pole rings. Information can easily be lost between these rings (or plane ledges), which are typically ~ 20 Å apart, and thus atomic detail is not achieved in these areas. Third, the specimen must be able to achieve and sustain a high electrical field at its surface and therefore must be a good electrical conductor. This confines most studies to metals, although not all (e.g., Al, Mg, Zn, Ti) can withstand the high electrostatic forces on the tip during operation. Fourth, crystallographic information such as orientation relationships may be found after detailed analysis but are often more readily available from techniques involving diffraction.

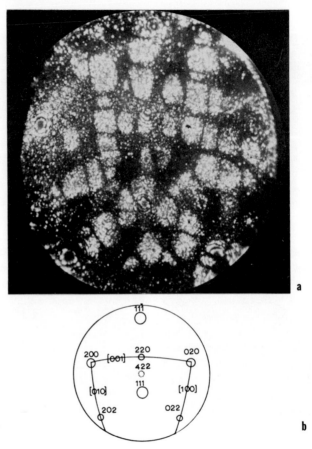

Fig. 16. Field-ion micrograph (a) and corresponding stereographic map (b) of Ni–14% Ti aged to peak hardness (500 hr at 600°C). The bright regions are γ' precipitates which are revealed in a characteristic distribution, aligned periodically in $\langle 100 \rangle$ directions (see Sinclair *et al.*, 1974).

Nevertheless, if a problem in materials science requires a solution at the atomic level, a well-chosen experiment by FIM is often superior to conventional methods. It is the only tool which can currently image, as a matter of routine, individual point defects in metals. This capability has proved useful in studying the distribution of vacancies and interstitials after radiation damage (O'Connor and Ralph, 1972; Robinson *et al.*, 1973) and in determining atomic arrangements in dilute alloys (Gold and Machlin, 1968), in order–disorder systems (Berg *et al.*, 1973), and in ordered precipitates (Sinclair *et al.*, 1974) (e.g., Fig. 15). The images of nondilute alloys (e.g., complex steels) are difficult to interpret due to the disruption of the characteristic ring pattern by the behavior of the various atomic species both to field evaporation and to ionization.

Some useful high-resolution information has been obtained about dislocations by FIM (e.g., Smith et al., 1969; Taunt and Ralph, 1974a; Seidman and Burke, 1974), but so few dislocations appear in an image (one every few specimens) that reliable statistics are rare and it is more convenient to scan thin areas in TEM and image in weak-beam or high-order bright-field conditions for high-resolution detail (Section III,C). Grain boundary topography may be obtained at a resolution of about 5Å and the technique appears quite promising in this field. Finding a suitable grain boundary is tedious and may be aided by high-voltage TEM (Smith et al., 1974), but very little of the grain boundary area can be mapped in detail (e.g., ∼ 1000 × 500 Å). The statistical sample may always be questioned when the analysis is extrapolated to explain bulk behavior (Murr, 1975), but a quite detailed picture of boundaries can be built up. In analysis of three-dimensional defects, the method is particularly useful where there is a high defect number density (e.g., Fig. 16). FIM can often yield information about homogeneous reactions [e.g., homogeneous nucleation (Driver and Papazian, 1973), spinodal decomposition (Sinclair et al., 1974) and ordering (Taunt and Ralph, 1974b)] which may be lacking in conventional methods due to lack of resolution or overlapping of fine detail (FIM is a surface technique). Chemical data can also be derived about small regions by use of an atom probe FIM (Muller et al., 1968) in which individual atoms can be identified.

Thus the technique is extremely useful in obtaining atomic detail about radiation damage, grain boundaries, and phase transformation processes. Its role would best appear to be complementary to the other techniques to provide specialist detail which is not available from the alternative methods.

G. Methods of Local Chemical Composition Determination

The localized segregation of the constituents of a material will affect its properties to an extent depending on the scale of the effect and the property under consideration. For instance, segregation or clustering at the atomic level will affect electrical properties to a larger degree than it would fracture toughness. Conversely, grain boundary segregation and precipitation commonly affects toughness but need not be considered for electrical devices made from large single crystals. Thus, chemical investigation on a wide range of scale is desired by the materials scientist. Amongst the various investigative techniques for direct defect observation described above, modifications of the basic approach also allows local chemical composition to be established in known areas and these methods will be described briefly in this section.

1. Optical Methods

The contrast obtained in polished and etched specimens can often indicate composition differences on a scale within the limit of the optical microscope. It is not possible, however, to establish quantitatively the degree of segregation nor to identify particular phases which may be present without recourse to other techniques. Thus, optical microscopy may be used as an indicator but not when detailed information is required.

2. SEM Methods

The large amount of space available near the specimen in the scanning electron mode allows the positioning of detectors suitable for identifying characteristic emitted radiation. The incident electron beam excites electrons in the atoms of the specimen to higher energy states. On subsequent decay to lower energies either characteristic electromagnetic radiation (x rays) or characteristic (Auger) electrons are emitted. The signals can be processed to determine either quantitatively the amount of various species in the area examined or to yield a picture of the distribution of a particular species.

The characteristic x rays may be analyzed for wavelength using a crystal spectrometer arrangement or directly for energy using an energy dispersion analyzer (Belk, 1974; Philibert and Tixier, 1975). The former has superior characterizing resolution although with recent technological advances in detector systems (e.g., lithium-drifted silicon diode detectors) this advantage is less marked and the two are comparative at high x-ray energies (~ 1-Å wavelength) (Gedcke, 1974; Kandiah, 1975). The latter is superior in efficiency of detection and may be preferable for scanning microscopy applications. Elements next to each other in the periodic table can generally be distinguished, but the analysis of lighter elements is still difficult. For quantitative analysis it is preferable to obtain, from suitable standard specimens, the variation of characteristic x-ray intensity with material content near the range of interest, and, for optimum quantitative data, purpose-built electron probe microanalysis equipment should be used.

The practical spatial resolution is set by the size of the region from which x rays are emitted, which depends on the probe diameter, incident electron energy, and the constitution of the region from which excitation occurs. A sufficiently high-intensity x-ray signal must be produced so that in commercial instruments probe diameters are not usually smaller than 1 μm. The x rays also originate from a depth of this order so that volumes of $\sim 10^{-12}$ cm^3 may typically be analyzed. The analysis of small particles should be treated cautiously as penetration of the beam through the particle to the underlying matrix may cause errors in the composition determination.

Auger electrons are most efficiently excited by low-energy incident electrons (e.g., 1–5 kV) compared with the higher energies favorable for x-ray excitation. The penetration of the electron probe is low and energy loss of the Auger electrons on exit from the sample restricts analysis to surface layers of the specimen ($\sim$10-Å depth). The spatial resolution is similar to the other electron probe techniques (MacDonald, 1975) and thus the volume analyzed is smaller than for x-ray detection. This is advantageous for studying the composition at surfaces, e.g., fracture surfaces may be analyzed for the presence of deleterious constituents which may be in too low a concentration for detection by other means. A three-dimensional analysis can be performed by stripping the original surface by ion sputtering, although this will eventually lead to a rough surface. The efficiency of Auger electron production does not fall off with atomic number as rapidly as x-ray production so that this method is superior for analyzing light elements. One disadvantage is the necessity (and expense) of extremely good vacuum at the specimen in order to ensure minimal surface contamination so that the true surface is being examined.

3. TRANSMISSION ELECTRON METHODS

For higher resolution analysis it is necessary to utilize transmission methods. However, the incorporation of detectors for emitted radiation in TEM is considerably more difficult due to lack of space near the specimen and very few TEM instruments take advantage of the characteristic radiations. The commercially available electron microscope with microanalyzer (EMMA–4) possesses an optimum resolution of about 0.1 μm. Alternatively, it is possible to study the transmitted electron beam to establish the spectrum of energies due to interaction of electrons with the elements of the crystal through which the beam has passed. The energy loss is dependent upon the species with which interaction has occurred and thus can be used for chemical analysis. The spatial resolution can be reduced to $\sim$100 Å by suitable choice of slits to select the transmitted electrons. An example of a comparison of electron velocity analysis with an EMMA analysis of the same specimen area is shown in Fig. 17. A disadvantage of electron energy or velocity analysis is that only low atomic number elements have narrow energy loss peaks and are suitable for such analysis (e.g., Al, Mg, Be). Thus superior resolution is gained at the expense of being able to study only a few "model" systems. Also, an energy loss analyzer for TEM is not widely available and the technique is confined to a few prototype instruments.

Conversely, STEM instruments can readily be designed to include provision for both x-ray and energy loss analysis. Furthermore, the inelastically scattered electrons may be separated from the unscattered beam and utilized in image formation. As beam spreading is also considerably smaller than for conventional SEM it seems that improved resolution for analysis will be possible with STEM. The versatility of STEM indicates that it could be an exciting development in the microanalysis field.

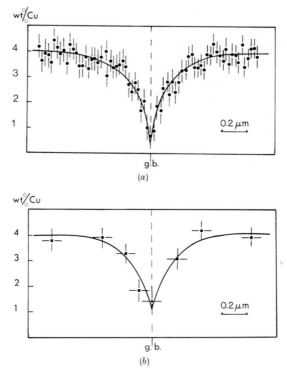

Fig. 17. A comparison of the determination of the Cu distribution near a grain boundary in an aged Al–Cu alloy by electron velocity analysis (a) and electron microscopy with microanalysis, EMMA (b) (Courtesy of Doig *et al.*, 1975.)

4. DIRECT ATOM ANALYSIS

Atoms removed from the surface of a specimen may be subsequently analyzed by a mass spectrometer for identification. This is the principle of secondary emission analysis, whereby ions are removed by bombardment of the sample with a beam of positive ions (Castaing, 1975). The resultant ion signal may be directly characterized, or focused, and a particular element chosen to form an image showing the distribution of this species. The resolution is ~ 0.5 μm, slightly superior to conventional electron probe methods, and in principle it is possible to analyze for any element or isotope.

The atom-probe field-ion microscope is capable of analyzing single atoms chosen from a FIM image by timing the flight of the ion between specimen and detector. The composition of extremely tiny regions can be obtained [e.g., Goodman *et al.*, (1973) found the composition of Cu-rich G.P. zones during the initial stages of precipitation in an Fe–Cu alloy]. As the statistical sample is normally small ($\sim 10^3$–10^4 atoms analyzed) the experiment must be carefully chosen if the results are to be projected as representative of the bulk of the material, but for studies of very fine-scale segregation phenomena in metals this is obviously an extremely powerful technique.

IV. Lattice and Structure Imaging

Current transmission electron microscopes possess the capability of 2-Å line resolution and 3.5-Å point-to-point resolution. Thus by suitable choice of imaging conditions it is possible to image directly the crystalline lattice itself. An increasing amount of research has been devoted over the past five years to establishing the conditions under which lattice images can be satisfactorily and readily interpreted in terms of the real lattice of the specimen. The advances which have been made and some applications of the technique will be illustrated in this section.

Two types of image are taken at present: one-dimensional *lattice* fringe *images* and two-dimensional lattice *structure images*. In the one-dimensional image, electron reflections perpendicular to the desired imaging planes are strongly excited by suitably tilting the specimen and information is derived from only one set of atomic planes in the crystal. This technique provides useful information about metals and other close-packed structures where it is difficult to obtain representative two-dimensional images. The structure image may be taken of more complex materials with larger unit cell dimensions. The interpretable resolution limit is only ∼3.5-Å point-to-point at present [set by the combination of incident beam divergence and spherical aberration of the objective lens (O'Keefe and Sanders, 1975)] and thus this approach is only appropriate for structures with larger periodicities.

The experimental variables which affect the lattice image are now well established (Allpress and Sanders, 1973; Lynch *et al.*, 1975; Sinclair *et al.*, 1975). They are as follows.

i. *Number of electron beams allowed through the objective aperture to form the image.* For satisfactory structure images a minimum number of about 20 reflections is thought necessary for periodic images (Lynch *et al.*, 1975). As the size of the objective aperture and so the number of chosen beams increases, the spherical aberration of the objective lens causes too great a phase shift with respect to the transmitted beam and eventually loss of resolution is encountered. An objective aperture of 40–50-μm diameter is typically employed. Many beams from all directions in the crystal are used in image formation, with the diffraction pattern normally set close to a low-index zone orientation. At appropriate microscope settings the image represents a projection of the electron density in the specimen and may be interpreted accordingly (Cowley and Iijima, 1972; O'Keefe, 1973). In fringe images only two or three beams are utilized, although for metals it is interesting that this requires a similar aperture size to that used for complex structure imaging.

ii. *Orientation of the specimen.* As outlined above, a systematic row of reflections is used for lattice images, whereas a complete zone is utilized in structure imaging. It is important to be very close to an exact Bragg reflecting condition for the former (Hirsch *et al.*, 1965; Sinclair *et al.*, 1975) and to the exact zone axis in the latter case (Iijima, 1971, 1973; Cowley and Iijima, 1972).

iii. *Objective lens setting.* The lattice image is essentially a phase contrast image formed by interaction of electron waves with various amplitudes and phases. Because of the spherical aberration of the objective lens electron beams off the optical axis of the microscope suffer a phase delay with respect to the central beam which may be counteracted by defocus of the objective lens from the true Gaussian focus plane. The amount of defocus [optimum defocus condition (Scherzer, 1949)] depends on the type of imaging and, for each microscope, on the spherical aberration coefficient. For the JEM-100B used by Iijima the optimum defocus for structure imaging is 900 Å (Cowley and Iijima, 1972). For the Philips EM301 used by the present author for superlattice imaging in Cu_3Au the value is 300 Å underfocus (Sinclair *et al.*, 1975). The exact setting must be found for each case by appropriate calculations or in practice by choosing the most suitable image from a through-focus series of micrographs.

iv. *Specimen thickness.* By increasing the thickness of a TEM specimen the intensity of transmitted and diffracted electron beams is affected. Thus, certain thicknesses are optimum for suitable images and these may be found by the appropriate calculations or by inspection of the image of wedge-shaped specimens. For structure images foils $\leqslant$100-Å thick are preferred, although an identical image may be obtained at much greater thicknesses (e.g., in $H-Nb_2O_5$ the image at 500–800-Å thickness is identical to that in the thin sub-100 Å area (Fejes *et al.*, 1973). For two-beam fringe images, the optimum thicknesses occur at $\xi_g/4$, $3\xi_g/4$, $5\xi_g/4$, etc. (ξ_g is the extinction distance of the reflection) with the first being the best (Phillips, 1971). It is desirable, therefore, to have thin high-quality foils for lattice imaging which may be prepared from bulk in the normal way (see Section III,C) or, for metals, prepared to the optimum thickness by vapor deposition techniques (Sinclair *et al.*, 1975).

Various other considerations are also important. The mechanical, electrical, and thermal stability of the microscope must be extremely good to allow 2-Å resolution, especially since exposure times up to 60 sec may be required. Specimen drift (e.g., due to heating or charging in the intense electron beam) must be minimized and the material should not degrade in the beam (this is a serious problem in imaging organic materials at high resolution). The contamination rate should be low. The objective aperture should be carefully positioned to reduce any effect it may have on the electron beams and it should preferably be of the thin-film type which suffers low contamination. It is also extremely important to correct the astigmatism of the objective lens as carefully as possible. This should be done on the area under examination and should be recorrected after any changes of specimen position, orientation, or even objective aperture location.

Finally, it should be emphasized that the interpretation of the images and the determination of optimum experimental conditions is best approached by theoretical image simulation. For structure images phase grating multi-slice calculations are preferred, whereas for the simpler fringe images the Bloch wave method of dynamical electron diffraction theory may be utilized. In this way ambiguous or incorrect interpretation may be avoided.

A. Ceramics

Fringe imaging was applied to examine defects in ceramics before the higher resolution structure imaging was attempted (e.g., Allpress and Sanders, 1972). Information was obtained about planar defects in the $Nb_2O_5-WO_3$ and $Nb_2O_5-TiO_2$ systems which are present to accommodate nonstoichiometry from crystallographically simpler structures. Elsewhere fringe images have been used in studies of phase transformations in ortho-pyroxene (Vander Sande and Kohlstedt, 1974; Champness and Lorimer, 1974) and defects in hexagonal ferrites (Van Landuyt *et al.*, 1973, 1974). However, the most important work on ceramics has come from structure imaging.

At the appropriate underfocus the structure image can be interpreted in terms of a projection of the charge density of the material (Cowley and Iijima, 1972). Thus in metallic oxides the projection of metal atom rows appears dark on the image while rows of oxygen atoms appear light. Metal atoms too close to one another for resolution appear as darker regions than those which can be resolved by the microscope. The image is essentially a charge-density map of the structure. An example of this correlation between the electron structure image of $Ti_2Nb_{10}O_{29}$ and the projection of the structure as established by x-ray diffraction is shown in Fig. 18a,b and that for a silicate (beryl) in Fig. 18d. Other materials so imaged include various Ti, Nb, and W oxides (Iijima and Allpress, 1973, 1974), silicates (Buseck and Iijima, 1974), wüstite (Iijima, 1974), perovskite polytypes (Hutchison and Jacobson, 1975), and tournaline (Iijima *et al.*, 1973). The correlation between the structure and projected images indicates that structure determination itself is possible

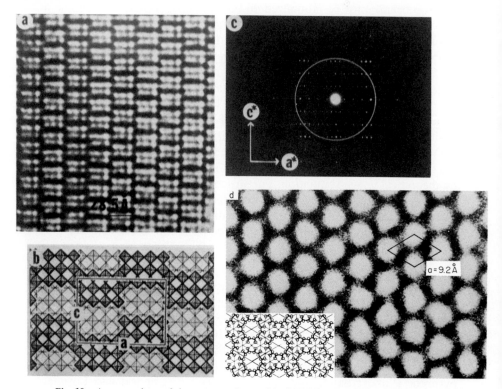

Fig. 18. A comparison of the structure image (a) of $Ti_2Nb_{10}O_{29}$ with a projection of the structure (b). The unit cell indicated in the latter has dimensions $a = 28.5$ Å, $c = 20.5$ Å and is composed of two layers (dark and light lines) of metal atoms in oxygen tetrahedra (shaded squares). The corresponding electron diffraction pattern is shown in (c) with a circle indicating the size of the objective aperture. (Courtesy of Cowley and Iijima, 1972.) (d) A similar comparison for the structure image of beryl. (Courtesy of Buseck and Iijima, 1974.)

if a reasonable basis of the structure is known. Examples whereby ambiguities of x-ray structures were resolved in various tetragonal tungsten bronze type materials in the Nb_2O_5–WO_3 system have been described by Iijima and Allpress (1973, 1974).

The advantage of using a direct technique such as microscopy to view defects is that their nature and distribution can be directly observed on a picture. The further advantage of structure and fringe imaging is that the atomic arrangement at the defect can be seen. An example of the effect of a planar defect on the atomic arrangement in H–Nb_2O_5 (Iijima, 1973) is shown in Fig. 19. The change of stacking sequence is quite clear and a shift of the fault plane is also revealed at the atomic level. Point defects in $Nb_{12}O_{29}$ and $Nb_{22}O_{54}$ (Iijima *et al.*, 1973, 1974) are shown in Fig. 20a,c. A model for the

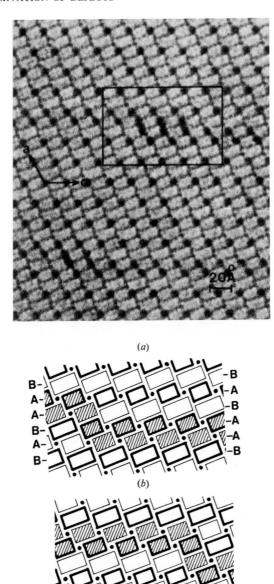

(a)

(b)

(c)

Fig. 19.   (a) Structure image of $H-Nb_2O_5$ showing displacement defects. The unusual black dots in the overlapping region of two defects (in the rectangular box) can be explained by a superposition of the two different arrangements of structural blocks shown in (b) and (c). (Courtesy of Iijima, 1973.)

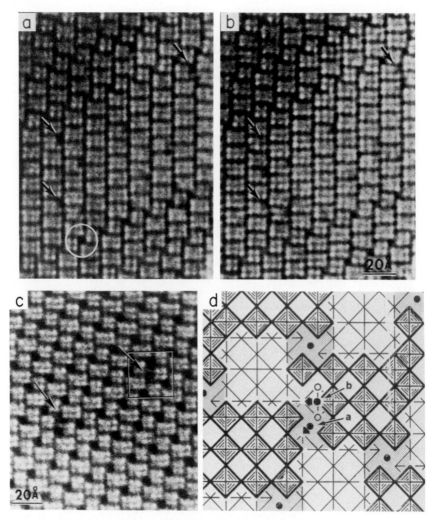

Fig. 20. Structure images showing point defects in $Nb_{12}O_{29}$ (a) and $Nb_{22}O_{54}$ (c). After a few minutes of electron-beam irradiation of the former, the black dots representing the defects can be seen to have disappeared (b). A proposed model for the origin of the black dots in $Nb_{22}O_{54}$ is shown in (d) (Courtesy of Iijima *et al.*, 1974.)

defect based on local changes of metal atom positioning is shown in Fig. 20d for $Nb_{22}O_{54}$. It is interesting that after a few minutes of irradiation under the electron beam these defects can be made to disappear (Fig. 20b), although in some cases they reappear in different parts of the image (Iijima *et al.*, 1974). This indicates that the dark spots are very likely to be point defects which can be induced to diffuse by beam heating or irradiation. Thus the structure imaging technique is capable of imaging even point defects in materials and gives much unequivocal data on the character of defects in ceramics.

B. Metals

In metallurgy, in which the crystal structures are simpler, much interest has centered around application of the method to defect and phase transformation studies. Considerable excitement was aroused by the first direct lattice fringe image of a dislocation whereby a terminating fringe was surrounded by fringes distorted from their natural position (e.g., Fig. 21a, from Phillips and Hugo, 1970). It was supposed that plane distortions could be directly measured from the image. However, Cockayne et al., (1971) showed that, since very few dislocations lie parallel to the imaging planes, the inclination of the dislocation line will strongly contribute to the observed fringe positions. From simulated images it was learned that there is not in general a one-to-one correspondence between fringe and atom plane positions in such circumstances. It can be seen, for instance, in Fig. 21a that the fringe distortions are not symmetrical with respect to the terminating fringe with greater disruption of the image on the right, whereas the atomic plane distortions should be symmetrical about an isolated dislocation. Thus one must be extremely cautious in choosing faults for analysis by lattice imaging. Care must also be taken over other electron optical effects such as fringe bending at the edge of thin foils, the causes of which are outlined by Hirsch et al. (1965). Nevertheless, it is becoming clear that, as in structure images, if the fault lies parallel to the imaging planes then fringe images may be associated with atomic plane positions [for a dislocation this may be ensured by a perfectly symmetrical lattice fringe pattern about the terminating fringe (Cockayne, 1975) as in Fig. 21b] and that lattice imaging may be utilized for studying defects in appropriate circumstances.

The superior information available from lattice imaging of alloys over conventional imaging modes is illustrated by Fig. 22 which compares the lattice image and superlattice dark-field image of an antiphase boundary (APB) in ordered Ni_4Mo. In the latter the boundary is revealed as a black line on a white background, in some regions ~ 20-Å thick. The translation at the APB must be determined by a series of various dark-field micrographs. However, the lattice image can be taken as representing the positions of the Mo planes of atoms, which occur every fifth plane in the superlattice with the other planes being all Ni. The continuity of the fringes (atomic planes) is shown to within an atomic diameter of the fault. From calculations of the fringe visibility with degree of long-range order it is established that little change in degree of order occurs up to the APB. It can be seen directly that the left-hand side is shifted upwards by the two-fifths of the superlattice

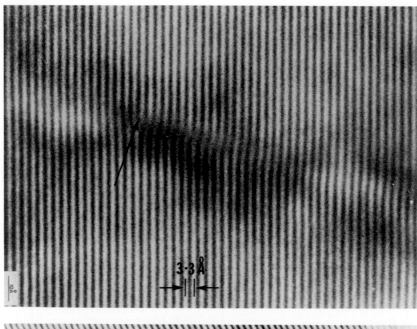

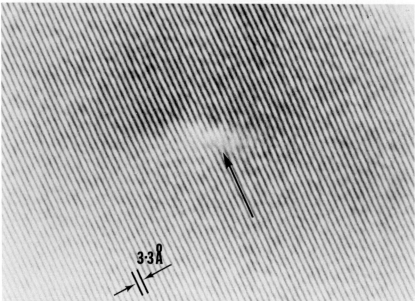

Fig. 21. (a) A lattice fringe image of a dislocation in germanium. A terminating fringe is clearly shown but the asymmetric distortion of fringes indicates that the dislocation line is not parallel to the imaging planes and that fringe positions may not be interpreted as representing directly the lattice plane positions. (Courtesy of Phillips and Hugo, 1970.) (b) Lattice image of an edge dislocation in germanium, with symmetrical fringe distortion about the terminating fringe indicating that the dislocation line is parallel to the imaging planes. The fringe distortions in this case should represent lattice plane distortions in the crystal. (Courtesy of Phillips and Wagner, 1973.)

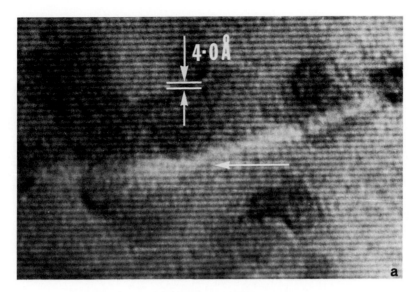

Fig. 22. A comparison of the lattice image (a) and conventional superlattice dark-field image (b) of an antiphase boundary in ordered Ni_4Mo. Information in the lattice image is given concerning the positions of atomic planes right up to the boundary, the degree of order in the vicinity of the boundary, the nature of the fault and any local composition changes (as in the position indicated). Such detailed information is not available in the dark-field image.

spacing (i.e., two fundamental lattice planes) with respect to the right, and that there is a local change of spacing (and hence composition) where the boundary becomes parallel to the imaging planes (arrowed). Considerably more information is present in the lattice image which is not available from the conventional micrograph.

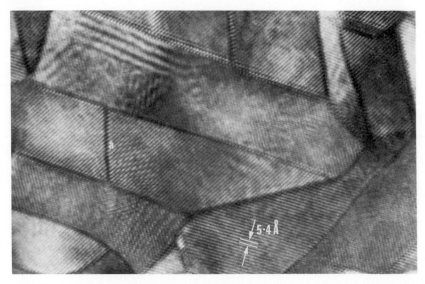

Fig. 23. A lattice image showing the distribution of domains in ordered Mg_3Cd. It can be seen that rotational domain boundaries and translational antiphase boundaries (as at **a**) are smooth at an atomic level in this alloy. (Courtesy of J. Dutkiewicz, 1975, private communication.)

A lattice image of fully ordered Mg_3Cd is shown in Fig. 23, in which the domain structure and several faults are clearly revealed. The orientation of each domain is directly shown and the very flat crystallographic nature of the domain interfaces is illustrated. A further comparison of the lattice and dark-field image is shown in Fig. 24 for partially ordered Mg_3Cd. The bright areas in the superlattice dark-field image represent ordered regions, the dark regions are disordered. In the lattice image this is revealed by differences in fringe spacings with double periodicity in the ordered material (Rocher and Dutkiewicz, 1976; Sinclair and Dutkiewicz, 1975). In some regions ribbons of ordered material are present in which the interface shows a slight bending. The lattice image reveals why this occurs (Fig. 25). There are discrete unit cell high steps at the interface which are not revealed in the dark-field image. This suggests that the mechanism of ordering in this alloy is one of movement of these steps across the interface, and this would also account for the flat domain boundaries found in the fully ordered alloy (Fig. 23). Again it can be appreciated that superior information is available in the lattice image.

To date this technique has been used to study dislocations and twins in various metals and semiconductors (Phillips, 1971), radiation damage in copper (Howe and Rainville, 1972), precipitation and single atom layer G.P. zone formation in Al–Cu (Parsons *et al.*, 1970; Phillips, 1973, 1975), and Cu–Be (Phillips and Tanner, 1973). The interest of our group has been in

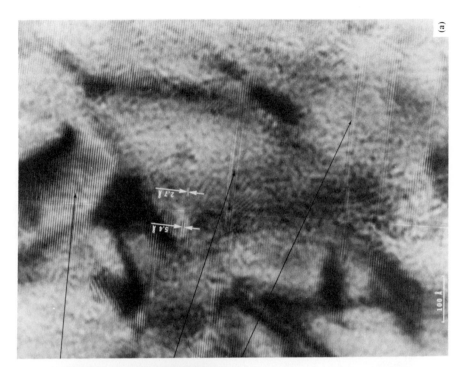

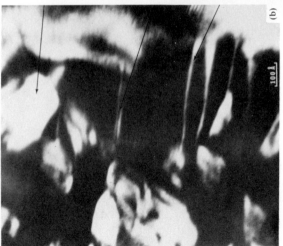

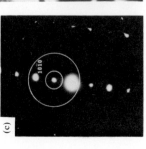

Fig. 24. A comparison of the lattice image (a) and superlattice dark-field image (b) of partially ordered Mg_3Cd. Ordered regions possess twice the fringe periodicity of disordered regions in the lattice image. An exact correspondence between the two is clear, with more detailed information present in the former (see Fig. 25). The corresponding diffraction pattern is shown in (c) with the smaller circle showing the objective aperture used for the dark-field image (b) and the larger circle that for the lattice image (a). (Courtesy of J. Dutkiewicz, 1975, private communication.)

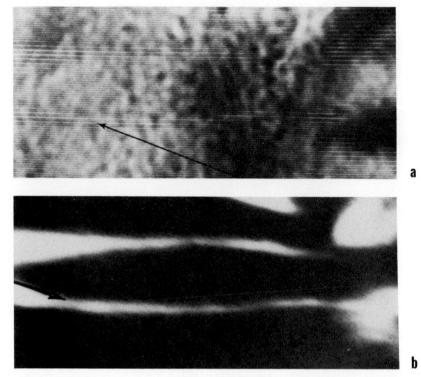

Fig. 25. A detail of the lattice (a) and dark-field (b) images shown in Fig. 24. The dark-field image reveals thin ribbons of ordered material which show some curvature. The lattice image indicates that this is due to unit cell high steps in the interface between ordered and disordered material. In addition to the higher resolution, information about the local degree of order can also be obtained from the lattice image but not from the dark-field image. (Courtesy of J. Dutkiewicz, 1975, private communication.)

studying short-range order (Sinclair and Thomas, 1975), ordering reactions and defects in ordered alloys (Sinclair *et al.*, 1975; Sinclair and Dutkiewicz, 1975; Dutkiewicz and Thomas, 1975), and spinodal decomposition processes (Gronsky *et al.*, 1975; Sinclair *et al.*, 1976). As in studies of ceramics the technique is superior to conventional imaging for yielding information at the atomic level.

C. *Amorphous Materials*

Interpretation of the structural nature of amorphous materials has been ambiguous since it was found that the diffraction pattern showed diffuse peaks at well-defined angular positions. Renewed interest in this problem has come from lattice imaging of such materials. Rudee and Howie (1972) formed images using the transmitted beam plus part of the first diffuse ring and found regions ~ 14 Å in diameter, in amorphous Si and Ge films, possessing regular fringes. These they interpreted as small crystallites existing in the predominantly irregular structure. However, more recent experiments (Herd and Chaudhari, 1974) on a range of amorphous films has indicated that their result was an imaging artifact arising from lens astigmatism on tilting the electron illumination and that the extent of coherently scattering regions is $\lesssim 6$ Å. It was concluded that the random network model is more appropriate for describing amorphous structures. This latter conclusion has also been drawn from independent microdiffraction experiments on specimen areas ~ 30 Å in diameter using a STEM (Geiss, 1975b). It may be noted that these experiments illustrate the care that must be taken over the interpretation of lattice images.

V. Summary and Conclusions

A wide range of investigative methods are available for studying and characterizing defects in materials. These range from direct visual observation down to the resolution of individual atoms by transmission electron microscopy or field-ion microscopy. Most structure characterization laboratories are equipped with optical, SEM, and TEM facilities and these offer the capability of studying each of the various crystal defects (cf. Fig. 1). These techniques should be regarded as complementary rather than directly competitive and the approach taken by the investigator should be carefully considered according to the problem to be solved.

For future trends it seems that direct lattice imaging by TEM will prove extremely useful for obtaining information about fine-scale detail of defects which is not currently available. Further direct conventional information about microstructure and defect characteristics is also available at lower resolution on the same instrument. The scanning transmission electron microscope also appears to be a considerably more versatile instrument than the conventional transmission microscope and if the cost becomes competitive much basic research may well be transferred to this mode. In any event the materials scientist has a wide scope in characterizing the defects which affect the properties of materials.

ACKNOWLEDGMENTS

Financial support for this work has been provided by the National Science Foundation and the Energy Research and Development Administration. The author would like to acknowledge the encouragement given by Professor G. Thomas during preparation of this chapter and, for

helpful and critical discussion, Dr. D. R. Clarke and the other members of the electron microscopy group in the Materials Science Department at Berkeley. Thanks are also due to the authors,
cited in the figure captions, who kindly provided micrographs for illustration.

References

Allpress, J. G., and Sanders, J. V. (1972). *In* "Electron Microscopy and Structure of Materials"
(G. Thomas *et al.*, eds.), p. 134. Univ. of California Press, Berkeley, California.

Allpress, J. G., and Sanders, J. V. (1973). *J. Appl. Crystallogr.* **6**, 165.

Amelinckx, S., Gevers, R., Remaut, G., and Van Landuyt, J. (1970). "Modern Diffraction and
Imaging Techniques in Material Science." North-Holland Publ., Amsterdam.

Armstrong, R. W., and Wu, C. C. (1973). *In* "Microstructural Analysis: Tools and Techniques"
(J. L. McCall and W. M. Mueller, eds.), p. 169. Plenum Press, New York.

Authier, A. (1970). *In* "Modern Diffraction and Imaging Techniques in Material Science"
(S. Amelinckx *et al.*, eds.), p. 481. North-Holland Publ., Amsterdam.

Banbury, J. R., Drummond, I. W., and Ray, I. L. F. (1975). *Proc. Ann. Conf., 33rd, EMSA,
Las Vegas* p. 112.

Belk, J. A. (1974). *In* "Quantitative Scanning Electron Microscopy" (D. B. Holt *et al.*, eds.),
p. 389. Academic Press, New York.

Berg, H., Tsong, T. T., and Cohen, J. B. (1973). *Acta. Metall.* **21**, 1589.

Booker, G. R. (1970). "Scanning Electron Microscopy" (*Proc. Ann. SEM Symp., 3rd, Chicago*),
p. 489. IIT Res. Inst., Chicago, Illinois.

Bouchard, M., Livak, R. J., and Thomas G. (1972). *Surface Sci.* **31**, 275.

Broers, A. N. (1969). *Rev. Sci. Instrum.* **40**, 1040.

Brown, L. M. (1971). *In* "Electron Microscopy in Material Science" (U. Valdré, ed.), p. 360.
Academic Press, New York.

Buseck, P. R., and Iijima, S. (1974). *Am. Mineral.* **59**, 1.

Butler, E. P., and Swann, P. R. (1975). *In* "Physical Aspects of Electron Microscopy and
Microbeam Analysis" (B. M. Siegel and D. R. Beaman, eds.), p. 129. Wiley, New York.

Castaing, R. (1975). *In* "Physical Aspects of Electron Microscopy and Microbeam Analysis"
(B. M. Siegel and D. R. Beaman, eds.), p. 355. Wiley, New York.

Champness, P. E., and Lorimer, G. W. (1974). *Phil. Mag.* **30**, 357.

Clarke, D. R. (1971). *Phil. Mag.* **24**, 973.

Cockayne, D. J. H. (1974). *J. Phys.* **35**, C7-12.

Cockayne, D. J. H. (1975). *Proc. Ann. Conf. EMSA 33rd, Las Vegas* p. 2.

Cockayne, D. J. H., and Vitek, V. (1974). *Phys. Status Solidi (b)* **65**, 751.

Cockayne, D. J. H., Ray, I. L. F., and Whelan, M. J. (1969). *Phil. Mag.* **20**, 1265.

Cockayne, D. J. H., Parsons, J. R., and Hoelke, C. W. (1971). *Phil. Mag.* **24**, 139.

Cowley, J. M. (1969). *Appl. Phys. Lett.* **15**, 58.

Cowley, J. M. (1975). *In* "Physical Aspects of Electron Microscopy and Microbeam Analysis"
(B. M. Siegel and D. R. Beaman, eds.), p. 17. Wiley, New York.

Cowley, J. M., and Iijima, S. (1972). *Z. Naturforsch.* **27a**, 445.

Crewe, A. V. (1974). *In* "Quantitative Scanning Electron Microscopy" (D. B. Holt *et al.*,
eds.), p. 65. Academic Press, New York.

Crewe, A. V., Wall, J., and Welter, L. M. (1968). *Rev. Sci. Instrum.* **40**, 241.

Crewe, A. V., Langmore, J. P., and Isaacson, M. S. (1975). *In* "Physical Aspects of Electron
Microscopy and Microbeam Analysis" (B. M. Siegel and D. R. Beaman, eds.), p. 47.
Wiley, New York.

Dingley, D. J., and Steeds, J. W. (1974). *In* "Quantitative Scanning Electron Microscopy"
(D. B. Holt *et al.*, eds.), p. 487. Academic Press, New York.

Doig, P., Edington, J. W., and Jacobs, M. H. (1975). *Phil. Mag.* **31**, 285.

Driver, J. H., and Papazian, J. M. (1973). *Acta Metall.* **21**, 1139.

Dutkiewicz, J., and Thomas, G. (1975). *Metall. Trans.* **6A**, 1919.

Fejes, P. L., Iijima, S., and Cowley, J. M. (1973). *Acta Crystallogr.* **A29**, 70.

Flower, H. M., and Swann, P. R. (1974). *Acta Metall.* **22**, 1339.

Gedcke, D. A. (1974). *In* "Quantitative Scanning Electron Microscopy" (D. B. Holt *et al.*, eds.), p. 403. Academic Press, New York.

Geiss, R. H. (1975a). *Appl. Phys. Lett.* **27**, 174.

Geiss R. H. (1975b). *Proc. Ann. Conf. EMSA, 33rd Las Vegas* p. 218.

Glauert, A. M. (1972). "Practical Methods in Electron Microscopy," Vol. 1. North-Holland Publ., Amsterdam.

Glauert, A. M. (1974). "Practical Methods in Electron Microscopy," Vol. 2. North-Holland Publ., Amsterdam.

Gold, E., and Machlin, E. S. (1968). *Phil. Mag.* **18**, 453.

Goodman, S. R., Brenner, S. S., and Low, J. R. (1973). *Metall. Trans.* **4**, 2371.

Goringe, M. J., Hewat, E. A., Humphreys, C. J., and Thomas, G. (1972). *Proc. Eur. Congr. Electron Microsc., 5th* p. 538.

Gronsky, R. (1974). M.S. Thesis, Univ. of California, Berkeley.

Gronsky, R., and Thomas, G. (1975). *Acta Metall.* **23**, 1163.

Gronsky, R., Okada, M., Sinclair, R., and Thomas, G. (1975). *Proc. Ann. Conf. EMSA, 33rd, Las Vegas* p. 22.

Head, A. K., Humble, P., Clareborough, L. M., Morton, A. J., and Forwood, C. T. (1973). "Computed Electron Micrographs and Defect Identification." North-Holland Publ., Amsterdam.

Hearle, J. W. S. (1972). *In* "The Use of the Scanning Electron Microscope" (J. W. S. Hearle *et al.* eds.), p. 1. Pergamon, Oxford.

Hearle, J. W. S., Sparrow, J. T., and Cross, P. M. (1972). "The Use of the Scanning Electron Microscope." Pergamon, Oxford.

Herd, S. R., and Chaudhari, P. (1974). *Phys. Status Solidi (a)* **26**, 627.

Hirsch, P. B., Howie, A., Nicholson, R. B., Pashley, D. W., and Whelan, M. J. (1965). "Electron Microscopy of Thin Crystals." Butterworths, London.

Holt, D. B. (1974). *In* "Quantitative Scanning Electron Microscopy" (D. B. Holt *et al.*, eds.), p. 335. Academic Press, New York.

Holt, D. B., Muir, M. D., Grant, P. R., and Boswarva, I. M. (1974). "Quantitative Scanning Electron Microscopy." Academic Press, New York.

Howe, L. M., and Rainville, M. (1972). *Radiat. Effects* **16**, 203.

Hutchison, J. L., and Jacobson, A. J. (1975). *Acta Crystallogr.* **B31**, 1442.

Iijima, S. (1971). *J. Appl. Phys.* **42**, 5891.

Iijima, S. (1973). *Acta Crystallogr.* **A29**, 18.

Iijima, S. (1974). *Proc. Ann. Conf. EMSA, 32nd, St. Louis* p. 352.

Iijima, S., and Allpress, J. G. (1973). *J. Solid State Chem.* **7**, 94.

Iijima, S., and Allpress, J. G. (1974). *Acta Crystallogr.* **A30**, 22.

Iijima, S., Cowley, J. M., and Donnay, G. (1973). *Tschermaks Mineral, Petrogr. Mitt.* **20**, 216.

Iijima, S., Kimura, S., and Goto, M. (1974). *Acta Crystallogr.* **A29**, 632; (1974). *ibid.* **A30**, 251.

Johari, O. (1968–1975). "Scanning Electron Microscopy" (*Proc. Ann. SEM Symp., 1st-8th, Chicago* IIT Res. Inst., Chicago, Illinois.

Johnson, V. E. (1975). "Scanning Electron Microscopy" (*Proc. 8th, Ann. SEM Symp., Chicago* p. 763. IIT Res. Inst. Chicago, Illinois.

Joy, D. C. (1974). *In* "Quantitative Scanning Electron Microscopy" (D. B. Holt *et al.*), p. 131. Academic Press, New York.

Joy, D. C., Thompson, M. N., Booker, G. R., and Andersen, W. H. J. (1974). *Phys. Status Solidi (a)* **21**, Kl.

Kandiah, K. (1975). *In* "Physical Aspects of Electron Microscopy and Microbeam Analysis" (B. M. Siegel and D. R. Beaman, eds.), p. 395. Wiley, New York.

Laidler, J. J., and Mastel, B. (1975). *In* "Physical Aspects of Electron Microscopy and Microbeam Analysis" (B. M. Siegel and D. R. Beaman, eds.), p. 103. Wiley, New York.

Lang, A. R. (1970). *In* "Modern Diffraction and Imaging Techniques in Material Science" (S. Amelinckx *et al.*, eds.), p. 407. North-Holland Publ., Amsterdam.

Lifshin, E., Morris, W. G., and Bolon, R. B. (1969). *J. Met.* **21**, 1.

Lynch, D. F., Moodie, A. F., and O'Keefe, M. A. (1975). *Acta Crystallogr.* **A31**, 300.

MacDonald, N. C. (1975). *In* "Physical Aspects of Electron Microscopy and Microbeam Analysis" (B. M. Siegel and D. R. Beaman, eds.), p. 431. Wiley, New York.

Makin, M. J. (1971). *Jernkonterets Ann.* **155**, 509.

Mishra, R. K. (1976). Ph.D. Thesis, Univ. of California, Berkeley.

Muir, M. D. (1974). *In* "Quantitative Scanning Electron Microscopy" (D. B. Holt *et al.*, eds.), p. 3. Academic Press, New York.

Muller, E. W., and Tsong, T. T. (1971). "Field-ion Microscopy–Principles and Applications." Elsevier, Amsterdam.

Muller, E. W., Panitz, J., and McLane, S. B. (1968). *Rev. Sci. Instrum.* **39**, 83.

Murr, L. E. (1975). *In* "Physical Aspects of Electron Microscopy and Microbeam Analysis" (B. M. Siegel and D. R. Beaman, eds.), p. 163. Wiley, New York.

Narasimha Rao, B. V. (1975). M.S. Thesis, Univ. of California, Berkeley.

Narasimha Rao, B. V., and Thomas, G. (1975). *Mater. Sci. Eng.* **20**, 195.

Oatley, C. W. (1972). "The Scanning Electron Microscope." Cambridge Univ. Press, London and New York.

O'Connor, G. P., and Ralph, B. (1972). *Phil. Mag.* **26**, 113, 129.

O'Keefe, M. A. (1973). *Acta Crystallogr.* **A29**, 389.

O'Keefe, M. A., and Sanders, J. V. (1975). *Acta Crystallogr.* **A31**, 307.

Parsons, J. R., Rainville, M., and Hoelke, C. W. (1970). *Phil. Mag.* **21**, 1105.

Philibert, J., and Tixier, R. (1975). *In* "Physical Aspects of Electron Microscopy and Microbeam Analysis," (B. M. Siegel and D. R. Beaman, eds.), p. 333. Wiley, New York.

Phillips, V. A. (1967). *Tech. Met. Res.* **1**, 25.

Phillips, V. A. (1971). "Modern Metallographic Techniques and Their Application." Wiley, New York.

Phillips, V. A. (1973). *Acta Metall.* **21**, 219.

Phillips, V. A. (1975). *Acta Metall.* **23**, 751.

Phillips, V. A., and Hugo, J. A. (1970). *Acta Metall.* **18**, 123.

Phillips, V. A., and Tanner, L. E. (1973). *Acta Metall.* **21**, 441.

Phillips, V. A. and Wagner, R. (1973). *J. Appl. Phys.* **44**, 4252.

Robinson, J. T., Wilson, K. L., and Seidman, D. N. (1973). *Phil. Mag.* **27**, 1417.

Rocher, A., and Dutkiewicz, J. (1976). (Submitted to *Acta Cryst.*, Lawrence Berkeley Laboratory Report No. 3559).

Rudee, M. L., and Howie, A. (1972). *Phil. Mag.* **25**, 1001.

Scherzer, O. (1949). *J. Appl. Phys.* **20**, 20.

Seidman, D. N., and Burke, J. J. (1974). *Acta Metall.* **22**, 1301.

Seshan, K. (1975). Ph.D. Thesis, Univ. of California, Berkeley.

Seshan, K., and Washburn, J. (1972). *J. Radiat. Effects* **14**, 267.

Siegel, B. M., and Beaman, D. R. (1975). "Physical Aspects of Electron Microscopy and Microbeam Analysis." Wiley, New York.

Sinclair, R., and Dutkiewicz, J. (1975). *Proc. Ann. Conf. EMSA, 33rd, Las Vegas* p. 10.

Sinclair, R., and Thomas, G. (1975). *J. Appl. Crystallogr.* **8**, 206.

Sinclair, R., Ralph B., and Leake, J. A. (1973). *Phil. Mag.* **28**, 1111.

Sinclair, R., Leake, J. A., and Ralph, B. (1974). *Phys. Status Solidi (a)* **26**, 285.

Sinclair, R., Schneider, K., and Thomas, G. (1975). *Acta Metall.* **23**, 873.

Sinclair, R., Gronsky, R., and Thomas, G. (1976). *Acta Metall.* **24**, 789.

Smith, D. A., Page, T. F., and Ralph, B. (1969). *Phil. Mag.* **19**, 231.

Smith, G. D. W., Smith, D. A., Taylor, G. S., Goringe, M. J., and Easterling, K. (1974). *In* "High Voltage Electron Microscopy" (P. R. Swann *et al.*, eds.), p. 240. Academic Press, New York.

Stern, R. M., Ichinokawa, T., Takashima, S., Hashimoto, H., and Kimoto, S. (1972). *Phil. Mag.* **26**, 1495.

Swann, P. R., Humphreys, C. J., and Goringe, M. J. (1974). "High Voltage Electron Microscopy." Academic Press, New York.

Taunt, R. J., and Ralph, B. (1974a). *Phil. Mag.* **30**, 1379.

Taunt, R. J., and Ralph, B. (1974b). *Phys. Status Solidi (a)* **24**, 207.

Thomas, G. (1973). *In* "The Science of Materials Used in Advanced Technology" (E. R. Parker and U. Colombo, eds.), p. 35. Wiley, New York.

Thomas, G., Fulrath, R. M., and Fisher, R. M. (1972). "Electron Microscopy and Structure of Materials." Univ. of California Press, Berkeley, California.

Thompson, M. G. R., (1975). *In* "Physical Aspects of Electron Microscopy and Microbeam Analysis" (B. M. Siegel and D. R. Beaman, eds.), p. 29. Wiley, New York.

Thornton, P. R. (1968). "Scanning Electron Microscopy." Chapman and Hall, London.

van der Biest, O. O., and Thomas, G. (1974). *Phys. Status Solidi (a)* **24**, 65.

van der Biest, O. O., Butler, E. P., and Thomas, G. (1975). *Proc. Ann. Conf. EMSA, 33rd, Las Vegas* p. 36.

Vander Sande, J. B., and Kohlstedt, D. L. (1974). *Phil. Mag.* **29**, 1041.

Van Landuyt, J., Amelinckx, S., Kohn, J. A., and Eckart, D. W. (1973). *Mater. Res. Bull.* **8**, 1173.

Van Landuyt, J., Amelinckx, S., Kohn, J. A., and Eckart, D. W. (1974). *J. Solid State Chem.* **9**, 103.

Veneklasen, L. H. (1975). *In* "Physical Aspects of Electron Microscopy and Microbeam Analysis" (B. M. Siegel and D. R. Beaman, eds.), p. 315. Wiley, New York.

Wu, C. C., and Armstrong, R. W. (1975). *Phys. Status Solidi (a)* **29**, 259.

Yakowitz, H. (1974). *In* "Quantitive Scanning Electron Microscopy" (D. B. Holt *et al.*, eds.), p. 451. Academic Press, New York.

Young, F. W., Baldwin, T. O., and Shirill, F. A. (1967). *In* "Lattice Defects and their Interaction" (R. R. Hisiguti, ed.). Gordon and Breach, New York.

Crystal Defects in Silicon Integrated Circuits—Their Cause and Effect

C. M. MELLIAR-SMITH

Bell Laboratories
Murray Hill, New Jersey

I. Introduction

In the twenty-five years since the first transistor was demonstrated, semiconductor devices have developed from relatively simple discrete units into extremely complicated integrated circuits of the type shown in Fig. 1 containing many thousands of transistors on a single chip. This development has been made possible by remarkable advances in the production of the very high-quality silicon crystals from which the devices are made. Indeed to the best of the author's knowledge no other manmade product makes more stringent requirements on its starting material (in terms of chemical purity and crystalline perfection) than silicon semiconductor devices. In addition, not only must the starting crystal be produced with high quality, but this quality has to be maintained throughout the many processing steps required

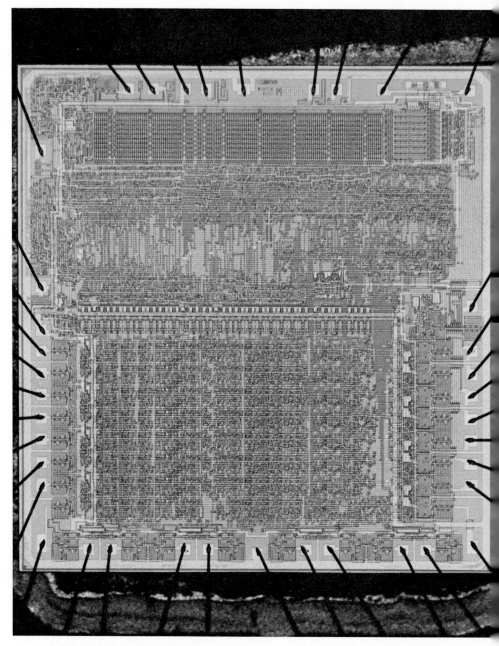

Fig. 1. Optical photomicrograph of Motorola M6800 microprocessor chip. This NMOS device contains approximately 5400 transistors and measures 0.210 × 0.217 inches.

in the fabrication of a large integrated circuit. This chapter is designed to review how this crystalline quality is achieved in crystal growth and maintained during subsequent processing; how crystal defects are introduced; their effect on the performance of the devices, and finally where possible how they can be prevented or rendered electrically inactive and thus harmless. The chapter is broken into two main sections—crystal growth and process induced problems. This rather neat classification of defects and processing steps is somewhat deceptive, however, as many of the different processes interact to form the finally observed defect. For example, the formation of point defect clusters during crystal growth may not become apparent until they nucleate stacking faults when the wafer is subsequently oxidized. It is this interaction between differing processes and defects which has caused the occasional discrepancies that have appeared in the literature. Unless careful control is maintained over all aspects of the sample preparation, significant irregularities can occur both in experimental results from the laboratory or in device performance and yield from a production line. In addition, different types of devices often require differing emphases on materials properties. Bipolar devices rely primarily on a high-quality epitaxial film for device performance with the starting crystal being less important. In contrast metal/oxide/semiconductor (MOS) devices often do not need the epitaxial film for their fabrication and in consequence the top few microns of the silicon substrate itself is particularly important.

Looking back over the past twenty-five years, it is obvious that the materials science involved has undergone a continuous and rapid evolution which is continuing today, even though many of the major problems have been largely solved. The advent of very large integrated circuits, typically 5×5 mm, but in extreme cases as large as 16×20 mm (Sequin *et al.*, 1975), calls for a continuing reduction in electrically active defects if the device yield is to be reasonable. The recent requirements of more complex processing, for example, ion implantation with its concurrent radiation damage problems or the heteroepitaxial deposition of single-crystal silicon onto insulating substrates, have resulted in the need for further materials study. In addition, the rapid increase in wafer size from around one inch in the late nineteen-sixties to three inches or larger has prompted further studies in both crystal growth and wafer processing.

One assumption which might be made from these comments is that all crystal defects in silicon are deleterious. Looking at the overall picture, however, this is not necessarily true (Lawrence, 1973); one type of defect gettering another and more detrimental defect. For example, the minority carrier lifetime in silicon is generally lower in dislocation-free silicon than in silicon containing several thousand dislocations per square centimeter. At this dislocation level, the dislocations effectively getter point defects, which in dislocation-free material, act as recombination centers and cause low lifetimes. A continuous and on-going study of silicon and its processing is required to ensure that the device requirements of the future will be met.

II. Bulk Crystal Imperfections

The development of single-crystal growth techniques, which allowed the production of semiconducting substrates of excellent purity and crystalline perfection, has provided the materials basis for the solid-state electronics industry. The importance of high-quality crystal as a starting point for semiconductor device fabrication and subsequent performance cannot be overstated.

Most of the single-crystal silicon is produced using one of three techniques, namely Czochralski (Teal and Buehler, 1952), float zone (Theuerer, 1956), or crucibleless growth (Dash, 1960). The details of these techniques have been well covered in the literature and will not be discussed in this review. However, as each technique has a slightly differing influence on purity and crystal perfection a brief description of each and its particular problems is given below.

Czochralski growth (Fig. 2), whereby a small seed crystal is slowly pulled from a molten melt initiating single-crystal growth at the solid/liquid interface, is the most widely used process. Large crystals, well in excess of 10 cm in diameter, may be economically grown and doping the crystal can be

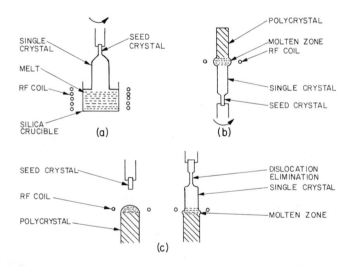

*.   (a) Czochralski crystal growth; (b) float zone crystal growth; (c) pedestral crystal growth with dislocation elimination.

achieved by adding a controlled amount of dopant to the melt. Low-resistivity silicon doped with boron at high concentrations is somewhat easier to obtain than the use of phosphorus and arsenic which have a higher vapor pressure. However, dopant evaporation from the melt can be controlled by the use of suitable gas ambients or the controlled addition of dopant to the melt during crystal pulling.

The major impurities present in Czochralski grown silicon are usually oxygen and carbon, the former derived from reaction with the silica crucibles and the latter either from the graphite susceptors used to heat the crucibles or from carbon contamination in the polycrystalline silicon source material. The effect of these impurities on microstructure and device performance is discussed below. It is possible to reduce the contamination level of oxygen and carbon in Czochralski grown silicon by suitable care in growth conditions and by replacing the graphite susceptor with molybdenum (Benson, 1969). However, where device performance is dependent on very low concentration of impurities, such as high-voltage devices, float zone or a cold crucible technique is used.

The float zone process, shown in Fig. 2, and developed from zone refining, makes use of surface tension to hold a narrow molten zone between the single crystal and polycrystalline portions of the source rod. As the molten zone does not have to be restrained by a crucible wall it is generally oxygen free.

A significant advance in the technique of crystal growth was made by Dash (1959, 1960), who was able to improve the crystal perfection by using a variation of both the Czochralski and float zone techniques (Fig. 2). Initially the process resembles Czochralski growth whereby the withdrawal of a seed crystal is used to initiate single crystal growth from the melt. However, to reduce oxygen contamination the melt is retained as a puddle at the top of the source rod and as a consequence is "crucible free." By using a narrow seed crystal and a high initial rate of pulling, the dislocations in the first stages of crystal growth are grown out. Subsequently, the rate of growth is reduced and the crystal diameter increases. When the molten zone reaches the full diameter of the source rod a process similar to float zone growth is obtained. The large dislocation-free crystal is resistant to the generation of new dislocations due to their high energy of formation (Cottrell, 1953) and remains dislocation free even if subject to large temperature fluctuations (Dash, 1959).

This use of a narrow seed to grow dislocation free crystals has been extended to crucible Czochralski growth with equally effective results, although oxygen contamination is correspondingly higher than the crucible-free techniques.

A variety of other specialized single crystal growth techniques for silicon have been developed, including vapor growth (Sandmann, 1969), vapor/ liquid/solid growth (Wagner and Doherty, 1966) web dendrite (Dermatis and Faust, 1963) and the ribbon process, (Groundy and Beatman, 1966) but will not be discussed in this review as they have yet to be used in significant amounts for device fabrication.

A. Dislocations

At first sight a section on the generation of dislocations in melt-grown semiconductor crystals might seem rather superfluous coming immediately after a description of a technique which allows crystals to be grown essentially dislocation free. However, now that this advancement in crystal growth has been made it has led to a better understanding of the interaction between dislocations and other lattice defects, particularly in the control of vacancies and interstitial impurities. In consequence a clear understanding of the mechanisms of dislocation occurrence in semiconductor material is required to allow the growth of crystals with the optimum crystal lattice, be it dislocation free or not, for device fabrication. The ability to grow dislocation-free silicon, germanium, and gallium arsenide is the result of a fortunate combination of favorable properties associated with these materials; specifically, a low dislocation mobility to reduce their growth from the seed into the crystal, and a low coefficient of thermal expansion which, coupled with the high thermal conductivity of these materials, reduces the thermal stresses present in the growing crystal.

Dash (1960) in his study of the growth of dislocation free crystals found few major sources for dislocation formation in the crystal pulled from the melt.

1. Propagation of dislocations present in the seed into the crystal growth area.
2. Multiplication due to high stress levels in the crystals usually due to thermal shock.
3. Generation of dislocations from surface damage on the seed crystal.
4. Poor epitaxy if the melt does not wet the seed.

As described in the previous section, the technique of initially pulling a narrow crystal at high rate is widely used to prevent any dislocations in the seed crystal from propagating far into the bulk crystal. Dash (1960) has postulated that if a high concentration of vacancies is present at the initial stages of growth, then the rate of dislocation climb will be sufficient to grow out many of the dislocations in the seed crystal. The rapid pulling of the crystal in the early stages of growth provides the required vacancy supersaturation due to the rapid cooling of the narrow crystal which does not couple well to the rf heating coils commonly used. The total removal of dislocations by climb only applies to edge dislocations. Screw dislocations are not moved but rather, if the vacancy concentration is high enough, forced into a helix.

In addition, crystal orientation will play a part in minimizing the number of plastically generated dislocations in the seed propagating down into the growing crystal. The dislocations progagate along the {111} planes and in consequence the larger the angle between these planes and the growth axis then the faster the dislocations will move to the surface. These angles are relatively large for ⟨100⟩ and ⟨111⟩ crystal axes (35° and 19°, respectively) and during the initial growth period when the crystal is narrow dislocations are fairly rapidly grown to the surface, occuring, for example, if twinning due to poor epitaxy is present. This is not the case for crystals grown with a ⟨100⟩ or ⟨112⟩ axis which have {111} glide planes parallel to the growth axis. As a result long dislocations can move undeviated from the seed into the crystal and it is more difficult to grow dislocation-free crystals using this technique.

Once the crystal growth has proceeded to the stage of being dislocation free it is much less susceptible to thermal shock effects due to the very high stress levels required to initiate new dislocations. In addition, the vacancy saturation required to accelerate dislocation climb is no longer needed and indeed is undesirable if vacancy agglomeration is to be prevented. In consequence the crystal growth can rapidly expand to the larger diameters required for device fabrication.

The presence of surface damage or contamination by precipitation will cause dislocation loops to propagate through the crystal if it is subjected to thermal shock. The velocity of a dislocation in silicon is given by the expression (Schonherr, 1968).

$$v \alpha \tau \exp - Q/kT$$

where τ is the applied stress and Q is the activation energy for formation of a kink pair on the dislocation. For silicon and germanium this has a relatively high value of between 1.5 and 2.0 eV and the velocity is low when compared to other materials. Patel and Freeland (1967) have shown (Fig. 3) that n-type doping above 10^{18} cm^{-3} markedly raises the dislocation velocity due to the interaction of kinks with charged dislocation acceptor sites (Glaenzer and Jordan, 1968). Although it is difficult to postulate the applied stress gradient as the dislocations moved into the crystal, the dislocation velocities shown in Fig. 3 suggest that prolonged high-temperature treatments can result in dislocation growth at significant distances into the crystal.

In a similar manner any surface damage on the seed crystal will result in dislocation formation during crystal growth. While the technique of fast initial growth will result in a reduction of dislocation propagation into the crystal, careful handling and a surface chemical etch (1 part 50% HF solution,

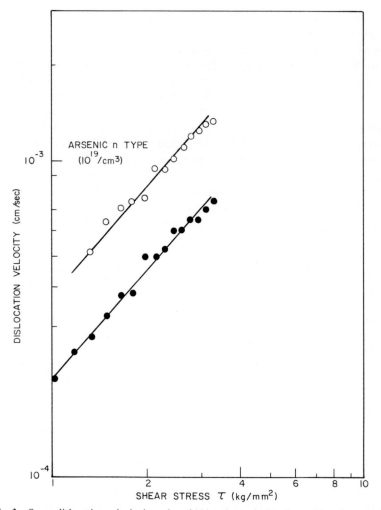

Fig. 3. Screw dislocation velocity in undoped (●) and arsenic doped (○) silicon ($T = 800°C$) (from Patel and Freeland, 1967).

3 parts 70% HNO_3 solution, and 6 parts glacial acetic acid for 15 min) (Dash, 1959) will remove the seed surface damage.

If the seed is contaminated or at too low a temperature, epitaxy will be affected and a large number of dislocations will be introduced into the crystal during subsequent growth. The same effect would occur if a solid impurity particle (such as SiO_2 or SiC) was present at the interface of the melt and the crystal at anytime during crystal growth.

Investigations of the electrical effects of dislocations are complicated by the presence of any precipitation which tends to occur preferentially at dislocations. In consequence, considerable care must be taken to ensure that the dislocations under study are "clean" or free of precipitation. Under these conditions dislocations are relatively inert (compared to dislocations on which precipitation has occurred), even when present in very high concentrations ($> 10^7$ cm^{-2}). Small changes in conductivity (Yu et al., 1967; Glaenzer and Jordan, 1969b), minority carrier lifetime (Glaenzer and Jordan, 1969a; Lawrence, 1968a; Glinchuk et al., 1966; Kurtz et al., 1956; Lemke, 1965; Noack, 1969; Leamy et al., 1975), noise (Yu et al., 1967), and reverse current and transistor gain (Lawrence, 1968a) have been observed due to dislocations.

Clean dislocations exhibit acceptor like properties in n-type silicon and donor properties in p-type. The effect on the energy band diagram for n-type silicon is shown in Fig. 4, in conjunction with a schematic diagram of the charge distribution around a dislocation. In p-type material the energy bands are bent downwards near a dislocation. Early work by Shockley (1953) and Read (1954) suggested that the dangling bonds at a dislocation with an edge component would act as acceptors in a diamond lattice semiconductor. However, screw dislocations which have no dangling bonds exhibit similar electrical properties and Heine (1966) proposed that the distortion in the bandgap is caused by the local decrease in atomic density at the dislocation.

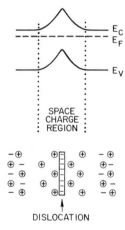

Fig. 4. Schematic diagrams of the effect of a single dislocation on the energy band and change in charge distribution in n-type silicon.

Read (1954) first showed that dislocations can have an effect on the conductivity of germanium, the effect being anisotropic with the dislocation direction. This effect has been further investigated by Glaenzer and Jordan (1969b) for silicon and the results for typical samples are shown in Fig. 5. The samples were cut from dislocation-free or float zone material containing

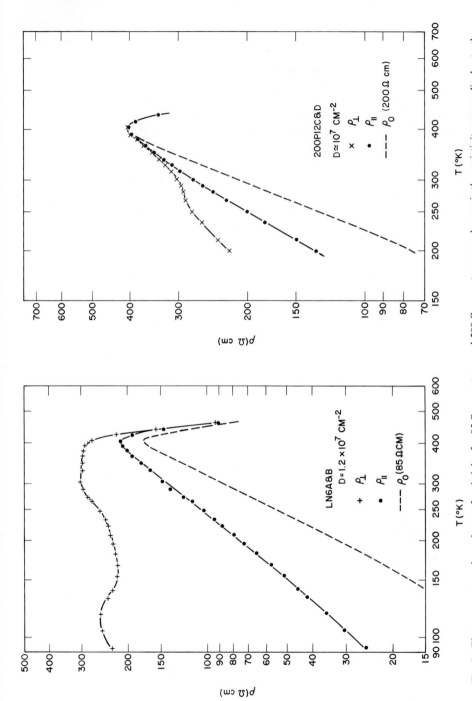

Fig. 5. The temperature dependence of resistivity for 85 Ω cm, *n*-type and 200 Ω cm, *p*-type samples. $\rho_\perp$ is the resistivity perpendicular to the dislocation array. $\rho_\parallel$ the resistivity parallel to the dislocation array, and ρ_0 the resistivity prior to dislocation formation (from Glaenzer and Jordan, 1969b).

10^4 dislocations/cm^2 and further parallel arrays of edge dislocations were introduced by plastic deformation at 750°C in vacuum (Green *et al.*, 1968). The anisotropy of the resistivity data shown in Fig. 5 demonstrates the existence of space charge cylinders surrounding the dislocation as shown schematically in Fig. 4.

Minority carrier lifetime is an important parameter in device performance, being directly related to transistor gain (Grove, 1967). The effect of dislocations has been studied by various workers and a composite of their data is shown in Fig. 6, and considering the sample preparation variables involved the agreement between the groups of data is excellent. At low dislocation levels the point defects in the crystal act as recombination centers which reduce the carrier lifetime. Increasing dislocation levels tends to getter the point defects, increasing the lifetime until the dislocation density becomes so high that dislocation-dominated recombination becomes important. This dependence of minority carrier lifetime is an excellent example of the fact that the best electrical properties may not necessarily be achieved in dislocation-free silicon.

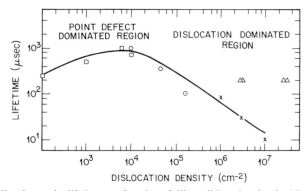

Fig. 6. Minority carrier lifetime as a function of silicon dislocation density (○, from Lemke, 1965; □, from Noack, 1969; ×, from Kurtz *et al.*, 1956; ΔΔ, from Glaenzer and Jordan, 1969a).

Lawrence (1968a) has investigated the effect of clean dislocations on transistor characteristics and found it to be relatively slight. Using (111) *n*-type silicon substrates gettered with phosphorus glass to prevent metal precipitation, Lawrence introduced up to 5×10^7 dislocations/cm^2 by rapidly removing the wafer from a 1270°C furnace. These dislocated wafers were used to fabricate both narrow (0.5 μm) and wide (4.5 μm) basewidth transistors by standard thermal diffusions and the electrical characteristics of these devices are shown in Table I and Fig. 7. The reverse leakage current introduced by the dislocations is relatively small (250 pA) at 10 V and much less than samples contaminated by copper precipitates. Small changes in

TABLE I

TRANSISTOR CHARACTERISTICS WITH HIGH AND LOW
DISLOCATION DENSITY[a]

	Low dislocation $(<10^3 \text{ cm}^{-2})$	High dislocation $(>5 \times 10^7 \text{ cm}^{-2})$
I_{BC} at 150 V (nA)	3.0	6.7
Gain β ($I_B = 100 \ \mu A$)		
Narrow base	274	201
Wide base	3.2	2.9
Storage time		
Narrow base (μsec)	1.66	0.49
Wide base (μsec)	1.10	0.63

[a] From Lawrence (1968a).

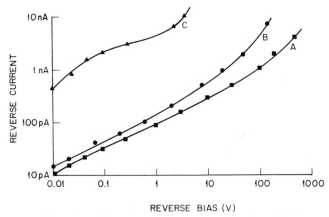

Fig. 7. Reverse bias electrical characteristics of p–n junctions with differing amounts of crystalline disorder. (A) $<10^3$ dislocations cm^{-2}; (B) $>5 \times 10^7$ dislocations cm^{-2}; (C) copper percipitation (from Lawrence, 1968a).

transistor gain and storage time were also observed with the narrow base-width devices being more susceptible. However, the high dislocation density did not result in catastrophic degradation of the transistors.

B. Defects

The presence of microdefects and inhomogeneities in silicon and germanium crystals has been observed for many years. However, with the advent of low dislocation silicon and the concurrent decrease in interaction between dislocations and defects (both vacancy and impurity related), the latter have become more readily observed and assumed a greater significance in recent years.

A variety of techniques have been developed to display the presence of defects, particularly in silicon, including surface etching solutions (Sirtl and Adler, 1961; Plaskett, 1965; deKock, 1970; d'Aragona, 1971; Ravi and Varker, 1973; Abe et al., 1973), metal decoration and x-ray topography (deKock, 1970, 1971; deKock et al., 1973), and electron microscopy (Ravi and Varker, 1973, 1974a; Bernewitz et al., 1974; Grienauer et al., 1974). The surface etching technique, in conjunction with crystal sectioning, has received the widest use. This technique, however, is largely restricted to locating the position of the defect and does little to determine the nature of the problem. As a result a myriad of literature has been published on a variety of etching effects, the terminology for which is often very confusing. The situation has recently improved somewhat with surface etching being used to position other analytical techniques such as electron microscopy (Ravi and Varker, 1973; Bernewitz et al., 1974; Grienauer et al., 1974) and infrared spectroscopy (Abe et al., 1973), and this has led to structure and composition related disgnostics of the defects.

The effect of Sirtl etching (Sirtl and Adler, 1961) (1 part 33 wt% CrO_3 with water, 1 part HF) on a dislocation-free silicon wafer is shown in Fig. 8. A series of circular regions contrasted by the etching treatment, appear across the wafer. At higher magnification two distinct surface morphologies are present, one a continuous layer structure and the other a circular pattern made up of discrete etch pits. For clarity the former are defined as *striations* and the latter as *swirls*. As can be seen from Fig. 8 both have a very similar form and curvature.

Striations are impurity inhomogeneities associated with small changes in the growth environments. While not strictly microstructure, in the sense that precipitation has not occurred and the crystal although strained is not defective, striations will be briefly discussed here to clarify their separation from the defects causing *swirls*. The segregation of solute atoms between the melt and the crystal is very dependent on the growth conditions at the growth interface. Nonuniform conditions can result from variable heating across the melt or facet growth of the crystal which can lead to localized supercooling. The solute inhomogeneities which result from these nonuniform growth conditions may consist of intentional dopant additions or impurity atoms such as carbon and oxygen. Due to the rotation of the crystal, the effect is to build up a spiral pattern which shows up as the striations depicted in Fig. 8.

The first indications of a defect-related striated structure (i.e., a *swirl*) in silicon were published by Plaskett (1965). He extended previous work on impurity striations in undoped Czochralski silicon to oxygen-free float zone

Fig. 8. Sirtl etched wafer showing impurity distribution striations and etch pits caused by swirl defects (from Ravi and Varker, 1973).

material. By so doing he was able to reduce the effect of oxygen segregation which had been proposed as the major cause for striated structures in Czochralski crystals. The structure was observed by copper decoration followed by transmission infrared microscopy. No precipitates were observed in the outer 2 mm of the wafers of the crystal. However, when viewed longitudinally, they showed stria whose separation d was given by the relationship

$$d = f/\omega$$

where f is the growth rate and ω the crystal rotation rate.

More recently, swirls have been exposed by Sirtl etching and a variety of surface etching effects exposed. Several workers (Sirtl and Adler, 1961; Plaskett, 1965; deKock, 1970; d'Aragona, 1971; Ravi and Varker, 1973, Abe *et al.*, 1973) have detected shallow, triangular shaped etch pits in (111) silicon. In addition if the Sirtl etch is kept cold during the etching Bernewitz (Bernewitz and Mayer, 1973) has shown that the pit formation is preceeded by a hillock at the same location. In (100) silicon, Sirtl etching shows hillocks rather than pits (d'Aragona, 1971; Ravi and Varker, 1973). deKock (deKock, 1970, 1971; deKock *et al.*, 1974) has observed two types of etch pit of differing size, A clusters and B clusters. The latter were much smaller and only detectable by carbon replica transmission electron microscopy. Surface etching shows that the defects occur in layers with many multiple and overlaping pits occurring to display the depth distribution. The concentration (deKock, 1971) of the A defects has been reported at 10^7 cm^{-3} in the swirl layers with the B defects present with a concentration several orders of magnitude higher. B defects are also present out to the edge of the wafer unlike A defects.

deKock has developed two techniques to reduce the defect concentration. The first experiment (deKock, 1971) involved the growth of the crystal in an argon, 10% hydrogen atmosphere. Hydrogen is thought to interact with the point defects in the silicon and by so doing will reduce the formation of the A and B swirl-type defects. The hydrogen, due to its high diffusivity in silicon, is effective when added to the growth atmosphere. The rate of defect formation is dependent on the crystal pulling rate as shown in Fig. 9 falling to zero at growth rates exceeding 3 mm/min. However, one disadvantage to this technique is that hydrogen precipitation in the crystal can occur. This results in an increased brittleness which makes wafer breakage during later processing more frequent.

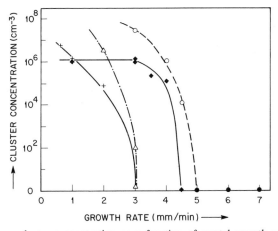

Fig. 9. Vacancy-cluster concentration as a function of crystal growth rate and growth atmosphere (○, B cluster, argon atmosphere; ◆ A cluster, argon atmosphere; △ B cluster, argon–10% hydrogen atmosphere; + A culster, argon–10% hydrogen atmosphere). (From de Kock, 1971 and de Kock *et al.*, 1973.)

The alternative approach (deKock *et al.*, 1973) for preventing swirl defects was to reduce the concentration of vacancies and oxygen introduced into the crystal during growth by increasing the growth rate. At first impression it would seem that this would be detrimental as less time would be available for defect diffusion. However, the process of "remelt" (Morizane *et al.*, 1967) is very important in defect inclusion which is more dependent on the microscopic growth rate at any point of space and time than the overall growth rate. It has been shown by Morizane *et al.* (1967) that the growth rate V at any point of a crystal in an assymetric thermal environment is given by

$$V = V_0(1 - \alpha \cos 2\pi Rt)$$

and

$$\alpha = 2\pi \, \Delta T R / V_0 G$$

where V_0 is the bulk growth rate, R the rotation rate, G the temperature gradient at the growth interface, and ΔT the temperature variation for one rotation. If α is greater than one, the growth rate will periodically become negative and this is associated with the crystal remelting. As a result, localized excessive variations in the growth rate will occur which, it is postulated, cause defects to be incorporated into the crystal (Webb, 1962). By growing the crystal sufficiently fast (with α proportional to V_0^{-1}) remelting can be prevented (Witt *et al.*, 1973). In addition, if R and ΔT are reduced, keeping the other conditions constant, swirl formation should be reduced. This was found to be the case for A clusters but not for B clusters. This suggests that somewhat different mechanisms may be involved for their formation. To illuminate this problem, deKock *et al.*, (1974) separated the effects of pulling speed and cooling rate by a series of quenching and variable growth rate experiments. In the former series of experiments the crystals were pulled at 1–5 mm/min and then the crystal and melt were separated very quickly to quench the lower part of the crystal. X-ray transmission topographs of longitudinal sections from crystals grown in this way showed that the B clusters were present closest to the melt/crystal interface with the A clusters further back up the crystal. The results of the experiments where the cooling rate and the growth rate were independently altered are shown in Table II. These results showed that in addition to remelt, the crystal cooling rate was also a major influence on the defect formation. Crystals grown fast (experiments 5 and 6) but cooled slowly showed defects while crystals grown slowly but cooled fast (experiment 8) showed none.

From these observations deKock concluded that the B clusters are formed first by vacancy condensation into vacancy–oxygen complex nuclei. The defect may be either a void or a superlattice of vacancies in the form of a regular array of Si═Si double bonds in the cubic lattice. As the defects become larger on further cooling they become unstable and collapse, re-

TABLE II[a]

THE INFLUENCE OF GROWTH RATE AND COOLING RATE ON VACANCY
CLUSTER FORMATION

Experiment	Growth rate (mm/min)	Cooling rate from the melting point according to growth rate (mm/min)	Cluster formation
4	5 (no remelt)	5	None
5	5 (no remelt)	3	B clusters
6	5 (no remelt)	2	A and B clusters
7	3	3	A and B clusters
8	3	5	None

[a] DeKock et al. (1974).

sulting in vacancy dislocation loops. Recent investigations using transmission electron microscopy (Bernewitz et al., 1974; Grienauer et al., 1974; Petroff and deKock, 1975) have provided additional information on the nature of the defects, suggesting that interstitial silicon atoms are probably more important than vacancies. The A clusters consist of large, perfect dislocation loops (1–10 μ) or a cluster of dislocation loops preferentially elongated along $\langle 110 \rangle$ directions. The dislocations, which were interstitial type without exception (Foll et al., 1975), generally lie on $\{111\}$ planes with a Burgers vector of $a/2\langle 110 \rangle$. In addition, Grienauer et al. observed small interstitial loops (500 Å) accumulated along the large loops and Petroff and deKock observed the presence of coherent or semicoherent precipitates unless the crystals were quenched, in which case no precipitation was observed. The latter workers also analyzed the character of the loops by the Burgers vector method of Maher and Eyre (1971) and found them to be interstitial rather than vacancy type. Only the latter type would be expected if the defect collapse model described in the previous paragraph was occurring, and the recent electron microscopy observations that interstitial defects are present suggests that the defect formation mechanism is rather more complex. In practical terms these swirl dislocation loops can cause additional problems in large diameter crystals in which the compressive stress caused by surface cooling can cause these loops to multiply via a Frank–Read source. This propagation can result in a significant dislocation density in crystals grown to be dislocation free (deKock, 1974). The B clusters were found to be of two types, both showing small semicoherent or coherent precipitates, the B_1 type having an anisotropic strain field and B_2 type a spherical strain field, the latter defect occuring an order of magnitude less frequently.

In addition to this work, impurity precipitation at swirl defects has been reported by Ravi and Varker (1973) who found platelet precipitates with a distinct superlattice in (111) silicon while in (100) silicon the precipitates were noncrystallographic and identified as α-cristobalite (SiO_2). The precipitation of oxide in this latter case suggests relatively high oxygen concentrations in the float zone silicon crystals used for the experiments. Carbon has also been found in the location of swirl defects by Abe *et al.*, (1973). By using a narrow (1 × 4 mm) beam infrared (16.5 μ) absorption technique (Newman and Willis, 1965), a distinct correlation between carbon concentration and surface etch topography was obtained. In addition, silicon crystals with a high carbon content showed a high incidence of circular etch patterns.

C. Precipitates

The rate of impurity precipitation in a crystal is related to the impurity concentration, the temperature of the crystal, and to some extent the number of crystal imperfections such as dislocations which can accelerate precipitation. For this to occur the impurity concentration must be sufficiently high that supersaturation occurs at a temperature at which the impurity atoms have significant diffusion rates.

In semiconductor silicon the major impurities are oxygen and carbon with lesser amounts of metallic impurities which are usually derived from the polycrystalline source material. With concentrations in the 10^{+13} cm^{-3} range or less, metal precipitation in bulk-crystal silicon is not significant. However, the combination of metallic contamination and process-induced imperfections can result in metallic precipitates during device processing. These precipitates, if located near p–n junctions can have a major effect on device characteristics and are discussed in Section III,B.

Oxygen and carbon concentrations depend on the crystal growth conditions. Oxygen contamination is a major problem in Czochralski silicon due to reaction with the crucible walls. The problem has been aggravated by the increased use of larger diameter crystals which require rapid rotation to ensure uniform dopant distribution. The rapid rotation, however, also accelerates the rate of oxygen dissolution from the crucible. The use of vacuum ambients during growth reduces oxygen contamination to some extent by lowering the oxygen partial pressure and also by encouraging oxygen removal as SiO from the surface by evaporation. Typical oxygen concentrations in Czochralski crystals lie in the mid 10^{16} to mid 10^{18} cm^{-3} range. Generally the larger the crystal diameter the higher the oxygen content because more rapid rotation is required to ensure dopant uniformity. Carbon is present in lower concentrations, the graphite heater blocks being generally considered as the major contamination source.

Float zone and pedestal techniques show significantly lower oxygen concentrations (10^{15}–10^{16} cm^{-3}) due to the lack of crucible contamination. The

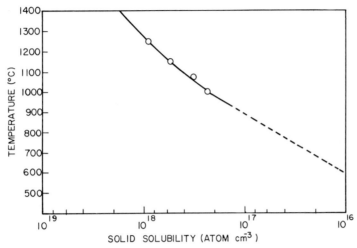

Fig. 10. Solid solubility of oxygen in silicon as a function of temperature (from Trumbore, 1960).

solubility curve for oxygen as a function of temperature is shown in Fig. 10 (Trumbore, 1960). From this it can be seen that supersaturation of oxygen in float zone silicon occurs at less than 600°C. At this temperature the diffusivity of oxygen is sufficiently low ($D < 10^{-20}$ cm²/sec, Grove, 1967) that aggregation will not occur. For carbon the figures for concentration and diffusivity are even lower and in consequence float zone and pedestal silicon is generally free of precipitation.

1. Oxide Precipitates

The effect of oxygen present in silicon crystals due to reaction with the quartz crucibles has been investigated for several years (Kaiser, 1957; Kaiser and Keck, 1957; Kaiser and Breslin 1958; Kaiser *et al.*, 1958; Lederhandler and Patel, 1957; Hrostowski and Kaiser, 1959; Patel, 1964, 1973; Patrick, 1970; Joshi and Howard, 1970). Fuller *et al.* found that the donor concentration in silicon increased on annealing at 450°C, rising to a maximum at approximately 100 h and then decreasing again (Fuller and Logan, 1957). This donor concentration was shown to be dependent on the oxygen concentration (Logan and Peters, 1957; Fuller and Logan, 1957; Kaiser *et al.*, 1958) and could be reduced by annealing in excess of 1000°C, at which temperature the oxygen aggregates into a second phase which can be detected by Tyndall light scattering (Kaiser, 1957). Kaiser *et al.* (1958) suggested a mechanism by which the oxygen on heat treatment formed a series of SiO_x aggregates until finally polymeric SiO_2 is formed. The principal donor species was $[SiO_4]$.

The precipitation rate is very dependent on heat treatment and, thus, processing subsequent to crystal growth. Below 450°C, precipitation is not observed (Patrick, 1970). However, local aggregation at this temperature must be postulated to explain the drop in donor concentration observed by Fuller and Logan (1957) and the changes observed by Patrick (1970) in the infrared spectrum. Temperatures above 1000°C cause precipitation to an increasing degree as observed by monitoring the dissolved oxygen using a variety of techniques (Fuller and Logan, 1957; Patrick, 1970; Patel, 1973). Patel (Patel and Batterman 1963; Patel, 1973) has used x-ray techniques to measure both the change in dissolved oxygen and to observe the precipitates. The results are shown in Figs. 11 and 12. Extending annealing at this temperature causes the precipitates to coalesce into larger groups up to 0.25 mm in diameter. Annealing at temperatures above 1300°C resulted in a decrease in the size and number of precipitates due to redissolution into the silicon. The time and temperature scales shown in Fig. 11 are well within the range of silicon integrated circuit processing and the effects of precipitation must be considered.

Shockley (1961) has postulated that the electrical characteristics of a $p–n$ junction are degraded if SiO_2 precipitation occurs in the space-charge region. Chynoweth and McKay (1956) have shown that avalanche breakdown in reverse bias $p–n$ junctions does not occur uniformly but at specific locations in the diodes. The local breakdown spots, called microplasmas, are made visible by the emission of light arising from the recombination of energetic

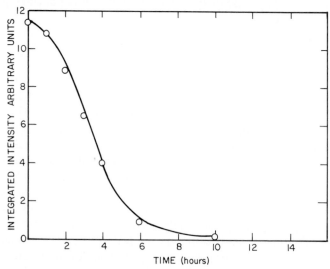

Fig. 11. X-ray anomalous transmission vs time at 1000°C for Czochralski silicon crystal containing 7.5×10^{17} cm^{-3} dissolved oxygen. (from Patel, 1973).

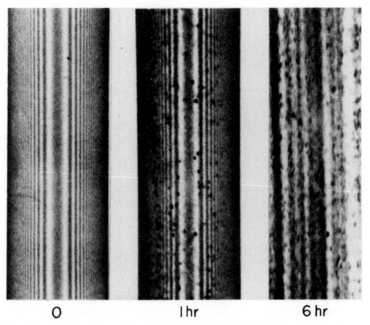

Fig. 12. X-ray section topographs after 0, 1, and 6 hr at 1000°C. These section topographs show the effect of oxide precipitation on the crystal perfection. The well-developed fringes indicate that the crystal is initially highly perfect. After 1 hr at 1000°C, dark spots have appeared. The spots appear to lie in bands, but Pendellosung fringes are still relatively clear. However, after 6 hr at 1000°C the fringes have completely disappeared and bands of defects are clearly seen. These defects are too small to resolve by x-ray topography. The size of the dark spots range up to about 10 μm. (From Patel, 1973.)

electrons and holes. Chynoweth and Pearson (1958) also observed that microplasmas tended to form where the space-charge region of the diode was intersected by dislocations, although only a small fraction of the dislocations caused light emission. Impurity segregation at these dislocations was considered as a possible cause for the microplasma by both Chynoweth and Pearson and also Shockley (1961), who suggested that the imperfection had to have two features to explain the observed effects. First, since the voltage across the junction only changes by a small amount (Senitzky and Moll, 1958) when the microplasma switches on, the imperfection produces a local electric field disturbance over a region in which the voltage drop is about 1 v, i.e., a few hundred angstroms. Second, to explain the characteristic "on–off" current behavior of the plasmas Shockley assumed that the imperfection had to contain traps which could immobilize high densities of charge. Based on this data, Shockley suggested that precipitation of SiO_2 in the space-charge region would explain the observed electrical effects the precipitation preferentially occurring at dislocations (Kikuchi and Tachikawa, 1960). The precipitate, due to its lower dielectric constant than silicon would disturb the local electric field as shown in Fig. 13. This field increase would cause the multiplication factor α to rise sufficiently so that the requirement for microplasma formation

$$\int_0^w \alpha[F(x)]\, dx > 1$$

would be met (Shockley, 1961).

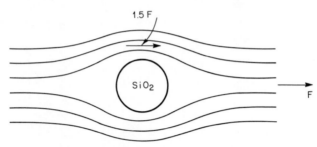

Fig. 13. The influence of a low dielectric constant SiO_2 precipitate on increasing the electrical field in the space charge region of a $p-n$ junction (from Shockley, 1961).

Batavin and coworkers (1967) provided some experimental verification for Shockley's theory by observing light emission from diodes fabricated from high-oxygen (8×10^{17} cm^{-3}) silicon substrates. Subjecting these samples to a 1050°C anneal to precipitate the SiO_2, they observed no light emission until the samples had been annealed for 7 hr. Thereafter, as the annealing time increased, the dimension and brightness of the microplasmas increased and their density decreased in accordance with an $N \sim E^{1/2}$ relation. The precipitates will only be effective in causing microplasmas if they terminate in the space charge region, in this case 1–2 μm wide. If the precipitate is outside, or totally penetrates the space-charge region, it will merely form a dielectric channel which has no effect. In consequence the observed annealing kinetics of microplasma formation can be qualitatively explained in terms of SiO_2 precipitate growth.

Interstitial oxygen which has precipitated as SiO_2 has also been associated with undesirable leakage currents in $p-n$ junctions (Batavin, 1970). Using dislocation-free silicon with an initial oxygen concentration of 6–8 $\times$ 10^{17} cm^{-3} and long annealing times at 1000°C, Batavin observed the effect of oxygen precipitation on leakage current as shown in Fig. 14. The observation that the junction started to harden up again after 50% of the oxygen had precipitated was explained, as described in the previous paragraph, on the growth of the precipitates to dimensions similar to the width of the space-

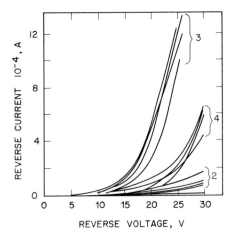

Fig. 14. The effect of oxygen precipitation on the reverse bias characteristics of a p–n junction: (1) no SiO_2 precipitation, (2) 12% SiO_2 precipitation, (3) 50% SiO_2 precipitation, (4) 100% SiO_2 precipitation (from Batavin, 1970).

charge region. In terms of device processing the time–temperature scales involved in oxygen precipitation are clearly important. Batavin observed a 50–70 hr incubation period at 1000°C before junction degradation (allowing the fabrication of his junctions by thermal diffusion without precipitation). However, the data available in the literature on the precipitation kinetics are very variable and reports of significant precipitation at much shorter times have been made (Patrick, 1970; Lederhandler and Patel, 1957; Pearce, 1974). It might be suspected that the dislocation density of the differing samples might be responsible; however, there is little agreement in experiments designed to examine this effect. Shul'pina *et al.* (1967) and Patrick (1970) found that dislocations occurring during crystal growth tended to retard oxygen precipitation, while Pearce (1974) found a marked increase in the precipitation rate if the substrate was plastically deformed during annealing. Until a more definitive picture is available, the interstitial oxygen content of the silicon crystal should be carefully monitored, particularly as the wafer diameter is increased.

Junction degradation is often masked in silicon integrated circuits because the active areas are located in an epitaxial layer deposited onto the bulk crystal substrate. However, the quality of the epilayer itself is very dependent on the amount of oxygen precipitation in the substrate. Precipitation in the substrate results in dislocation networks and stacking faults (Patel, 1964) due to the stress set up in the lattice due to the 10% increased specific volume of SiO_2 over silicon. Patrick (1970) has shown that these lattice imperfections degrade the quality of the epitaxial layer. The data are presented in Table III. Wafers 1, 2, 5, 6, 7, and 10 which were either oxygen free or with the oxygen in solution provided relatively defect-free epitaxy. Wafers 3, 4, 8, and 9 had precipitation damage which caused defective epitaxy. Wafers 3 and 4 were

TABLE III

STACKING FAULT AND ETCH PIT DENSITIES
AFTER EPITAXIAL DEPOSITION[a]

Wafer	State of oxygen before deposition	Stacking faults (cm^{-2})	Etch pits (cm^{-2})
1	Dissolved	140	350
2	Dissolved	100	950
3	Precipitated	9100	250,000
4	Precipitated	1200	240,000
5	Oxygen-free	260	820
6	Dissolved	400	970
7	Dissolved	270	4,000
8	Precipitated	4700	1,200,000
9	Precipitated	9300	740,000
10	Oxygen-free	930	690
3'	Dissolved precipitates	2700	1,400,000
4'	Dissolved precipitates	1300	100,000

[a] Patrick (1970).

annealed at 1300°C for 60 min to dissolve the precipitates. However, sufficient residual damage remained to prevent much improvement. Sylwestrowicz (1962) and Patel (1964), using yield point measurements, have shown that recovery occurs at 1350°C suggesting that precipitate dissolution at this temperature might be effective in improving the epitaxial quality.

The presence of dislocations in the silicon has been shown to accelerate the precipitation rate (Lederhandler and Patel, 1957). The dislocations act as nucleation sites for precipitation and light scattering experiments of deformed samples, the dislocations in which are generally oriented in one direction, suggests that the precipitates have a similar orientation.

2. CARBIDE PRECIPITATES

Carbon precipitation has been less studied than oxygen although present as a supersaturated impurity in silicon with typical concentrations of around 3×10^{15}–3×10^{17} cm^{-3} depending on growth technique (Nozaki et al., 1970). Precipitation as βSiC has been observed (Newman and Wakefield, 1962) in Czochralski silicon containing 10^{18} atoms/cm^3. The precipitates, which were large enough to be observed by optical microscopy, were detected in crystals heated in the temperature range 1000–1250°C and slowly cooled. Above 1250°C precipitation was lighter and confined to dislocations. In addition to precipitation, large dislocation loops on $\langle 111 \rangle$ planes were observed and postulated to be due to the vacancy generation which occurs when the substitutional sited carbon is precipitated.

In general, the precipitation rate of carbon is very slow and appears to require some form of nucleation, possibly oxide precipitates (Bean and Newman, 1971) as it is not discovered in oxygen-free silicon. When precipitation does occur it has been observed to degrade the reverse bias characteristics of $p-n$ junctions (Akiyama et al., 1973).

In extreme cases where the carbon content of the melt is very high, the liquid can reach a silicon/carbon eutectic (Voltmer and Padovani, 1973). At this point, with a melt composition containing 1×10^{19} cm^{-3} of carbon, two-phase freezing will occur producing silicon carbide inclusion into the silicon crystal. The resulting lattice strain will generate dislocations which degrade the crystal perfection.

III. Process-Induced Imperfections

The fabrication of an integrated circuit from the silicon single crystal involves many steps. Initially the crystal has to be cut and polished into wafers, referred to in this section as wafer shaping operations, followed by a series of fabrication steps to construct the planar active devices which constitute an integrated circuit. For convenience in this section the defect producing steps have been broken down to thermal, epitaxial, oxidation, and diffusion processes. The latter three processes also involve high temperatures (1000°C plus) and the dislocation and precipitation processes discussed under thermally induced defects can equally well occur during the epitaxy, oxidation, or diffusion. Finally, ion implantation is discussed. The combination of these processes include all the fabrication steps which can cause crystal damage which is likely to be detrimental to device performance. The final processing steps such as metallization, separation, bonding, and packaging are all completed at low temperature, and any crystal damage, for example, due to the separation procedure or beneath the wirebonding pads, should not propagate into the active area of the device.

A. Wafer Shaping Operations

Wafer shaping procedures are taken to include all the processes involved in transforming the crystal boule into a large number of polished wafers ready for subsequent processing. The processes are all largely mechanical and include grinding the boule to the correct diameter, sawing the boule into wafers, and finally lapping and polishing one or both of the wafer surfaces. In general, the mechanical and abrading processes used result in crystal damage which can be divided into two classes; macroscopic material damage (chipping and scratching) and microscopic crystal imperfections such as dislocations. The use of progressively finer polishing materials reduces the former, while the surface layers containing the dislocations can be removed by chemical etching.

The degree of surface perfection required varies widely for differing devices (Soper, 1970). A simple diode may only require the etching of the sawn wafer while the silicon targets used in vidicon image tubes must be shaped to a high degree of accuracy, polished flat, parallel, and crystallographically damage free.

The effects of mechanical treatment have been observed using a variety of techniques. Metallographic (Thomas, 1963; Pugh and Samuels, 1964) techniques such as angle lapping and chemical etching reveal a depth profile of defects to optical microscopy, and transmission electron microscopy (Stickler and Booker, 1962, 1963; Thomas, 1963; Schwuttke, 1974) has been used to observe and directly characterize the damage. The effect of damage on the crystal etch rate has also been used to determine the average depth of the damage layer (Buck, 1960; Stickler and Faust, 1966). In addition, nondestructive techniques include x-ray topography (Turner et al., 1968; Joshi and Howard, 1970; Schwuttke, 1973), optical reflectance spectroscopy (Sell and MacRae, 1970), ion scattering (Buck and Meek, 1970), and a variety of indirect electrical measurements (Buck, 1960; Taloni and Rogers, 1970). These latter techniques indicate evidence of damage but provide little data on the microscopic nature of the damage.

Schwuttke (1974) has used a combination of experimental techniques including transmission electron microscopy and x-ray topography in conjunction with MOS capacitance relaxation studies to investigate the effect of residual wafer shaping damage on device performance.

In dealing with the damage caused by the wafer shaping process we will mainly deal with the damage which remains towards the end of the process, as much of the macroscopic damage can be removed by suitable polishing and etching techniques. However, for completeness, the damage introduced during each step sill be briefly considered.

The use of automatic handling equipment requires wafers of a set diameter and to achieve this the crystal boule is centerless ground. An orientation flat is generally ground at this stage of the process. Unless very severe abrasion is used or a high temperature allowed to develop, the grinding damage is restricted to the periphery of the slice. As such, these will not cause device problems, but peripheral damage acts as a source of dislocations which can spread into the wafer during subsequent high-temperature processing and if severe enough can reduce the mechancial strength of the slice and increase breakage losses.

Surface damage is very dependent on the sawing operation. Sawing the boule into slices is most commonly an abrasive cutting process using a diamond cutting wheel. Other less damaging cutting techniques such as chemical (Madin and Asher, 1950; Sheff, 1967) and spark cutting (Livshits, 1960) are available but not much used as they do not produce flat parallel sided wafers. The depth profile of diamond wheel saw damage has been investigated (Joshi and Howard, 1970; Schwuttke, 1974) using etching and x-ray topography. This work showed two damage profiles, "uniform" damage which covered the whole wafer and "nonuniform" damage which was concentrated on the leading edge towards the saw cut. The relative depth of the damage is shown for each type in Fig. 15 as a function of sawing distance down the boule and the surface of the wafer being cut. The most important fact which can be drawn from this data is that, although the damage type, the face of wafer being cut, and the length of the boule are all variables, damage penetrates 30 ± 10 μm into the wafer under all conditions. This damage is removed by lapping and polishing with progressively finer abrasives.

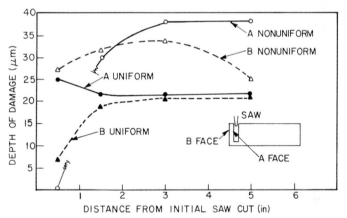

Fig. 15. (A) Schematic flow diagram of the sequential processing required for a T^2L epitaxial bipolar transistor. (B) Schematic flow diagram of the sequential processing required for a silicon gate NMOS transistor.

Lapping is used to remove the saw marks and to produce flat and parallel surfaces on the wafer. Its application, however, can further degrade the wafer by driving saw damage, such as dislocations; deeper into the wafer (Whitten *et al.*, 1964; Benson and Steward, 1968). To prevent this effect it is preferable to use a chemical etch to remove most of the saw damaged surface layer, followed by the most gentle abrasive technique compatible with the time available and the surface finish required (Buck and Meek,

1970). For optimum results this step can proceed directly to the polishing operation, obviating the need for lapping altogether. If lapping cannot be eliminated it should be followed by a careful cleaning procedure to ensure that any abrasive particles and contaminants are removed from the wafer surface (Martin, 1969) or subsequent high-temperature processing results in degraded device properties.

The microscopic damage caused by lapping silicon has been investigated using optical (Thomas, 1963; Pugh and Samuels, 1964) and electron microscopy (Stickler and Booker, 1962, 1963) and has been found to be due to the formation of surface cracks with dislocation arrays extending beyond and below the crack. The dislocation arrays are probably formed as "dislocation cracks", first proposed by Allen (1957) in indium antimonide, whereby a high local stress causes a crack to form which immediately closes. Any resulting atomic mismatch results in a wall of dislocations. Subsequent annealing of the lapped samples propagates dislocations from the initial networks into the surrounding regions (Stickler and Booker, 1963). The damage depth is dependent on the size of the abrasive particles, as shown in Fig. 16. It is particularly deep for fixed abrasive particles (e.g., silicon carbide papers) as opposed to similar sized, but free (e.g., diamond powder) abrasives.

The extension of abrasive materials to smaller diameters is referred to as

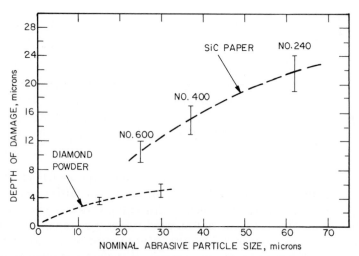

Fig. 16. Damage profile for abraded silicon showing damage depth as a function of abrasion treatment (from Stickler and Booker, 1963).

polishing. As polishing is often the final stage in the mechanical wafer shaping operation any damage induced at this point is of considerable interest. The use of diamond or alumina powder, although providing an optically perfect surface, results in surface scratches which can be observed by electron microscopy (Stickler and Booker, 1963) or by enhancement etching (Pugh and Samuels, 1964). Stickler and Booker found that in addition to the shallow grooves single dislocations were present at very high densities (up to $10^{10}/cm^2$). For diamond powder (0.25 μm diameter) the dislocation penetrate about 2000 Å into the slice and are probably caused by very high localized strain caused by the rolling action of the abrasive particles. With the use of Linde B (alumina) the depth of damage was generally less.

In addition to the purely mechanical abrasion of diamond and alumina polishing a mechanical/chemical removal procedure commonly referred to as Syton (trade name, Monsanto) polishing is often used. In this processs a colloidal suspension of silica gel in sodium hydroxide is used as the polishing medium. The polishing action of the colloidal suspension of silica is not well understood but experimental data suggest that a chemical removal process is important (Holland, 1964). Kaller (1959) has suggested that the polishing action results in a layer of silica gel on the wafer surface, this layer being continually removed by a chemical reaction with the polishing agent. This technique provides an almost damage-free surface although a thin oxide/contamination film remains on the surface after polishing (Adams and Kaiser, 1970). The superiority of Syton polishing over alumina powder was well displayed in the data of Buck and Meek (1970), shown in Fig. 17, using 100-keV He$^+$ ion scattering. The amount of backscattering in the 62–63-keV region (silicon peak) is dependent on the degree of crystalline disorder at the silicon surface. As shown in Fig. 17 the surface polished using Linde B (alumina) shows significantly higher backscattering yield at the silicon peak than either the Syton polished specimen or the chemically etched sample used as a standard. In addition, scattering at higher energies shows heavy metal contamination (especially Fe at 78 keV) for the Linde B polished sample.

Crystal damage and contamination introduced into the wafer by lapping and polishing the *back* surface of the wafer can be propagated through the wafer and reduce the device yield (Wang *et al.*, 1970). The percentage yield of good chips, using a NAND/NOR gate device as a test vehicle, fabricated from wafers with various back surface treatments is shown in Table IV.

The lapped devices were showed to have the lowest yield and when dynamically tested at temperatures from -55 to $+125°C$ were characterized by soft junctions and high output leakage currents. These results may well have been due to diffusing contamination effects following inadequate cleaning after lapping. Similar effects on MOS devices have been detected

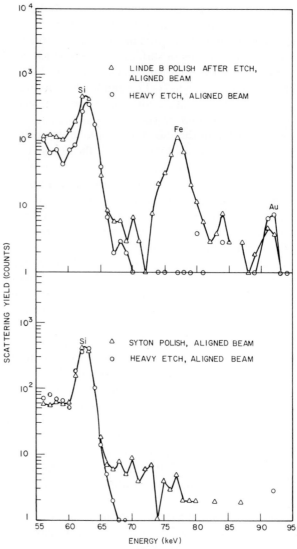

Fig. 17. Scattering yield versus energy of He^+ ions backscattered from etched, Linde B and Syton polished silicon surfaces (from Buck and Meek, 1970).

by Nigh (H. E. Nigh, private communication). X-ray transmission topographs of mechanically and chemically back-polished wafers showed the device yield to be very dependent on slip dislocation density, the lower the density the higher the device yield.

TABLE IV

Yield of Circuits as a Function of Back Surface Treatment
of the Wafer[a]

Back surface treatment	Total circuits tested	Number of acceptable circuits	Yield (%)
Chemical polish	1675	530	31.6
Mechanical–chemical polish	1474	648	44.0
Lapped	4108	1026	25.0

[a] Wang et al. (1970).

B. Thermally Induced Defects

It is not easy to separate defects caused by thermal treatments from other process-induced defects. Many of these processes such as epitaxy, diffusion, and oxidation involve the use of high temperatures and thermally induced defects can be introduced into the wafer in addition to or in conjunction with the specific defects linked with that process. However, two defect processes—namely, thermally induced dislocations and metal precipitation—are described in this section because they can occur throughout semiconductor processing and continuous vigilance is required to prevent their formation.

1. Thermally Induced Dislocations

Large temperature variations across a silicon slice can set up thermal stress sufficient to generate slip and cause dislocations. This can occur if adequate care is not taken in heating and cooling the wafer or if the thermal environment caused by the substrate fixturing is nonuniform. Star-shaped dislocations similar to those observed in thermally shocked crystal boules are observed by Matukura and Mirua (1963) and Henderson (1964) and subsequently Fairfield and Schwuttke (1966) reported the importance of thermal shock in dislocation generation. The formation of thermal shock induced dislocations has also been observed by Jungbluth and Wang (1965) during epitaxial deposition. These dislocations are important in device performance and the comparison of yield maps with slip patterns clearly show (Plantinga, 1969) that transistor problems occur in the region of sliplines. Thermally induced dislocations (in contrast to misfit dislocations) run from the front surface of the wafer to the back and as a consequence any precipitation at these dislocations will be particularly favorable in terms of shorting or softening junctions. The dislocations are also affected by the nonuniform stresses induced by the growth or deposition of thin films on the silicon surface (Hu, 1975a).

Processing which leads to thermally induced dislocations can also cause wafer warpage. When the stress in the wafer exceeds the yield point of silicon, dislocations form and plastic deformation occurs to partially relieve the stress. However, when the wafers have cooled to room temperature a reversed stress distribution which cannot be relieved by plastic deformation, will be present. This stress causes the wafer to buckle, destroying its planar shape and leading to photolithography problems in subsequent pattern delineation steps.

Morizane and Gleim (1969) and Hu (1973a) have studied the thermal environment which can lead to slip in slices during processing. Slip is particularly severe if the wafers are held in vertical rows as the centers of the wafers tend to be shielded from radiation heating by their neighbors and because the slice is locally heat sinked to the carrier. More uniform heating will be achieved if the slices are layed horizontally on a flat plate, ensuring uniform

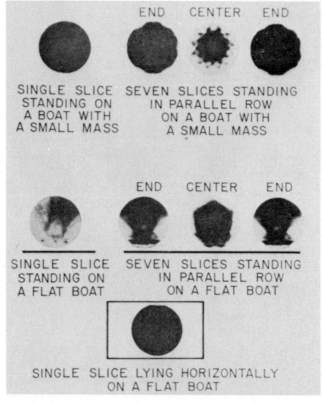

Fig. 18. Slip patterns resulting from type of boat used and the loading position of the slice (from Morizare and Gleim, 1969).

radiant heating and thermal contact. This effect is shown in slip patterns (Fig. 18) from the work of Morizane and Gleim (1969) for slices which were were pushed into a 1200°C furnace. However the throughput requirements of the semiconductor industry are such that the higher capacity of vertically stacked wafer systems makes their use obligatory and significant studies have been made to prevent slip formation under these conditions.

The dislocation structures produced by thermal stress have been investigated by Hu (1973a). Using Sirtl etching and copper decoration after the wafers had been rapidly pulled from an 1100°C furnace, arrays of $\langle 110 \rangle$ 60° type dislocations were observed on (111) planes as shown in Fig. 19. The dislocation sources appeared to be irregularities on the surface, the dislocation starting as a triangular loop of a screw segment and a $\langle 110 \rangle$ 60° dislocation segment. If the 60° segment propogates, a new segment is created breaking away from the source to form a trapezoid, the process repeating itself to form a new cycle of dislocation generation. A row of wafers cools more rapidly on the outside of the slice and in consequence the lattice tends to contract near the edge. The stress set up by this contraction can be relieved by the generation of dislocations, the density of which is given by the approximate expression (Billig, 1956)

$$\rho = \frac{\alpha}{bt} \frac{\partial T}{\partial r}$$

where α is the coefficient of thermal expansion, b the Burgers vector, t is a unit vector in the tangential direction, and $\partial T/\partial r$ is the temperature gradient. Assuming a typical temperature gradient figure of 120°C/cm, Hu's calculations give an expected dislocation density of about 10^4 cm^{-2}, which is close to that actually observed. Repeated thermal cycling causes this value to increase by one to two orders of magnitude and the formation of dislocation tangles.

One way of preventing thermal shock to the wafers is to control the rate of temperature rise either by ramping the furnace temperature and/or by slowly moving the wafers into or out of the furnace. In determining the correct "load" temperature before ramping the furnace, the prehistory of the wafer must be known. A dislocation-free wafer has a threshold temperature of 1100°C (Porter, 1970; Porter et al., 1972) but Patel and Chaudhuri (1963) have shown that the yield stress can be lowered by as much as 40% by dislocation densities on the order of 10^3 cm^{-2}, and in consequence lower load temperatures may be considered advisable (Schwuttke, 1974).

A highly effective alternative is to use a quartz covered boat to hold the wafers, an example of which is shown in Fig. 20 (Porter, 1970; Porter et al., 1972; Schwuttke, 1974). Using this technique vertically stacked wafers can

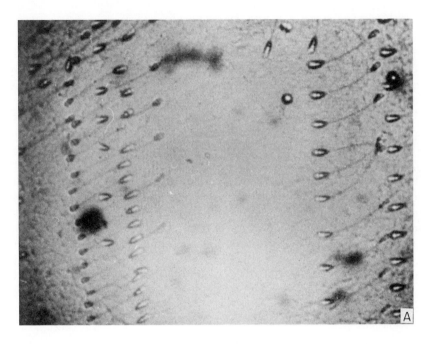

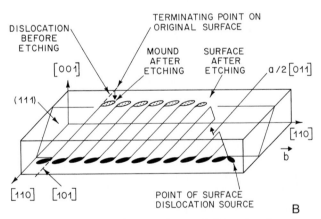

Fig. 19. (A) ir photomicrograph of copper decorated dislocations in a thermally stressed silicon wafer. The larvalike figures attached to both ends of each dislocation are etch mounds on opposite surfaces of the wafer. (B) Structure representation of the dominant type of dislocations and the mechanism of generation by thermal stress. (Figures from Hu, 1973.)

be loaded and unloaded directly at 1200°C without introducing significant slip, the boat geometry acting to prevent excessive temperature variations across the wafer. In addition, wafer warpage is significantly reduced, as shown in Fig. 21 (Schwuttke, 1974).

Fig. 20. Quartz boat and cover (from Schwuttke, 1974).

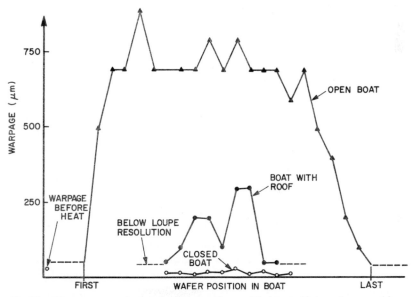

Fig. 21. Warpage measurements for 2 inch wafers rapidly inserted into and removed from a 1200°C furnace. (from Schwuttke, 1974).

2. METAL PRECIPITATION AND GETTERING

The high diffusivity (Fig. 22A) and steep solid solubility function with temperature (Fig. 22B) make transition metal impurities in silicon very susceptible to precipitation after any thermal treatment. Generally, the precipitation will occur preferentially at lattice deformation sites such as defects

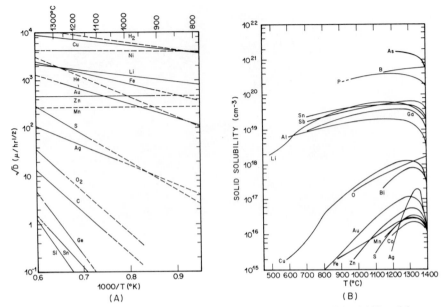

Fig. 22. (A) Diffusivity of miscellaneous impurities in silicon. (B) Solid solubility of dopant and impurity atoms in silicon. (From Grove, 1967.)

and this explains the often observed fact that one lattice defect will degrade a semiconductor device while a similar adjacent defect has no effect on the device performance. The preferential precipitation at lattice strain is the basis for "gettering," the most effective cure to this problem. This technique involves generating lattice damage generally on the back surface of the wafer followed by a thermal treatment. At high temperature, the impurities diffuse to the back surface and are precipitated at the lattice damage and away from the active areas of the device located on the front side of the wafer.

Goetzberger and Shockley (1959, 1960) first showed that junctions treated with metals (Au, Cu, Mn, Fe) to cause precipitation showed rounded or "soft" reverse characteristics caused by excess current below the avalanche voltage. The excess current was related to the reverse voltage by the expression $I \propto V^N$, where $N = 3$–7. A typical example is shown in Fig. 23. Using a small voltage probe to plot the voltage contours over the area of the junction they were able to show that the excess current flowed through a single area of the diode. Later experiments using a light probe (Haitz et al., 1963) showed that the soft characteristics were caused by a field–sensitive carrier recombination multiplication phenomenon initiated at the high electric field at the tip of a metal precipitate which had extended into the space-charge region.

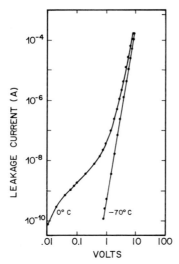

Fig. 23. Reverse IV characteristics of a "soft" junction containing copper precipitates (from Goetzberger and Shockley, 1959).

Lawrence (1965) extended these studies for copper and showed that the shape of the reverse IV characteristics was also dependent on the type of precipitate formed. The soft characteristics were observed when copper precipitated onto stacking faults—generally uniformly for small faults ($<5\ \mu m$)—and concentrated around the periphery, where the lattice strain is greatest, for the larger faults. In contrast, copper precipitation at voids resulted in arrays of separate small precipitates of 50–500 Å in size and loosely distributed along the $\{110\}$ silicon planes. These precipitates do not cause soft characteristics ($I \alpha V^N$) but rather a two orders of magnitude increase in reverse bias current. The characteristics for devices with these two types of precipitate are shown in Fig. 24.

The observation of Hall and Racette (1964) that interstitially dissolved copper acts as a donor while precipitated copper acts as a triple acceptor has been used by Das (1973) to explain qualitatively copper induced "thermal conversion" effects. The presence of this type of copper in dislocation free n-type silicon will tend to reduce n-type conductivity possibly beyond the point of compensation. Subsequent process-induced defects precipitate the copper to an acceptor state reversing the doping effect. Similar precipitation effects on deep depletion–capacitance experiments (Heiman, 1967) have also been observed.

The morphology of copper precipitation in silicon has been studied by Das (1973). Several authors (Rieger, 1964; Hu and Poponiak, 1972; Schwuttke, 1961; Fiermans and Vennik, 1965, 1967a,b; Nes and Washburn, 1972)

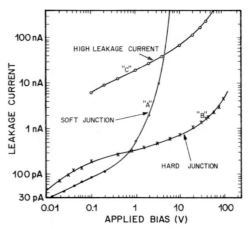

Fig. 24. Reverse IV characteristics of a junction containing (A) copper decorated stacking faults, (B) no defects, and (C) loosely packed {110} copper precipitates (from Lawrence, 1965).

have observed copper precipitation using relatively low-resolution techniques such as infrared microscopy and x-ray topography. A variety of "star" and "platelike" precipitates were observed which, when observed by Das using transmission electron microscopy, appeared as colonies or agglomerations of small spherical defects 50–500 Å in diameter. These defects were postulated to be covalently bonded copper–silicon structures with a diamond cubic lattice isomorphous with the silicon host lattice. This structure, which was identified by electron diffraction and Moiré spacing analysis, is not stable at room temperature and is thought to be stabilized by the strain energy of the precipitate/matrix system. However, increasing compressive stress ($A_{Si} < A_{Cu-Si}$) as the precipitates grow restricts the size of the precipitates to 500 Å.

A recent study of iron precipitation (Cullis and Katz, 1974) also showed that silicide formation is important, the precipitate structure being identified by electron microscopy as $\xi_a - FeSi_2$ (α-lebolite). As shown in Fig. 25 two defects were observed in poor devices, stacking faults (A) up to 7 μm in length and penetrating the (100) silicon lattice on inclined {111} planes and rod like defects penetrating the lattice along $\langle 101 \rangle$ directions up to 2 μm in depth. The rods occasionally occurred in isolation (B) but more commonly (C, D) in conjunction with irregular dislocation loops which contained small colonies of copper precipitates. The presence of this type of deep rod defect is clearly undesirable as $\xi_a FeSi_2$ exhibits metallic conduction and will short out any junction it penetrates.

It is also of considerable interest to review the precipitation kinetics which leads to large $FeSi_2$ rods but small, more numerous CuSi precipitates,

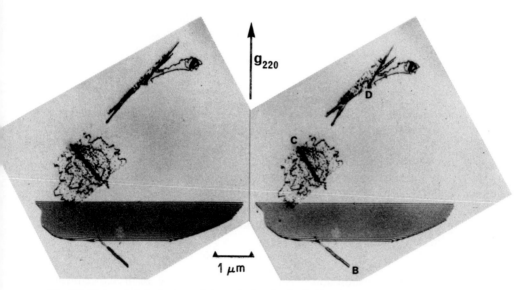

Fig. 25. Transmission electron micrograph showing inclined stacking fault (A) and $FeSi_2$ rod precipitates (B, C, D) (from Cullis and Katz, 1974).

under identical thermal treatments. Precipitation requires impurity diffusion, which can occur either by slow substitutional motion (I_s) or rapid interstitial motion (I_i), the equilibrium between these species being given by the expression (Frank and Turnbull, 1956)

$$I_s \underset{k_2}{\overset{k_1}{\rightleftharpoons}} I_i + V$$

where V is a lattice vacancy and k_1 and k_2 are rate constants. Under quasi-equilibrium conditions the behavior of this system can be derived (Penning, 1958; Seeger and Chik, 1968) to give a precipitation time constant τ, where

$$1/\tau = \tfrac{1}{2}\{(k_1 + k_2 C_i^e + k_v) \pm [(k_1 + k_2 C_i^e + K_v)^2 - 4k_v k_1]^{1/2}\}$$

with k_v the rate constant for vacancy annihilation and C_i^e the equilibrium impurity concentration. This expression can be reduced to

$$1/\tau \approx k_v(Cv^e/C_s^e)$$

assuming $k_1 + k_2 Cv^e \gg k_v$ and $C_s^e \gg Cv^e$, where Cv^e and C_s^e are the thermal equilibrium vacancy and substitutional impurity concentrations, respectively. For iron these conditions, where $C_s^e \gg C_i^e$ (Collins and Carlson, 1957), are met and using the expression from Penning (1959)

$$k_v = 2\,{}^2 D_v/t^2$$

where t is the effective wafer thickness and D_v the vacancy diffusion coefficient and from Kendall and DeVries (1969) the best estimates are

$$D_v = 1.4 \times 10^{-4} \exp(-0.33/kT) \quad cm^2 \, sec^{-1}$$

and

$$Cv^e = 6.5 \times 10^{29} \exp(-4.44/kT) \quad cm^{-3}$$

and a value of C_s^e (at 1050°C) of $8 \times 10^{15} \, cm^{-3}$ from Collins and Carlson (1957) one obtains values for τ in the region $\tau \approx 4 \times 10^3$ sec (for $t = 0.025$ mm, wafer thickness) to $\tau \approx 10$ sec (for $t = 10 \, \mu$, narrow surface zone). The correct value probably lies in between.

In contrast to iron, for copper $C_i^e \gg C_s^e$ (Hall and Racette, 1964; Meek and Seidel, 1975; Struthers, 1956) and the precipitation rate will be largely controlled by the interstitial diffusivity D_i. Under these conditions the precipitation rate at random dislocations may be estimated from the expression (Ham, 1959)

$$1/\tau \approx \beta N D_i \quad sec^{-1}$$

where β is a constant and N is the dislocation density. In the experiments of Cullis and Katz $N \approx 10^6 \, cm^{-2}$ (corresponding to Frank partial dislocations and dislocations at the rod defects) and using $D_i = 10^{-4} \, cm^2 \, sec^{-1}$ (Hall and Racette, 1964) and $\beta \approx 0.25$ (see Fig. 5, Ham, 1959), then $\tau \approx 10^{-2}$ sec.

From these calculations it is clear that all other things being equal, the rapid precipitation of copper, due to its lower τ value, would be expected to produce the multiplicity of small (50–500 Å) precipitates, whereas iron silicide is formed more slowly as larger more widely spaced rods.

The deleterious effects of metal contamination on silicon devices can be removed by one of a variety of processes under the overall title of gettering. Generally defined this process involves creating a sink for metallic impurities at some location on the wafer (generally the back surface, but it can be in areas of the front surface not used for active devices) and then annealing the wafer at a temperature sufficient that the metallic impurities diffuse to the sink and are preferentially precipitated at that region. A variety of gettering techniques have been used, including ion implantation (Masters et al., 1970; Buck et al., 1972, 1973; Hsieh et al., 1973; Seidel and Meek, 1973; Meek and Gibbon, 1974; Seidel et al., 1975a), glass layer doping (Goetzberger and Shockley, 1960; Ing et al., 1963; Mets, 1965; Murray and Kressel, 1967; Buck et al., 1968; Nakamura et al., 1968; Lambert and Reese, 1968; Lawrence, 1968b; Cagnina, 1969; Katz, 1974; Meek et al., 1975), and to a lesser extent back surface abrasion (Mets, 1965; Nakamura et al., 1968) and nickel plating (Goetzberger and Shockley, 1960; Lawrence, 1968b).

There are no exact time–temperature conditions for all gettering cycles— the optimum conditions depending on contaminating species, concentration, and the gettering technique used. For phosphorus glass gettering, annealing at 1000°C for 60 min or 1050°C for 30 min is usually sufficient. The efficiency of isochronal gettering as a function of temperature is shown in Fig. 26 (Lambert and Reese, 1968) for gold impurities along with the hole lifetime in the *n*-base region of a p^+-n diode gettered under the same conditions. A rapid decrease in the gold concentration in the wafer is observed for temperatures above 1000°C. This is not unexpected in view of the work of Seidel and Meek (1973) who showed that both gold and copper are largely

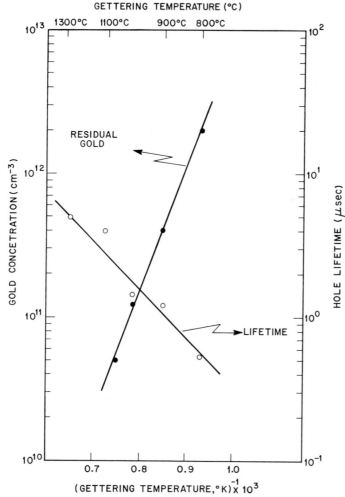

Fig. 26. Residual (ungettered) gold concentration in silicon wafer after 60-min isochronal phosphorus glass gettering treatment and the effect of the residual impurity on the hole lifetime in a diode gettered under the same conditions.

gettered onto substitutional sites in the silicon (not the glass overlayer) in the region of the phosphorus diffusion caused by the doping glass, rather than into the glass itself. The enhanced solubility of gold in n-doped silicon has been explained (Cagnina, 1969; Lambert and Reese, 1968) in terms of an ion pairing reaction of the type

$$Au_{sub} + e^- + P^+ \rightleftharpoons AuP$$

In consequence, the gettering efficiency is closely related to the phosphorus dopant profile in the back of the wafer and which is rather shallow unless drive in temperatures of above 1000°C are used. Lawrence (1968b) provided additional experimental evidence for this process by observing that copper precipitates are gettered much more efficiently by phosphorus glass in contact with the back of the wafer than if the doped glass is separated from the silicon by an additional undoped oxide layer. For iron contamination, Nakamura et al. (1968) found the reverse effect with the majority of the contaminant gettered into the glass rather than the doped silicon.

A direct comparison between phosphorus glass and ion implantation gettering has been made by Seidel et al. (1975a). In the latter case the copper and gold were found to occupy nonsubstitutional sites in contrast to their lattice position after phosphorus glass gettering. Thus, implantation gettering occurs by a different mechanism, related to the degree of crystallographic damage (see Section III,F) caused by the implantation (Hsieh et al., 1973; Seidel et al., 1975a), and because this damage is related to the implanted element, dose, and voltage and also to the subsequent annealing cycle, no one set of optimum conditions can be specified for gettering. Typical gettering conditions range in voltage from 50–250 keV at a dose of about 10^{15}–10^{16} cm^{-2}. The effectiveness of differing implanted elements in gettering gold impurities has been studied by Seidel et al. (1975a), and ranked in the order $Ar > O > P > Si > As > B$ using a 10^{16} cm^{-2} implant at 100 keV. This order is correlated with the disorder in the silicon after annealing which ranged from amorphous in the argon implanted samples to relatively little residual disorder for arsenic and boron.

The amount of gold gettered into the implanted layer was also found to depend on the annealing temperature, the results being shown in Fig. 27A along with similar data taken for phosphorus glass gettering. From this figure it can be seen that argon implantation gettering is more efficient for annealing temperatures below 1000°C. An example of its effective use, where the annealing temperatures were restricted to 900°C for design reasons, is in the gettering of the silicon photodiode array camera targets (Hsieh et al., 1973). These targets consist of a planar array of 800,000 p^+–n diodes which during operation are scanned by a low-voltage electron beam. Light incident on the back of the target will induce a photocurrent in the diodes and this is used to produce the image. Reverse current caused by defects in the

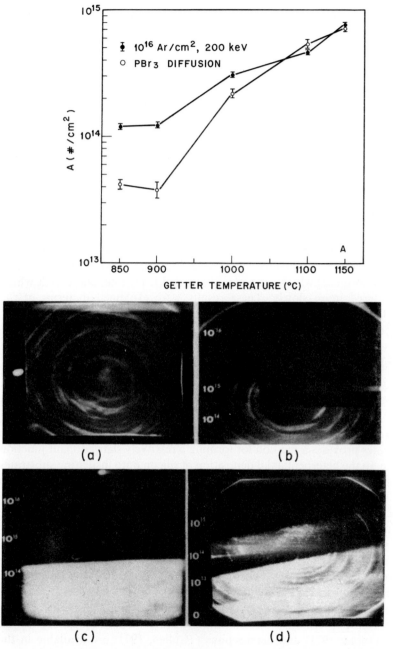

Fig. 27. (A) Gettered gold content (No./cm^2) for Ar implanted layers and phosphorus diffused layers as a function of the getter temperature. Sample thickness is ~200 μm; Au equilibrated at 1135°C (B) Twelve volt dark field video display of (a) phosphorus glass, (b) phosphorus implanted, (c) arsenic implanted, and (d) argon implanted targets gettered 30 min at 900°C.

silicon (as opposed to the impinging light) will cause spurious white spots to appear on the image. The requirements for satisfactory performance (reverse current per diode less than 5×10^{-14} A) were very severe and considerable care was taken to keep the number of process-induced defects in the silicon to a minimum (Buck *et al.*, 1968). However, defective diodes were observed in a cyclonic pattern in about 10% of the dislocation free float zone material used, as shown in Fig. 27B. This preferential impurity precipitation, probably related to the swirl patterns described previously, were resistant to phosphorus glass gettering at 900°C but were well gettered by implanted phosphorus, arsenic or argon. The relative efficiency of arsenic using a 900°C/30-min annealing cycle is in contrast to the work of Seidel *et al.* (1975a). This is probably due to the differing degree of recrystallization in the implanted silicon at the lower temperature.

Buck *et al.* (1973) have studied the rate at which transition metal impurities are gettered in ion-implanted silicon and find them to fall into two groups: copper and nickel which are gettered fast and iron, gold, and cobalt which are gettered more slowly. These results were explained in terms of the interstitial solubility and diffusivity of the impurities which are highest for copper and nickel.

C. Epitaxy

The ability to deposit epitaxial films onto both silicon and passive substrates (such as sapphire or spinel) has resulted in significant advances in device performance. In the former case, often referred to as homoepitaxy, the growth of epitaxial films with a dopant level different from that of the substrates provides a method for forming a relatively sharp dopant junction profile. In the heteroepitaxial deposition of silicon onto insulating substrates advantage is taken of the ability to physically isolate the transistors from each other by etching the epitaxial film down to the substrate, thereby obviating the need for junction isolation and reducing parasitic capacitances.

The epitaxial films are almost universally deposited by chemical vapor deposition at atmospheric pressure, either by the hydrogen reduction of chlorosilanes ($SiCl_xH_y$) or by the disproportionation of silane (SiH_4). Deposition at a temperature of between 1000 and 1300°C is usually used, frequently preceded by a gas-phase etching technique to remove any surface damage or contamination from the substrate. Suitable gas-phase dopants are also added to the reaction gas stream. Excellent reviews of homoepitaxy (Joyce, 1974; Gupta, 1971) and heteroepitaxy (Cullen, 1971) are available in the literature.

The crystallographic perfection of the epitaxial film is of great importance in achieving adequate electrical performance. Homoepitaxial films can now be obtained with a degree of perfection near to that of the substrate with dislocations, stacking faults, and various twin structures being the major imperfections. In heteroepitaxy, less perfection is possible due to the high density of dislocations caused by lattice and thermal expansion mismatchs between the film and substrate, resulting in poor minority carrier lifetimes. In consequence, there has been little development of a heteroepitaxial bipolar device technology, with the majority of effort being directed towards MOS devices where the more important electrical property, the surface mobility, is comparable to that measured in bulk silicon (Ronen and Micheletti, 1975).

1. HOMOEPITAXIAL MISFIT DISLOCATIONS

Misfit dislocations can occur close to the epitaxial/substrate interface if there is a marked difference in the dopant concentration of the substrate and the deposited film. The dislocations, called misfit dislocations, are caused by the expansion or contraction of the silicon lattice caused by the presence of the differing sized solute atoms. The effect is shown schematically for diffusions in Fig. 41, with the resulting edge dislocations lying parallel with the surface of the slice. Similar lattice contraction and expansion effects causing dislocation formation are observed in thermal diffusions and are dealt with in a later section.

Misfit dislocations, introduced by depositing low-doped boron epitaxial films onto highly doped boron (2×10^{18}–6×10^{19} cm^{-3}) substrates, have been studied by Sugita and coworkers (Sugita *et al.*, 1969, Tamura and Sugita, 1970, 1973) using x-ray topography and transmission electron microscopy. The effect of the boron doping on the silicon lattice is shown in Fig. 28 (Horn, 1955). The epitaxial films were deposited by the hydrogen reduction of silicon tetrachloride at 1140°C, onto [111] or [100] substrates which had been gas-phase etched just prior to deposition. Under these experimental conditions some redistribution of the substrate dopant onto the epitaxial film occurs and the dopant interface is not particularly sharp. In this respect the dopant profile and the resulting misfit dislocations strongly resemble those observed after thermal diffusions and which are described in a later section. The dislocations are observed in the regions of maximum concentration gradient which extends approximately one micron on either side of the epitaxial interface. Outside of this interfacial region both the substrate and the epitaxial film are relatively dislocation free.

The dislocation geometry was found to be dependent on the epitaxial film thickness. For thin films (3–5 μm) long straight dislocations were observed, oriented in $\langle 110 \rangle$ directions and lying nearly parallel to the film surface in the (111) plane. For thicker epitaxial films (>8 μm) more complex and higher density dislocation networks were observed. In addition to dislocations along $\langle 110 \rangle$ directions, additional ones with $\langle 211 \rangle$ orientations were also observed. These latter dislocations were caused by the intersection of two $\langle 110 \rangle$ dislocations which subsequently decompose into two threefold nodes joined by a third dislocation along the $\langle 211 \rangle$ direction. In addition, a small number of thicker samples showed irregular hexagonal arrays of $\langle 211 \rangle$ dislocations. The formation of these networks had been predicted by Holt (1966) as a major stress relief mechanism for heterojunctions in diamond lattices.

The density of misfit dislocations was found to be dependent on the dopant level difference between the film and substrates. Depositing a low-doped epitaxial film onto various conductivity substrates, no misfit dislocations were observed if the substrates were doped less than 5×10^{18} cm^{-13}. The dislocation density increased with the substrate doping reaching very high levels for 8×10^{19} cm^{-3} doped substrates at which the individual dislocations could not be resolved by x-ray topography. Quantitative dislocation density measurements made for intermediate misfit levels were somewhat less than those expected from the theoretical predictions derived from van der Merwe (1963), suggesting that a significant amount of the misfit is accommodated by elastic strain.

Sugita also observed that a critical epitaxial film thickness is required before misfit dislocations occur. The results are shown in Table V. The experimental observation of a critical film thickness confirms the van der Merwe model for misfit dislocation growth, which proposed that the critical value for misfit dislocation formation is related to film thickness and various physical parameters such as the shear moduli of the film and the substrate. For thin epitaxial films the misfit strain is absorbed elastically until the critical thickness is reached, at which point the misfit dislocations occur. The agreement between Sugita's data and the calculated values are good; for a substrate resistivity of 0.0035 Ω-cm (Table V) the calculated critical thickness is 1.0 μm compared with the observed value of 2.4–2.9 μm.

TABLE V

DEPENDENCE OF DISLOCATION DENSITY ON EPITAXIAL
FILM THICKNESS

Sample number	Substrate resistivity (Ω cm)	Thickness (μm)	Dislocation density
1	0.0031	0.3	0
2	0.0032	1.0	0
3	0.0035	1.5	0
4	0.0033	2.4	0
5	0.0036	2.9	Slight
6	0.0038	12.7	heavy

Nishizawa *et al.* (1975) have observed that misfit dislocations may be prevented by simultaneously doping the epitaxial layer with tin. This electrically inactive Group IV element has a larger atomic radius than silicon and will compensate for the lattice reduction caused by the smaller phosphorus or boron dopant atoms.

2. STACKING FAULTS

One of the more commonly observed defects in epitaxial silicon films are stacking faults. The faults generally initiate at a small interfacial growth region of film which is mismatched with the substrate. The faults subsequently grow along inclined {111} planes intersecting the surface along ⟨110⟩ directions giving rise to characteristic geometrical patterns after surface etching. The stacking faults can be extrinsic or intrinsic. The former may be considered as the introduction of an additional layer of atoms in the atom plane stacking sequence for example, ABCACBCABC while an intrinsic fault can form by the removal of an atom plane, ABCBCABC. The use of the simple stacking sequence for face-centered cubic lattice (abc) rather than the more complex stacking sequence (aa'bb'cc') for the diamond lattice of silicon is justified when considering epitaxial stacking faults (Hornstra, 1958). A variety of models have been proposed for the initiation of the stacking faults. Three models, the faulted deposited layer model of Booker and Stickler (1962), the mechanical displacement model of Finch *et al.* (1963), and the twinned deposited layer model of Mendelson (1965), are all basically similar. These models all propose an initial mismatch between the film and the substrate occurring at some imperfection on the substrate surface. A fourth model (Jaccodine, 1963) has suggested that the stacking faults are initiated by the collapse of a vacancy cluster caused by imperfections at the film/substrate interface. This model however has been critized for its inability to explain the occurrence of intrinsic stacking faults which are commonly observed, and may be more applicable to ion implantation/oxidation induced stacking faults discussed in Section III,D.

The growth of the stacking faults along the inclined {111} planes leads to an inverted polyhedra structure which is shown schematically in Fig. 29 and causes the characteristic etch formations at the surface. Various etching solutions, including Dash etch, iodine etch, and Sailer's solution have been used in conjunction with optical microscopy to display the stacking faults (Batsford and Thomas, 1962, 1963; Booker and Stickler, 1962; Charig *et al.*, 1962; Light, 1962; Quiesser *et al.*, 1962; Chu and Gavaler, 1963; Finch *et et al.*, 1963; Mendelson, 1964, 1966). The geometry of the polyhedra and the resulting etch figure is dependent on the substrate orientation. A comprehensive study of this dependence has been made by Mendelson, who observed equilateral triangles (111, 110, 221, 334), isosceles triangles (112), square (100), or trapezoid (114) etch figures depending on substrate orientation. From the geometry of Fig. 29 it is also clear that if the stacking faults are initiated at the substrate/film interface the size of the etch figure will have a direct geometrical relationship with the film thickness, and as the etching progresses the figures will become smaller and smaller until they vanish when the epitaxial film has been completely etched away. This is almost universally observed and provides strong evidence for interfacial initiation of the faults. Occasionally, etch figures of differing size are observed (Chu and Gavaler, 1963, Booker and Stickler, 1962), suggestive of nucleation during film growth. This presumably occurs due to contamination from the reactant gases or deposition equipment and is not very common.

In addition to the closed polyhedra shown in Fig. 29 a variety of other etch figures can occur. A variety of these are shown in Fig. 30 (Booker, 1964). The open figures occur when the stacking fault does not propagate on all of the {111} planes giving rise to partial geometrical figures and line defects. Line defects consisting of two close parallel dissimilar defects (Booker and Howie, 1963) and small prismatic defects can also occur in closed figures. Furthermore, complicated etch figures can occur when the faults start to overlap. This will occur if the stacking fault density is very high and/or the epitaxial film thick. The various types of interaction are shown in Fig. 30. Identical faults (i.e., both extrinsic or both intrinsic) generally annihilate when they meet, because, although each is mismatched with respect to the surrounding lattice, they match each other on meeting and hence cancel out. A less common interaction between similar stacking faults is shown in Fig. 30 where one of the faults flips over just before interaction to form an extrinsic/intrinsic pair defect. Where an extrinsic and intrinsic fault interact, the extrinsic fault changes to intrinsic but continues to propagate along its original {111} plane while the original intrinsic fault is annihilated. This can be understood if, as previously mentioned, the extrinsic fault is

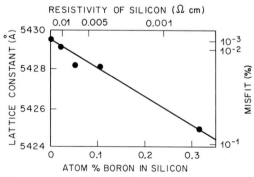

Fig. 28. Effect of boron concentration on silicon host lattice constants (from Horn, 1955).

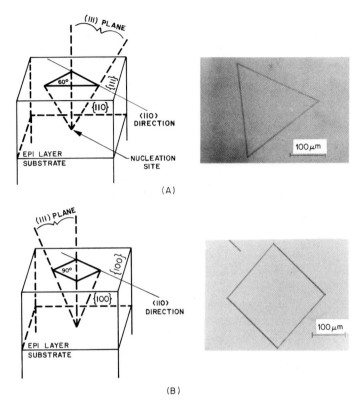

Fig. 29. (A) Geometry and optical micrograph after etching of stacking fault in epitaxial silicon film on {111} substrate. (B) Geometry and optical micrograph after etching of stacking fault in epitaxial silicon film on {110} substrate. (Optical micrographs from Chu and Gavaler, 1963.)

considered as two intrinsic faults. One of these faults is annihilated by the other intersecting fault, while the second continues to propagate.

The crystallography of epitaxial stacking faults have been carefully studied using transmission electron microscopy by Booker and coworkers (Booker and Stickler, 1962, 1963; Booker, 1964; Booker and Joyce, 1966). Booker observed that the stacking faults nucleate at growth areas mismatched with respect to the substrate and on {111} oriented substrates these areas could be considered as dislocation loops. The crystallography of the stacking fault depends on the loops, intrinsic faults growing from extrinsic loops and vice versa. Booker also observed that if the partial dislocation between the faults were to be the low-energy $a/2 \langle 110 \rangle$ variety then geometrical considerations would require that when two faults subtend an acute angle they be of the same type and when an obtuse angle of different type. This is shown schematically in Fig. 30 where in {111} silicon the stacking faults all are the same type, opposite to that of the nucleating dislocation loop, while in {100} silicon the faults are alternately intrinsic and extrinsic.

The effect of the epitaxial deposition conditions has a marked effect on the stacking fault density. Substrate orientation and preparation, reactant gas purity, and film growth rate have all been investigated.

The presence of dislocations in the substrate does not appear to be a favorable sight for stacking fault nucleation. Epitaxial films which are essentially free of stacking faults have been grown both by chemical vapor deposition (Finch et al., 1963; Li, 1966) and vacuum evaporation (Unvala and Booker, 1964) onto substrates with dislocation densities of up to 10^3 cm^{-2}. Further, Notis and Conrad (1964) found that if the stacking fault density in the epitaxial film was relatively high ($>10^3$ cm^{-2}), it had an inverse relationship to the substrate dislocation density, suggesting that the dislocations in some way retard the stacking fault nucleation. This data is somewhat in discrepancy with the suggestion by Mendelson (1966) that substrate dislocations caused by mechanical damage could be a source of stacking faults. It has been frequently observed that scratched substrates or mechanically polished samples which have not received a chemical polish to remove the work damage, show a high concentration of stacking faults, particularly linear faults, in the epitaxial films (Schwuttke, 1962; Chu and Gavaler, 1963; Finch et al., 1963; Mendelson, 1964), and Mendelson suggested that the substrate dislocations might split into partials propagating at 60° with a linear stacking fault between them.

Chemical impurities are an additional source of abnormal growth conditions which can lead to stacking faults. The two major sources of these impurities are the presence of cleaning stains or oxide residues on the substrates and impure reactant gases. Cleaning stains, mainly hydrocarbon residues, or even the water quality of the final rinse have been shown to be important in nucleating stacking faults (Mendelson, 1966; Batsford and Thomas, 1963; Finch et al., 1963; Handelman and Povilonis, 1964). Oxide residues can have a similar effect (Finch et al., 1963). Gas-phase etching with HCl or the highly reducing environment of the deposition conditions generally removes these problems. The effect of oxidizing impurities such as oxygen or water vapor in the reactant gas stream has been reported to reduce stacking fault densities (Haneta, 1964). However, in this case fault densities were initially extremely high (10^3–10^4 cm^{-2}) and still relatively high ($>10^2$ cm^{-2}) even after the addition of water vapor (0.1%) to the gas stream. The observed reduction is probably due to a gas-phase etching action of the oxygen or water vapor but its application to high-quality epitaxial films with stacking faults of less than 10 cm^{-2} is unclear.

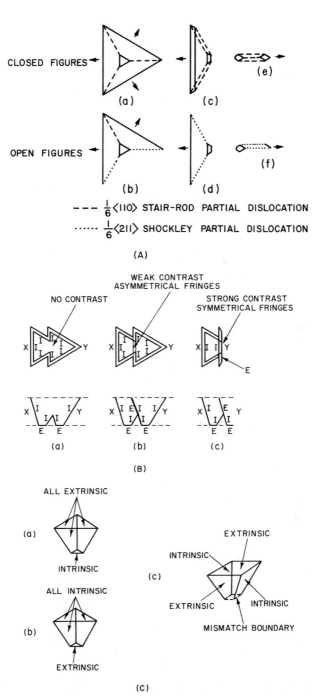

Fig. 30. (A) Various geometries of stacking fault defects in (111) autoepitaxial silicon (from Joyce, 1974). (B) Analysis of stacking fault interaction (from Booker, 1964). (C) Stacking fault crystallography in (111) substrates (a,b) and (100) substrates (c). (from Booker, 1964).

The effect of growth rate has been studied by Notis and Conrad (1964) who show that, under their experimental conditions, the fault density is proportional to the growth rate, rising from essentially no stacking faults for growth rates of around 1000 Å/min to nearly 10^4 cm^{-2} for rates of 1 μm/min. Other workers, however (Finch et al., 1963), while not investigating the growth rate effects, have been able to achieve similar low fault densities while still depositing the epitaxial film at the higher rates. For the low-pressure pyrolysis of silane (Booker et al., 1964) a somewhat different dependence on growth rate was observed, the stacking fault density decreasing with increasing growth rate although remaining very high even under the best conditions, while the overall perfection of the lattice tended to deteriorate, approaching polycrystallinity, at very high rates (17 μm/min).

Mendelson (1964) has observed that annealing the epitaxial films for several hours in hydrogen had the effect of removing almost all the polygon faults and most of the linear faults. This treatment also resulted in a slight surface etching of the sample ($\sim$ 1000 Å) which occurred in the early stages of the annealing treatment. Annealing in argon had no effect on the fault density so the stacking faults are thermally stable at 1250°C, which is reasonable since the partial dislocations can neither glide nor climb. Mendelson proposed that the hydrogen annealing removed the localized area of mismatch at the film/substrate interface which nucleated the stacking fault. In consequence, correct stacking recommences at this site and the fault collapses with the partial dislocations decomposing into Heidenreich–Shockley partials which can glide out of the film taking the stacking faults with them. The few remaining linear faults probably had a different source (dislocations in the substrate, for example) and in consequence resist the hydrogen anneal. During device processing it is often necessary to oxidize the silicon substrate (to mask subepitaxial diffusions, for example) prior to epitaxy. Pomerantz (1967) has observed that this can lead to stacking fault densities as high as 10^5 cm^{-2} in the epitaxial film. He also observed that if the silicon substrate was structure etched after the oxide had been removed but prior to epitaxial deposition the surface was populated with saucerlike etch pits. Split slice experiments further showed that these etch pits had a similar density and distribution on the substrate as the stacking faults in the subsequently deposited epitaxial film. Pomerantz found that both these defects could be prevented by generating dislocations on the back side of the slice (by abrasion or boron doping) prior to oxidation, and suggested that fast-diffusing metal impurities (Ni, Fe, Cu, Mn, and Au) precipitate at or near the surface during oxidation to nucleate the saucer-shaped defects or the stacking faults during subsequent epitaxial growth. Dislocation generation on the back of the slice would act as an effective getter for these impurities which would have a sufficient diffusivity to cross the wafer

(250 μm) in the two hours of oxidation at 1200°C. Further evidence that heavy metal precipitation can nucleate stacking faults is provided by the UHV sublimation epitaxy work of Thomas and Francombe (1967, 1971), who observed higher stacking fault densities in epitaxial films deposited onto low lifetime substrates than on long lifetime substrates.

3. TWINNED DEFECTS

There are two types of twinned defect observed in epitaxial films, often associated, namely tripyramidal defects and microtwin lamellae. Tripyramidal defects are only observed in (111) silicon films and appear as hillocks, ideally with a three-cornered star configuration as shown in Fig. 31. The hillocks are often less regular than this figure shows, with blunted segments or with one or two of the segments missing (Booker, 1965). Figure 31 also shows the associated linear defects which grow along inclined {111} planes and join up to form a closed triangle symmetrically arranged around the pyramidal defect.

These twinned defects have been observed by several authors (Tung, 1962; Miller et al., 1962; Batsford and Thomas, 1963; Chu and Gavaler 1963, 1964; Mendelson, 1964, 1967; Unvala and Booker, 1964; Booker, 1964, 1965; Booker and Joyce, 1966). The defects were found to occur when the substrates were contaminated with carbon in the form of β–SiC on the surface (Miller et al., 1963; Unvala and Booker, 1964; Booker and Joyce, 1966). The defects are nucleated at the substrate/film interface and increase in size as the epitaxial film grows. At the center of the defect is a small triangular mismatch area usually a few thousand angstroms across (Booker, 1965).

The crystallography of these defects has been comprehensively detailed by Booker (1965). The central triangular core is singly twinned with respect to the matrix about the (111) matrix plane. Each of the pyramidal segments (A, B, and C) are doubly twinned, initially about the (111) matrix plane and then about one of the inclined {111} planes. The differing segment orientations result in three types of interface. Between the core and segment is a "first-order twin join" and at the segment/segment interface a "second order twin join" as each segment is twinned about the matrix plane and separately twinned about differing inclined {111} planes. Kohn (1956, 1958) has observed both types of interface in melt-grown silicon crystals, showing that these interfaces are stable configurations in the diamond lattice. The segment/matrix interface is also a "second-order twin join" as shown by Chu and Gavaler (1964), who suggested a {1352}–{1352} interface. Although not a low index plane, it has a definite element of symmetry and represents a stable configuration. The stability of the twinned configurations explains the regular geometrical growth of the tripyramidal defects.

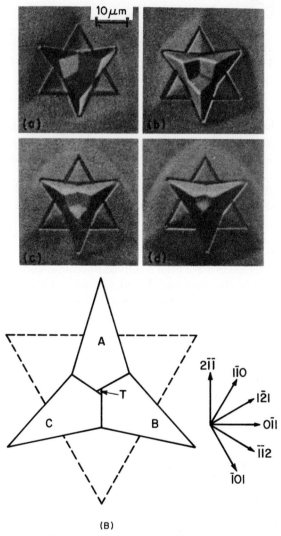

Fig. 31. (A) Tripyramidal defects and associated line defects (from Mendelson, 1967). (B) Crystallography of tripyramidal defect (full line) and associated line fault (broken line).

The line defects associated with the tripyramidal defects have an appearance similar to stacking faults. They are initiated at the substrate/film interface, grow along inclined {111} planes and intersect the surface along ⟨111⟩ directions. Booker (1965), using selected-area transmission electron diffraction and tilting the specimen, was able to show that the defects are microtwin lamellae singly twinned about one of the inclined (111) matrix planes.

The origin of twin defects in epitaxial silicon has been attributed to β–SiC contamination by several authors (Miller *et al.*, 1963; Unvala and Booker, 1964; Booker and Joyce, 1966). Booker (1965) has postulated that silicon growth centers nucleating on a silicon carbide platelet may be misoriented with respect to other growth centers nucleating on the silicon substrate surface. This mismatched growth center is the core triangle, and after it has grown to a certain size it twins again about the inclined {111} planes to produce the individual pyramidal segments. This subsequent twinning may occur spontaneously or at steps in the substrate. This nucleation theory is supported by later results obtained by Booker and Joyce (1966) who studied epitaxial nucleation of silicon by molecular beam techniques. When the substrate was contaminated with silicon carbide, 5% of the growth centers were singly twinned about the (111) matrix plane. These growth centers underwent further twinning or coalesced with other centers to produce more complex centers.

4. HETEROEPITAXY OF SILICON ON INSULATING SUBSTRATES

The heteroepitaxial deposition of silicon on nonconducting substrates has recently assumed a significant importance in semiconductor device fabrication (Ronen and Micheletti, 1975). The advantage of this technique is that by chemically etching the heteroepitaxial thin film the individual active devices of the integrated circuit can be separated into discrete, isolated areas of silicon on an insulating substrate. In consequence, junction isolation is obviated, parasitic capacitance and junction leakage reduced, and higher packing densities are achieved. This allows faster and lower power integrated circuits to be fabricated.

Heteroepitaxial silicon films have been deposited on sapphire (Al_2O_3), quartz (SiO_2), and spinel (MgO, Al_2O_3) and this work has been the subject of an extensive review by Cullen (1971). The majority of work has involved the use of sapphire substrates because spinel, although having some advantages over sapphire, has not until recently been commercially available with the required crystalline quality. Deposition on quartz has received little attention (Joyce *et al.*, 1965; DeLuca, 1969). The physical and chemical properties of sapphire and spinel which make them attractive as substrates are their refractory and inert nature and a close lattice spacing match to silicon.

The lattice mismatch between silicon and sapphire is dependent on the relative orientation between substrate and film. This epitaxial relationship has been studied by several groups (Manasevit and Simpson, 1964; Manasevit *et al.*, 1965; Nolder and Cadoff, 1965; Bicknell *et al.*, 1966) and the following relationships found $(100)_{Si} \| (1\bar{1}02)_{Al_2O_3}$, $(111)_{Si} \| (0001)_{Al_2O_3}$, $(111)_{Si}$ or $(100)_{Si} \| (11\bar{2}0)_{Al_2O_3}$, $(111)_{Si} \| (11\bar{2}4)_{Al_2O_3}$. These orientation relationships have been considered by Nolder and Cadoff (1965), who postulated that the silicon atoms substitute for aluminum at surface sites bonding with oxygen atoms of the sapphire to form the first layer of the epitaxial growth with a varying degree of misfit depending on orientation. For example, the mismatch of (100) silicon on ($1\bar{1}02$) sapphire is 12.5% and 4.2% in the $[11\bar{2}0]$ and $[1\bar{1}01]$ directions, respectively.

The lattice match between silicon and spinel is somewhat superior which is an advantage of this type of substrate. Spinel has a cubic lattice and the epitaxial silicon films have parallel orientation with the substrate $(111)_{Si}\|$ $(111)_{spinel}$, $(110)_{Si}\|(110)_{spinel}$, and $(100)_{Si}\|(100)_{spinel}$ (Manasevit and Forbes, 1966). Assuming the same epitaxial mechanism of silicon on sapphire, namely silicon substitution into a Mg^{2+} surface site, the lattice mismatch is less than 1% for stoichiometric spinel (MgO, Al_2O_3). However, even a relatively small lattice mismatch will result in a high dislocation density at the film/substrate interface. Mroczkowski *et al.* (1968) have measured the effect of mismatch on dislocation density in compound semiconductors using a meltback process which yields perfect homoepitaxy. For a lattice mismatch of only one percent, however, a dislocation density in excess of 10^8 cm^{-2} was measured. The perfection increases markedly if thicker films are grown and the use of thicker films and surface property devices to some extent mitigate this problem.

Spinel has an additional advantage over sapphire in that autodoping of aluminum from the substrate is reduced by an order of magnitude (Robinson and Dumin, 1968). With the increasing emphasis on reducing the carrier concentration below 10^{16} cm^{-3} the reduction of autodoping has received considerable attention. The major source is vapor transported Al_2O produced by the reaction of the substrate with either hydrogen

$$2H_2 + Al_2O_3 \rightarrow Al_2O + 2H_2O$$

or with the deposited silicon

$$2Si + Al_2O_3 \rightarrow Al_2O + 2SiO$$

These reactions are particularly severe at the early stages of epitaxy before complete surface coverage has been obtained, but the back surface of the substrates may act as a continuing source of autodoping during the remainder of the deposition (Mercier, 1968, 1970; Hart *et al.*, 1967). Attempts to reduce autodoping by reducing the temperature and hydrogen flowrates have led to the increased use of silane pyrolysis in a helium carrier gas (Mercier, 1970, 1971; Chiang and Looney, 1973).

The differences in thermal expansion coefficients of silicon ($4.2 \times 10^{-6}/°C$) and sapphire ($8.3 \times 10^{-6}/°C$) or spinel ($8.6 \times 10^{-6}/°C$) are sufficiently large that, on cooling from the deposition temperatures, the silicon films are in a state of compressive stress of 10^9–10^{10} dyn/cm^2 (Ang and Manasevit, 1965; Dumin, 1965; Schlötterer, 1968). These stress levels do not cause macro-

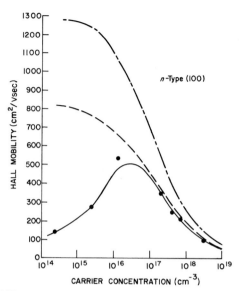

Fig. 32. Hall mobility as a function of carrier concentration for: ——— epitaxial silicon film on spinel (20-μ thick); ––– bulk silicon strained (extrapolation); ——–—— bulk silicon, unstrained (data, Irving, 1962) (from Schlötterer, 1968).

scopic physical damage such as cracks to the epitaxial film (Dumin, 1965), although dislocations may be formed to relieve the stress. The effect of stress on the Hall mobility has been investigated by Schlötterer (1968) and although significant in films with carrier concentrations below 10^{17} cm^{-3} the stress effects are largely masked by the more important mobility reductions due to defect scattering. This is shown in Fig. 32 where the Hall mobility is shown as a function of carrier concentration for unstrained and strained (8 × 10^9 dyn/cm^2) silicon and for epitaxial silicon on spinel. It is clear from this figure that below 10^{17} cm^{-3} the reduced mobility cannot be accounted from by residual stress, and there is considerable evidence that this effect is due to crystalline defects (Dumin and Robinson, 1968a; Schlötter, 1968). Indeed much of the data on the quality of silicon epitaxial films on silicon and sapphire have been inferred from electrical (carrier mobility, minority carrier lifetime) and optical (absorption, photoresponse) measurement rather than by direct observation.

Crystallographic defects have been detected using transmission electron microscopy, x-ray diffraction, and chemical etching (Bicknell *et al.*, 1966; Joyce *et al.*, 1965; Dumin and Robinson, 1968b; Manasevit *et al.*, 1965; Nolder *et al.*, 1965; Robinson and Wance, 1973; Robinson and Dumin, 1968; Wang *et al.*, 1969) and dislocations, stacking faults, and twinning defects observed, similar to those observed in homoepitaxial silicon, although in much higher densities. Dislocation densities of the order of 10^7 cm^{-2} have been commonly observed, and unless a better matching substrate material can be found there will be little improvement over this figure. Mismatch mechanisms may also explain the high density of stacking faults and microtwin lamellae (Robinson and Dumin, 1968; Nolder et al., 1965) that have been observed.

The defect density is found to decrease as the film thickness increases. This has been demonstrated using electrical measurements (Dumin and Robinson, 1968a,b; Norris, 1972) and by ion channeling studies (Picraux, 1972) which show ion scattering centers significantly higher than 10^5 cm^{-2}. The density of crystal imperfections was somewhat higher for (100) layers than for (111) layers. In both cases the crystalline quality improved by a factor of three on increasing the film thickness from 5000 Å to 2 μm.

Electrical measurements of minority carrier lifetime also show an improvement as the film thickness is increased. Using diffused diodes Dumin and Silver (1964) showed that lifetime increases with distance from the sapphire surface, reaching a maximum of several nanoseconds for a silicon film 15-μm thick. A similar result was obtained by Kranzer (1974) using gate controlled diodes. Significant improvements in lifetime have recently been achieved by using chloride gettering treatments (Robinson and Heiman, 1971; Ronen and Robinson, 1972) which are thought to remove heavy metal impurities as volatile chlorides or by introducing a heavy doped n^+ silicon layer between the active device silicon and the sapphire (Schroder and Rai-Choudhury, 1973; McGreivy and Viswanathan, 1974). In the latter case lifetimes in the microsecond range were reported.

The removal of work damage from the surface of the substrate is very important in ensuring improved heteroepitaxy. If residual polishing damage is present on the substrates, defects have been found to propagate into the epitaxial film causing device processing problems (Cullen, 1971). Any gross defects such as scratches or subgrains in the substrate are replicated in the epitaxial film (Wang et al., 1969). To ensure defect-free substrate surfaces various etching techniques have been developed (Filby and Nielsen, 1967). Sapphire is usually hydrogen etched at 1200–1300°C to remove about 1 μm below which, if the correct polishing procedures have been used, the substrate is damage free. Stoichiometric spinel reacts with hydrogen at high temperatures and is best prepared by etching in phosphoric acid (Aeschlimann et al., 1970) or molten borax (Robinson and Wance, 1973).

D. Oxidation-Induced Defects

The appearance of stacking faults in the silicon substrate after thermal oxidation has been studied by many authors. This interest is due to the detrimental effects of the defects, on device performance, causing soft junctions (Queisser and Goetzberger, 1963; Varker and Ravi, 1974) and emitter–collector shorts in transistors due to diffusion pipes at the defects (Barson *et al.*, 1969).

Stacking fault nuclei lie in the wafer prior to oxidation and this high-temperature treatment causes the faults to grow from these nuclei. Several causes of the nuclei have been observed including mechanical damage, crystal growth defects, and ion implantation damage. The growth of oxidation induced stacking faults caused by incipient growth defects in the silicon crystal usually only occurs in dislocation free, float zone material, having the same spiral distribution across the wafer as swirl type defects to which they are probably related. The other sources of the nuclei give a more random distribution.

For mechanically damaged samples stacking fault formation has been observed in {111} silicon (Thomas, 1963; Booker and Stickler, 1965; Jaccodine and Drum, 1966; Fisher and Amick, 1966; Sanders and Dobson, 1969; Ravi and Varker, 1974a), in {110} silicon (Booker and Tunstall, 1966; Fisher and Amick, 1966), and in {100} silicon (Quiesser and van Loon, 1964; Fisher and Amick, 1966; Mayer, 1970; Ravi and Varker, 1974a) and in {100} epitaxial silicon films (Drum and van Gelder, 1972; Hsieh and Maher, 1973). Stacking faults were first observed by Thomas (1963) in silicon wafers which had been air annealed, subsequent to diamond polishing. The inference that polishing damage was an important factor in nucleating the defects was confirmed by Fisher and Amick (1966) who observed a reduction by two orders of magnitude in defect density if the samples were chemically polished prior to oxidation. If web crystals, which should be totally free of mechanical damage were used, no defects were formed. Joshi (1966), however, showed that mechanical damage was not the only possible cause for stacking fault formation. If the samples were subjected to a deep chemical etch no defects were formed on subsequent oxidation unless an additional vacuum annealing step (1000°C or higher) was introduced before oxidation. Joshi postulated that this step caused oxygen precipitation, these precipitates nucleating the stacking faults. In addition, other processing steps prior to oxidation have been found to supply nuclei for stacking faults. Etching the silicon surface with HF prior to oxidation, was found by Drum and van Gelder (1972) to cause stacking faults in (100) silicon but rarely in (111) silicon, and Prussin (1974) and Seidel *et al.* (1975b) have observed oxidation induced stacking faults in wafers subject to ion implantation.

The oxidation-induced stacking faults have been observed using chemical etching (Drum and van Gelder, 1972; Fisher and Amick, 1966; Pomerantz, 1972; Queisser and van Loon, 1964) and transmission electron microscopy (Booker and Stickler, 1965; Booker and Tunstall, 1966; Jaccodine and Drum, 1966; Joshi, 1966; Lawrence, 1969; Ravi and Varker, 1974a; Sanders

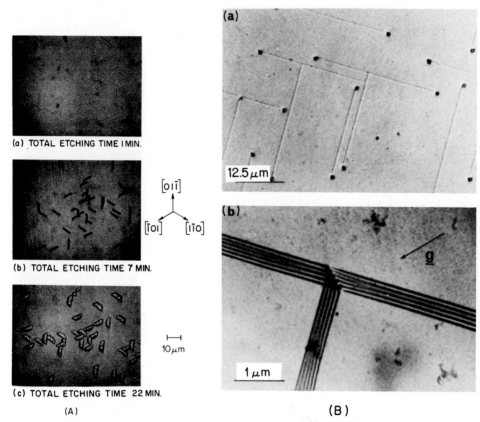

Fig. 33. (A) Line defect structure revealed by Sirtl etching of (111) surface wafer oxidized 18 hr at 1050°C in steam (from Fisher and Amick, 1966). (B) Optical interference (a) and bright field electron (b) micrographs of intersecting stacking faults in (100) surface silicon (from Hsieh and Maher, 1973).

and Dobson, 1969; Thomas, 1963). The defects are revealed by Sirtl etching the silicon wafer after the oxide layer has been removed. The line defect structure revealed this way on a (111) silicon wafer is shown in Fig. 33A (Fisher and Amick, 1966). The initial etching displays the defects as lines, aligned along the (110) directions where the ⟨111⟩ planes intersect the surface. Extending the etching time broadens, but does not lengthen, the lines and finally triangular dislocation etch pits become apparent at the ends of the lines. This data is consistent with the defects being two-dimensional stacking faults lying on inclined ⟨111⟩ planes and bounded by partial dislocations. Similar etching patterns are obtained for (100) and (110) silicon. The stacking faults differ from those found in epitaxial silicon which are usually closed polyhedra and not open line defects. As can be seen from

Fig. 33B (Hsieh and Maher, 1973) the line defects do not cross, rather one defect being bounded by the intersecting one to form a T etching pattern. This data is consistent with the fact that stacking faults do not readily propagate through other stacking faults lying on a different (111) plane.

Transmission electron microscopy has shown clearly that the defects are extrinsic stacking faults bounded by the silicon/oxide interface on one side and within the crystal by a semicircular Frank $(1/3) \langle 111 \rangle$ partial dislocation. Booker and Stickler (1965) detected the presence of an amorphous second phase, probably an oxide, giving rise to the postulation that the stacking faults form in a similar manner to that reported by Silcock and Tunstall (1964) for the precipitation of NbC in stainless steel. By this mechanism a second-phase platelet precipitate forms on the Frank partial dislocation on the side away from the stacking fault. The resulting crystallographic stress is relieved by the emission of vacancies from the dislocation which subsequently moves out passed the precipitate incorporating it into the stacking fault. Oxide precipitates were also observed by Joshi (1966) in the early stages of oxidation; however, they subsequently were converted to stacking faults and were considered to be more of a nuclei for the defect than an associated product.

An alternative mechanism for stacking fault movement based on vacancy transfer, not precipitation, was proposed by Sanders and Dobson (1969). Based on annealing experiments which showed that the stacking faults grew in oxygen but shrank in vacuum, they proposed that the growth of the stacking fault was dependent on the relative concentration of vacancies in the silicon and at the oxide or vacuum interface. In vacuum annealing, the interface concentration of vacancies is greater than the equilibrium vacancy concentration around the fault and in consequence the fault shrinks. In contrast, during oxidation, because vacancies are transported out of the silicon through the oxide film as part of the oxygen diffusion growth mechanism, the growing oxide interface acts as a sink for vacancies and the stacking faults grow.

The nucleation and growth of the stacking faults are very dependent on the processing parameters. On polished samples annealed in wet oxygen, defect densities up to 10^7 cm^{-2} have been reported (Fisher and Amick, 1966), while substrates chemically etched to remove damage are relatively free of stacking faults (Joshi, 1966). Defect densities below 200 cm^{-2} are reported for the oxidation of epitaxial films provided they are not exposed to HF. If nucleation of stacking faults occurs, the growth of the defects depends on the oxidation temperature, although the density is relatively temperature independent.

The kinetics of the stacking fault growth has been studied by several authors and recently, in detail by Hu (1975b) and Shiraki (1975). The effect of oxidation temperature on the observed size of the stacking faults is shown in Fig. 34 for several slice orientations (Hu, 1975b). Below 850°C stacking faults are not observed (Joshi, 1966; Lawrence, 1969). Above this temperature an increasing growth rate is observed with increasing temperature up to a temperature of 1150–1250°C depending on slice orientation. At this point the stacking faults start to shrink rapidly, disappearing completely at high enough temperatures (Thomas, 1963; Lawrence, 1969). In the growth region, for example at 1150°C, the stacking fault size L usually follows a power relationship with time t of the type

$$L = at^n + b$$

where n has been observed to lie between 0.8 (Fisher and Anick, 1966; Hu, 1975b) and 0.5 (Mayer, 1970).

Shiraki (1975) has observed similar results to those shown in Fig. 34. In addition, it was observed that the addition of between 0.1 and 0.5% HCl to

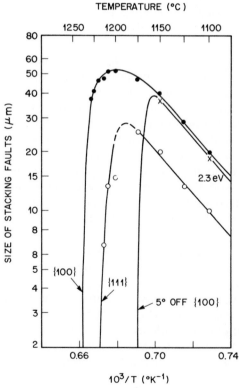

Fig. 34. The growth versus temperature of oxidation stacking faults in dry oxygen for 3 hr (from Hu, 1975b).

the dry oxidation environment caused a reduction in stacking fault growth, reducing the critical temperature for annihilation from the 1200°C observed by Hu for (100) silicon to 1100°C. This reduction was postulated to be due to the migration of silicon through the oxide, leaving vacancies in the lattice, followed by vapor-phase etching of the silicon at the oxide, oxygen/HCl interface.

It has been reported (Thomas, 1963) that stacking faults can also be prevented if highly doped (10^{20} cm^{-3}) silicon is used or the stacking faults annihilated if the faulted lattice is stressed by diffusing a high concentration of phosphorus into the wafer (Lawrence, 1969). This process, however, resulted in a very high 10^8 cm^{-2} dislocation density in the wafer, greater than would be expected from solute diffusion effects alone.

The most effective way to prevent stacking fault formation is to getter the wafer by introducing dislocations into the back side of the wafer prior to oxidation. Misfit dislocations introduced by high concentration boron (Pomerantz, 1967) or phosphorus (Rozgonyi et al., 1975) diffusions are very effective for this purpose or the deposition of a highly stressed Si_3N_4 layer to the back side of the wafer (Petroff et al., 1975) can be used if the diffusion step is undesirable.

For the case of ion implant induced nucleation centers, which are thermally unstable, a high-temperature annealing treatment in nitrogen (containing less than 0.1% O_2) prior to oxidation will prevent the formation of stacking faults (Prussin, 1974; Seidel et al., 1975b).

There is general agreement that stacking fault densities are much higher in wet oxidizing ambients than dry, regardless of the nucleating process (Queisser and van Loon, 1964; Lawrence, 1969; Drum and van Gelder, 1972). The cause for this enhancement has not been completely elucidated.

The nucleation and growth of stacking faults has recently been studied by Hsieh and Maher (1973). Using {100} epitaxial silicon which had been HF cleaned just prior to oxidation they followed the defect formation as a function of oxidation time (steam at 1050°C) by Sirtl etching and transmission electron microscopy. The Sirtl etching showed hillock formation in the early stages but longer oxidation times led to the distinct line and dislocation etch patterns shown in Fig. 33. Transmission electron microscopy showed the formation of needlelike precipitates (4×10^6 cm^{-2}), due to silicon/fluorine polymerization, lying along $\langle 110 \rangle$ directions at the silicon/oxide interface. After five minutes of oxidation the precipitates have been replaced by dislocation loop segments, again intersecting the surface along $\langle 110 \rangle$ directions. The dislocations are $(1/3) \langle 111 \rangle$ Frank partials, and subsequent oxidation causes them to climb into the silicon in some cases with a velocity sufficient that they are not annihilated by the following silicon/oxide interface.

The direct observation of the precipitate, dislocation, stacking fault sequence provides an experimental insight into the cause of oxidation induced defects. In the presence of surface damage, Lawrence (1969) has provided further detail on the processes of stacking fault formation. The presence of microgrooves in the silicon surface due to diamond polishing apparently retards the recovery of the deformed lattice by acting as pinning sites for dislocations. These dislocations subsequently form stacking fault boundaries which can grow by vacancy emission and the consumption of silicon at the oxide interface provides an effective sink for the emitted vacancies. The experiments of Joshi (1966) are also consistent with this overall model. In that work the initial annealing in vacuum is sufficient to cause oxide precipitation in the silicon crystal and the dislocations caused by the precipitates can act as nucleation points for stacking fault growth in a similar manner to the processes described above.

Recently, Ravi and Varker (1974) have observed stacking fault growth which is consistent with a continuous nucleation process. Using silicon substrates which displayed "swirl" type etch patterns (see Section II,B) they observed the formation of stacking faults of various lengths (in contrast to the relatively uniform stacking fault length generally observed) in the swirl bands. The swirl defects are thought to be point defect clusters and, as the silicon/oxide interface approaches these sites, an enhanced oxygen diffusion will occur. If this local enhanced diffusion of oxygen exceeds the solid solubility limit the oxygen interstitial atoms collapse into a Frank loop composed of an extrinsic stacking fault surrounded by a $(1/3) \langle 111 \rangle$ partial dislocation. This process, although a bulk nucleation process, is restricted to within 3000 Å of the oxide interface there being insufficient oxygen diffusion beneath this "near-surface" layer to cause supersaturation.

The degrading effects of stacking faults on the electrical characteristics of diodes have been carefully studied by Ravi et al. (1973) and Varker and Ravi (1974). Typical IV characteristics are shown in Fig. 35 from which it can be seen that the defective diodes have excessive reverse leakage current below the avalanche voltage. These characteristics were obtained for diodes produced by boron diffusing $\langle 111 \rangle$ oriented n-type epitaxial films to a junction depth of 4 μm. The diffusion mask was 5000 Å of steam oxide grown at 1100°C. Ravi and coworkers analyzed the stacking faults nondestructively in a scanning electron microscope using the electron beam induced current (EBIC) mode of operation. In this technique a signal is obtained by collecting the charge carriers generated in the diode by the electron beam as it sweeps over the sample. The presence of a crystal defect causes localized charge–carrier recombination, modulating the EBIC signal which can be used to display the defect on a cathode ray tube. Electrically active stacking faults (those faults which cause excess reverse leakage current)

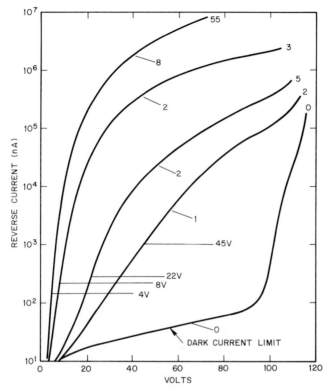

Fig. 35. IV characteristics of five diodes illustrating the effects of electrically active stacking faults on excess reverse bias current. The total number of faults is shown at the ends of the curves and the total number of electrically active faults is indicated at the center. Arrows indicate the minimum threshold voltage for the group of active faults in a given diode. (from Ravi *et al.*, 1973)

can be selectively displayed by applying a reverse bias to the diode in the SEM, an example of this type of display being shown in Fig. 36. This effect is attributed to an electrical interaction between the defect and the depletion field which causes a localized increase in the flow of charge carriers in the depletion region with the degree of enhancement related to the degree of electrical activity of the defect. The bias voltage required to cause current enhancement at an EASF is referred to as the "threshold voltage."

The bias EBIC displays clearly demonstrated that not all the stacking faults were electrically active confirming IV data for good diodes in which stacking faults were observed by etching techniques. The quality of the diode was determined by the stacking fault having the lowest threshold voltage, this EASF representing the "electrically weakest" point in the diode through

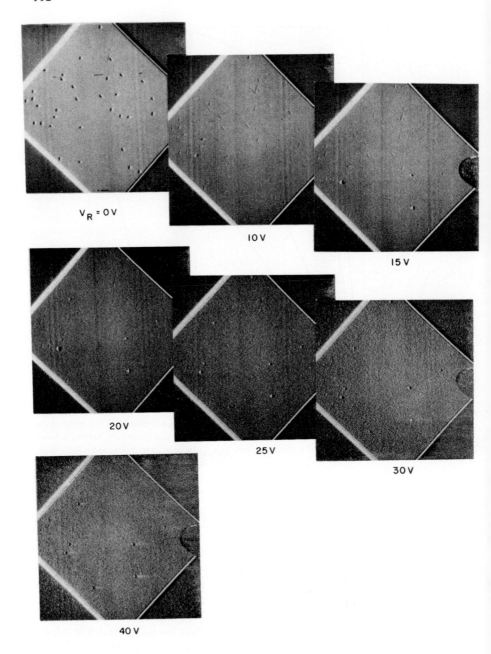

$V_R = 0 V$

10 V

15 V

20 V

25 V

30 V

40 V

Fig. 36. EBIC display of a diode with various reverse voltages illustrating the concept of electrically active faults and the threshold voltage of the faults (from Ravi *et al.*, 1973).

which most of the excess reverse current passed. This effect is graphically demonstrated in Fig. 35 which shows the IV characteristics for five diodes sampled from a single wafer. The reverse leakage current of the diodes correlates strongly with the minimum threshold voltage for EASF, less well for the number of EASF in the diode, and poorly with total number of active and inactive stacking faults.

Further analysis of the EASF showed that there was a strong correlation between the stacking fault length and the reverse diode conductance, the shorter the fault, the higher the conductance. Ravi and coworkers (1973) postulated two possible mechanisms for junction degradation to explain this size effect. Transmission electron microscopy showed evidence for second-phase precipitation of the stacking faults, the smaller faults decorated both in the fault zone and on the surrounding partial dislocation, while the larger stacking faults were decorated only on the partial dislocations. This precipitate decoration can weaken a diode by two possible mechanisms (shown schematically in Fig. 37); by retarding the boron diffusion causing junction contouring or by setting up strain fields in the crystal which are known to generate leakage currents in p–n junctions (Rinder, 1962). In

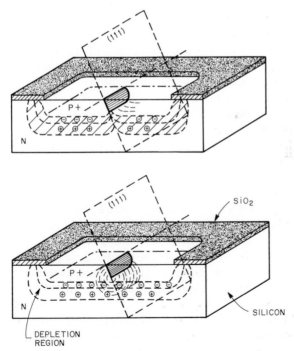

Fig. 37. Schematic sketches illustrating two possible direct interactions of stacking faults with a p–n junction (from Ravi *et al.*, 1973).

either case, the larger stacking faults, with the major portion of the precipitation located below the diode depletion region have less electrical effect than smaller stacking faults on which the precipitation is much more concentrated in the active region of the device.

The elemental composition of the precipitates was not analyzed but experimental evidence suggested that they could have been boron, silicon oxide, or metallic depending on processing and starting material variables. If the initial silicon wafers had a low oxygen content and were not processed at high temperature, the electrically active stacking faults could, to a large degree, be neutralized by a standard gettering cycle (Lawrence, 1968b) suggesting that fast diffusing metal impurities are involved. However, if the initial oxygen content is high these gettering procedures are much less effective, suggesting that a slow diffusing impurity such as oxygen or boron is responsible.

E. Diffusion-Induced Defects

The effect of high-temperature diffusion on the defect structure of silicon substrates has been the subject of considerable study. The majority of the work is covered in reviews by Willoughby (1968), Gibbons (1970), and Hu (1973b). Phosphorus, boron, and to a lesser extent arsenic have been the subjects of study. Misfit dislocations have been observed due to the difference in covalent radii of the dopant atoms and at higher concentrations the occurence of dopant precipitation. These defects have been studied by transmission electron microscopy, x-ray topography, and surface etching and have been correlated with anomalous diffusion effects which interfere with the fabrication or performance of electronic devices. The observation and difficulties of these diffusion effects which include "emitter dip," diffusion pipes, and the precipitation modification of electrically active dopant profiles, was the major stimulation for diffusion-induced defect studies in silicon.

The diffusion of a restricted area emitter can have a marked effect on the base diffusion profile below the emitter window. This effect was first observed by Miller (1959) and is most commonly observed as an enhanced diffusion into the silicon of the base/collector junction beneath an n-type emitter. This effected is commonly referred to as the emitter dip effect and an example is shown in Fig. 38. By restricting the fabrication of very narrow basewidths this effect has prevented the production of very high-frequency devices.

The presence of crystallographic defects in the silicon can also lead to enhanced diffusion effects. If these defects are present beneath the emitter, the emitter diffusion can "pipe" down the defect shorting out the transistor. This effect is shown schematically in Fig. 39 and is particularly important with the advent of large-scale integrated circuits with increasing total emitter area.

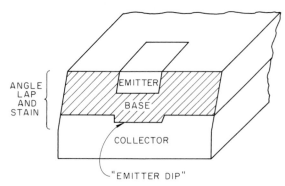

Fig. 38. Schematic diagram of the emitter dip effect in double diffused *npn* transistor.

Precipitation effects similar to those described in the previous section on bulk crystal properties, apply if the dopant concentration is above the solid solubility limit at temperatures high enough to allow diffusion and subsequent precipitation. The dopant precipitates, being nonsubstitutional, are not electrically active and are the cause for the modification of the diffusion profile from the expected (erfc) distribution shown in Fig. 40.

The detailed studies of these diffusion effects have led to a good understanding of the important factors involved and to a considerable extent have resulted in the development of procedures to prevent their occurrence.

1. DIFFUSION-INDUCED DISLOCATIONS

a. Dislocations in the diffused region. Diffusion-induced dislocations have been observed in silicon after phosphorus diffusions (Prussin, 1961; Schwuttke and Queisser, 1962; Lander *et al.*, 1963; Ino *et al.*, 1964; Jaccodine, 1964; Sukhodreva, 1964; Washburn *et al.*, 1964; Joshi, 1965; Joshi and Wilhelm, 1965; Joshi and Dash, 1966; Lawrence, 1966a; McDonald *et al.*, 1966; Rupprecht and Schwuttke, 1966; Schwuttke and Fairfield, 1966; Joshi *et al.*, 1967; Levine *et al.*, 1967a,b; Fairfield and Schwuttke, 1968; Yoshida *et al.*, 1968; Barson *et al.*, 1969; Ravi, 1972), after boron diffusions (Queisser, 1961; Prussin, 1961; Miller *et al.*, 1962; Schwuttke and Queisser, 1962; Schwuttke, 1963; Ino *et al.*, 1964; Jaccodine, 1964; Lawrence, 1966a; Levine *et al.*, 1967a; Ravi, 1973), and after arsenic diffusions (Joshi and Wilhelm, 1965; Dash and Joshi, 1970).

Phosphorus and boron, the most commonly used dopant elements, have smaller covalent radii (1.10 and 0.97 Å, respectively) than silicon (1.17 Å). The substitutional presence of these dopant elements in a silicon lattice will result in a contraction of the lattice constant, the reduction being dependent on the dopant concentration. In consequence when a planar diffusion of phosphorus or boron is made, the dopant concentration profile results in a corresponding profile of lattice spacing. If the lattice spacing gradient is

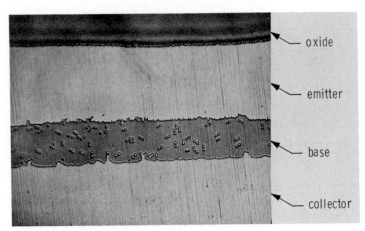

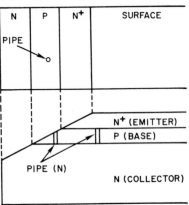

(A)

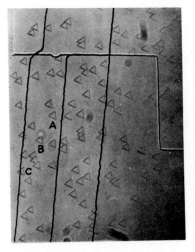 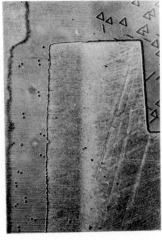

(B)

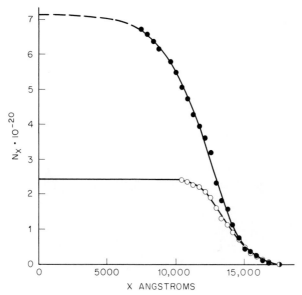

Fig. 40. Comparison of diffused phosphorus distribution by tracer measurement (●) (total phosphorus) and that obtained by resistivity measurements (○) (electrically active phosphorus).

very steep the resulting strain will result in the formation of dislocations in the diffused region. This effect is shown schematically in Fig. 41A. The resulting edge dislocations form networks lying parallel with the surface and are often associated with precipitation effects which also result in reducing the lattice stress. Where diffusion through oxide windows have been made the dislocations do not usually extend into the undiffused silicon, rather bending upwards to the surface (Schwuttke and Fairfield, 1966). The presence of dislocations extending under the oxide windows after diffusion, commonly referred to as emitter edge dislocations, are formed at the diffused/undiffused interface at the window edge. Emitter edge dislocations are discussed in later paragraphs. Arsenic, with a covalent radius 1.17 Å, very close to that of silicon, produces markedly fewer dislocations, mainly interstitial dislocation loops with Burgers vectors which do not lie on the diffusion plane (Dash and Joshi, 1970).

The experimental factors affecting the dislocation density caused by phosphorus diffusions have been extensively studied by Joshi and coworkers (Joshi and Wilhelm, 1965; Dash and Joshi, 1970). Deep diffusions, using (111)

Fig. 39. (A) Schematic diagram of diffusion pipes between emitter and collector and the microsectioning technique used for their observation. (B) Optical micrograph showing the correlation between stacking faults and diffusion pipes. (From Barson *et al.*, 1969.)

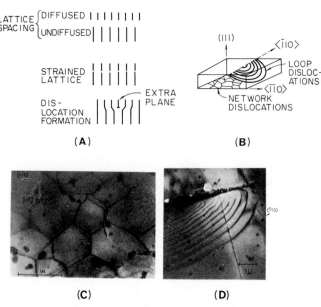

Fig. 41. (A) Schematic diagram of edge dislocation relief of lattice stress induced by diffusion. (B) Schematic diagram of dislocation networks and loops introduced into silicon by diffusion. (C) Electron micrograph of diffusion induced dislocation network (Joshi and Wilhelm, 1965). (D) Electron micrograph of diffusion induced dislocation loops (Lawrence, 1966a)

silicon and a surface dopant concentration of 10^{21} cm^{-3} resulted in a very high dislocation density (10^9 cm^{-2}) near the surface, of the type shown in Fig. 41C. Below 3000 Å, however, the dislocation network starts to break up into smaller patches and at still greater depth, 5000 to 15,000 Å, the networks disappear and only a few single dislocations remain. The experimental observation of a low dislocation density at the junction depth is of considerable importance, as their presence at this point would encourage precipitation which would degrade the junction. In addition to the network dislocations described above, Lawrence (1966a) observed an array of concentric loops lying on an inclined (111) plane and intersecting the surface along a $\langle 110 \rangle$ direction. The dislocation networks are shown in Fig. 41D. Until a junction depth of 3750 Å or greater was achieved no dislocations were observed at all, regardless of the surface dopant concentration, or the diffusion temperature. At a junction depth of 4250 Å they observed a few dislocation nodes while a junction depth of 5150 Å produced an extensive array of misfit dislocation nets. In (111) silicon these arrays were planar while in (110) silicon they were interconnected into a three-dimensional network.

A similar depth dependence study for boron diffusions was made by Levine et al. (1967a). Using (111) silicon, diffusion conditions to give a surface concentration of $2.5 \ 10^{20}$ cm^{-3}, and a junction depth of 5.5 μm, they observed no visible defect structure at the surface. The observation of dislocations started at 1 μm, reached a maximum between 3 and 6 μm and then dropped off to a low value at the junction depth. The maximum dislocation density coincided with the depth at which the dopant concentration gradient was steepest. The absence of dislocations at the surface was explained in terms of a relatively flat concentration gradient at this point due to the diffusion conditions. The dislocations were observed in three-dimensional networks which appeared to have moved into the crystal by a nonconservative or climb motion. In contrast, in (100) silicon a glide mechanism is more favorable for dislocation movement (Washburn et al., 1964). In this case an edge dislocation parallel to the surface can be formed when two gliding dislocations with inclined Burgers vectors intersect along a line that is parallel to the surface and at right angles to the Burgers vector of the edge dislocation. For (100) silicon the inclined glide plane is (111), the normal slip plane for silicon, and the process is therefore more favorable than in (111) or (110). In consequence dislocations may occur more deeply in (100) than (111) silicon.

The formation of dislocations is also dependent on the concentration of dopant, as shown by McDonald et al. (1966) and Dash and Joshi (1970). McDonald and coworkers, using phosphorus-doped (111) silicon, showed qualitatively that the density of slip patterns was dependent not on the surface concentration, but rather on the dopant profile. Using the conductivity of the slice as a measure of this they found the slip density became negligible at conductivities less than 1.2×10^{-3} Ω cm^{-1}. Dash and Joshi found a similar dependence but used the total amount of impurity (Q_{crit}) as an experimental parameter and finding it to be $\approx 1.1 \times 10^{15}$ cm^{-2}. This is close to the value for Q_{crit} of 3×10^{15} cm^{-2} calculated by Shockley (Queisser, 1961) for boron diffusions.

A number of models have been proposed to quantitatively explain the formation of diffusion-induced dislocations in silicon. An early model by Prussin (1961) suggested that the dislocation density was dependent on the surface concentration while that of Czaja (1966) included the dependence on the impurity concentration profile. However, the experimental work of both McDonald et al. (1966) and Dash and Joshi (1970) clearly demonstrated that the total amount of impurity (Q_{crit}) and the junction depth (x_j) were important considerations. The importance of Q_{crit} was alluded to by Shockley (see Queisser, 1961), but it was not until Dash and Joshi (1970) applied the van der Merwe (1964) model for interfacial misfit dislocation generation in epitaxial layers to the diffusion system that all the experimentally observed features could be explained. This model allows the calculation of the thickness dependence of the critical misfit (δ_c) before the spontaneous generation of dislocations. By reducing the bicrystal calculations to the case of diffusion induced strain in single crystal silicon Dash and Joshi obtained the expression

$$l_n 23.4\delta_c + 14.6\,(h/a)\,\delta_c = 0$$

where h/a is the depth (in units of lattice parameter) below which misfit dislocations occur. For an average dopant concentration of 3.6×10^{20} cm^{-3} the critical junction depth beyond which dislocations occur is 4000 Å. This figure has been reduced somewhat (to 2400 Å) by more refined calculations made by van der Merwe (1970) and Jesser and Kuhlmann-Wilsdorf (1967), but considering the assumptions made the agreement with the experimental data is excellent.

b. Emitter edge dislocations. When diffusions are made through an oxide window, dislocations have been observed to propagate over considerable distances, laterally under the oxide, into the undiffused portion of the wafer. This effect is commonly referred to as the "emitter edge" effect and has been investigated using surface etching, x-ray topography, and transmission electron microscopy (Sato and Arata, 1964; Ino et al., 1964; Blech et al., 1965; Lawrence, 1966a; Schwuttke and Fairfield 1966; Duffy et al., 1968; Fairfield and Schwuttke, 1968). Unlike the defect structure observed in the diffused region, these dislocations can spread out for several hundred microns and have been reported to penetrate up to 20 μm into the silicon (Fairfield and Schwuttke, 1968). In consequence, the dislocations can extend through the electrical junctions of the device and their presence has been found to cause a significant degradation of device performance (Lawrence, 1966a; Duffy et al., 1968).

The emitter edge effect is only observed in (111) silicon and the dislocations themselves lie in the (111) primary slip planes. The dislocations form either a network in areas of nonuniform stress where the strain field is unidirectional close to corners of the diffused region, or form as dislocation loops aligned to a general $\langle 211 \rangle$ direction which is perpendicular to the $\langle 110 \rangle$ stress vector.

The critical doping concentration in the diffused window required to cause edge dislocations has been the subject of some study. Lawrence (1966a) was unable to observe the generation of fresh dislocations outside the diffused region using phosphorus or boron diffusions up to 4.8×10^{16} cm^{-2}, although stress relieving movement of pre-existing dislocations occurred at about half this value. Fairfield and Schwuttke (1968) observed edge dislocations for boron at $Q \sim 1.3 \times 10^{17}$ cm^{-2} and, using lower diffusion temperatures (970°C for phosphorus as opposed to 1200°C for boron), a value of 2×10^{16} cm^{-2} for phosphorus. These workers also demonstrated that the phosphorus diffusion conditions could be used to control the edge dis-locations. The use of a nonoxidizing ambient and/or a high temperature (1150–1200°C) during the drive in cycle was found to prevent dislocation formation.

   c.   *Defects enhanced diffusion (The emitter-dip effect).*   Cooperative dif-fusion effects, that is the enhancement or decrease of the diffusion coefficient of one dopant by the introduction of a second, may be divided into two classes. The first concerns the effects of a static nature, such as electrical or elastic phenomena which are essentially short-range effects and only ob-served with close junction spacing (Hu and Schmidt, 1968). The second effect, which is caused by the dynamic movement of defects during high-temperature diffusions, occurs over a wide range. Several mechanisms have been postu-lated for the diffusion of common III and V group dopants in silicon (Seeger and Chik, 1968), but all involve the interaction of the dopant atom with defect structure in the host lattice (single or double vacancies or an extended interstitial group). In consequence, the introduction of crystal damage due to dopant-induced strain can cause anomalous diffusion effects. In this sec-tion only enhanced diffusion effects such as the "emitter dip" effect are described. However, a retardation effect has also been observed caused by dopant precipitation (and hence "removal" of active dopant) at dislocations caused by a subsequent diffusion.

   The emitter dip effect has been observed under a variety of experimental conditions and in both *pnp* and *npn* transistors by Lawrence (1966b, 1967). In the latter case the *npn* double diffused transistor was produced by diffusing boron ($Q = 1 \times 10^{15}$ cm^{-2}) into a phosphorus doped (111) silicon wafer to give a base/collector junction depth of 2.5 μm. The "emitter" junction was produced by diffusing phosphorus or boron (to produce a $p^{+}pn$ structure) to a junction depth of 2.3 μm. Following either of these "emitter" diffusions, the base/collector junction was found to have moved 0.5 μm further into the silicon. When the emitter diffusions were carried out under conditions which allowed the onset of emitter dip to be observed it was found that the n^{+} diffusion initiated the effect at basewidths (2 μm) twice that possible with the p^{+} diffusions. Transmission electron microscopy studies of the *npn* tran-sistor showed the presence of dislocations in the base region of a type similar to those observed by Joshi and Wilhelm (1965) for deep phosphorus diffu-sions except that their presence in the base region shows them to have pene-trated beyond the emitter junction, a fact not observed by Joshi and Wilhelm.

   The *pnp* transistors were produced by diffusing phosphorus at $Q = 6 \times 10^{14}$ cm^{-2} to a junction depth of 1.8 μm followed by a boron emitter diffusion. Anomalous movement of the base/collection junction under the emitter was observed when the emitter concentration exceeded a Q_{crit} of 2×10^{16} cm^{-2}, this figure being in good agreement with previous work on the boron concentration required to cause strain-induced dislocations (Lawrence, 1966b). The direction of movement of the base/collector junction was also found to depend on the phosphorus concentration in the base. If the concentration was below 4×10^{15} cm^{-2} emitter dip was observed. However, if it was above 4×10^{15} cm^{-2} the base/collector junction was observed to move towards the advancing emitter junction due to phosphorus precipitation on the emitter-induced dislocations.

Lawrence (1966b, 1967) was also able to reproduce enhanced and retarded diffusion in single n–p and p–n junctions by applying mechanical stress from a load applied to the sample through a sharp point. In both cases the anomalous diffusion occurred in the first few minutes during which the maximum plastic deformation occurred. From this data Lawrence postulated that vacancies generated by the moving dislocations were responsible for the enhanced diffusion. This postulation had been previously considered by Girifalco and Grimes (1959) and Ballufi and Ruoff (1963) to explain enhanced diffusion in metals and was developed in greater detail by Hu and Yeh (1969, 1970) for silicon.

Localized enhanced diffusion, forming "diffusion pipes" has been observed in silicon devices under certain conditions. Goetzberger and Stephens (1962) showed that phosphorus surface contamination can cause this effect, while Barson et al. (1969) demonstrated that diffusion pipes were often located at many of the stacking faults present in the epitaxial layer. With careful control of the processing environment and high-quality epitaxial layers, the incidence of diffusion pipes can be controlled. Barson et al. (1969) also showed that the use of a high-temperature (1150°C) gold diffusion from the back side of the wafer, used for lifetime control purposes, caused a significant increase in the diffusion pipe density. Lowering the gold diffusion temperature to 970°C prevented this effect.

2. Diffusion-Induced Precipitation

If diffusions are done under conditions whereby the impurity concentration exceeds the solid solubility, at a temperature where significant diffusion can occur, then the excess dopant atoms will tend to precipitate out of the silicon lattice. Early indications of this effect were observed by Tannenbaum (1961). Using conditions which produced high surface concentrations and shallow diffusions, significant differences in the dopant profiles were observed dependent on the use of electrical or radiotracer techniques. This discrepancy, shown in Fig. 40 led to the postulation that phosphorus precipitation had occurred during cool down, where the dopant concentration exceeded 2×10^{20} cm^{-3}, causing the electrical measurements, which would only detect the electrically active substitutional phosphorus, to be constant at this value. The radiotracer technique measuring the total concentration of phosphorus, including the precipitated material, gives corresponding higher values.

Precipitation effects in phosphorus diffused silicon have been observed by Schmidt and Stickler (1964), Kooi (1964), Joshi (1965), Joshi and Dash (1966, 1967), Jaccodine (1968), in double diffused structures by Levine *et al.* (1967b); in boron diffused silicon by Miller *et al.* (1962), Schwuttke (1962), Joshi and Dash (1967), and Dobson and Filby (1968); and in arsenic diffused silicon by Joshi (1965) and Joshi and Dash (1967).

Phosphorus precipitation in silicon has been studied in detail using transmission electron microscopy by Joshi and Dash (1966, 1967) for blanket diffusion and in double diffused *npn* structures by Levine *et al.* (1967b). The phosphorus precipitates were observed as platelets lying on (111) planes regardless of the orientation of the silicon matrix. An electron micrograph of precipitation in (111) silicon is shown in Fig. 42A. At dopant concentrations of less than 10^{21} cm^{-3} the precipitates were much smaller ($\approx \frac{1}{2}$ μm) than those shown for the higher concentrations in Fig. 42A. In the double diffused structures the precipitates were restricted to the top surface of the wafer where the dopant concentration was highest. The precipitates were often found to be associated with helical dislocations oriented along $\langle 110 \rangle$ directions, which in some regions reached a density of 10^9 cm^{-2}. Occasional stacking fault formation at the tips of the precipitates was also observed. The composition of the precipitates was investigated using electron diffraction. The results are somewhat inconclusive although Levine *et al.* (1967b)

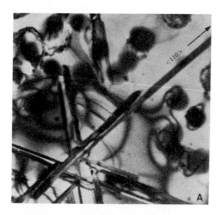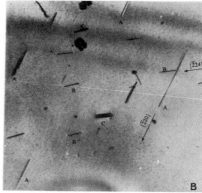

Fig. 42. (A) Electron micrograph of phosphorus precipitation in diffused (111) silicon. Regions of dark contrast are due to phosphorus rich clusters. Doping level $> 10^{21}$ cm^{-3} (from Joshi and Dash, 1967). (B) Electron micrograph of boron precipitation in diffused (110) silicon. Precipitates designated by letter A are platelets, by letter B, rods and by letter C, three-dimensional structures (from Joshi and Dash, 1967).

tentatively indexed a major fraction of the precipitates to have a base centered orthorhombic structure with $a = 6.3$, $b = 3.8$, and $c = 6.57$ Å. The precipitates were also observed to be of the vacancy type and of smaller specific volume than that of silicon. This information suggests that the precipitates are not similar to the SiP precipitates which have been observed by Schmidt and Stickler (1964).

Platelet precipitates (Fig. 42B) were also observed in those foils diffused with boron (Joshi and Dash, 1967; Dobson and Filby, 1968). Like the phosphorus precipitates described above they are of the vacancy type and lie on (111) silicon planes. Less commonly observed were one-dimensional rod precipitates lying along $\langle 211 \rangle$ directions and three-dimensional precipitates. The latter type were surrounded by mismatch boundary dislocations while the rod precipitates were invariably enveloped by single or double helices. Dobson and Filby (1967) extended their foil diffusion studies to planar diffusions into wafers and found surprisingly little precipitation. The only defect structure observed was occasional rod-type precipitates with helical misfit dislocations and stacking faults lying on (111) inclined planes. It was postulated that the foil experiments were more susceptible to compressive strain-induced precipitation due to the thermal expansion mismatch between the boron glass diffusion source and the silicon. The influence of the glass in the thicker silicon wafer samples would be very much less due to the constraining effect of the rest of the sample.

F. Ion Implantation

The use of a high-energy ion beam to implant dopant atoms into silicon has some important advantages over high-temperature diffusion doping. The dopant concentration and profiles can be accurately controlled by adjusting the ion beam conditions and the avoidance of the high temperatures required for diffusions restricts both the anisotropic "undercutting" of the dopant atoms beneath the diffusion mask or the movement of previously doped junctions. However, the impact of a high-energy ion beam results in significant crystal damage to the substrate and, unless removed by annealing, this will degrade the electrical properties of the implanted layer, a subject recently reviewed by Gibbons (1972) and Kimerling and Poate (1975).

The implanted atom looses energy to the substrate by two mechanisms, "nuclear collisions" in which the energy is transfered as translational energy to a substrate atom and "electronic collisions" which only involve the excitation of electrons and do not cause lattice atom displacement. Nuclear energy loss predominates for ions of high mass and low energy while electronic effects occur at higher energy and lower mass. The former process is the one of major concern in causing lattice damage and the energy transfer can be sufficiently large that not only is a substrate atom displaced, but itself becomes a secondary projectile called "knock-ons" which can cause further lattice damage by nuclear collisions. As a result of these energy loss processes the trajectory of an implanted atom is surrounded by a cluster of lattice damage. If the dose (implanted atoms per square centimeter) is sufficiently high then the damage clusters start to overlap and the substrate surface becomes totally disordered and approaches an amorphous state.

1. THE DISTRIBUTION OF DAMAGE IN THE SUBSTRATE

The depth distribution of the damage as a function of the experimental parameters has been the subject of both theoretical and experimental studies. A comprehensive review on the subject has been published by Gibbons (1972). A detailed survey of the techniques used to calculate the depth distribution of ion implantation damage are beyond the scope of this chapter. However, a brief discussion of the subject is in order to provide an insight into the damage processes involved and because the theoretical calculations have been refined to the point where excellent predictions of the damage distribution can be made. The calculations are based on determining the rate of nuclear energy loss dv/dx which is taken as a measure of the damage. The number of atoms (N_E) displaced is usually estimated by the relationship

$$N_E = \zeta \frac{dv}{dx} (2E_d)^{-1}$$

developed by Kinchin and Pease (1955) and modified by Sigmund (1969) whose ζ is a factor of about 0.8 and is dependent on the form of the scattering potential.

Monte Carlo calculations of the damage distribution have been made by Brice (1970b) and Pavlov et al. (1967). This technique is particularly suitable where the crystallinity of the substrate is important, for example in channelling implantation. A second approach to the calculation involves the formulation of LSS-type integrodifferential equations to determine the spatial distribution of the deposited energy. Potentially, the most accurate calculation, it has been used by Sigmund and Sanders (1967) to provide data on a wide variety of ion/target mass ratios. However, the mathematics of the model are such that at higher energies it looses accuracy and under these conditions an improvement is obtained using the calculation of Brice (1970b). Brice used a two part calculation, first determining the spatial distribution of the implanted atoms as they slowed down. Using this data and a knowledge of the interaction cross section and energy division between electronic and nuclear collisions, dv/dx could be calculated. The results were plotted as energy contours, an example of which is shown in Fig. 43. If a horizontal line is drawn through a given ion energy on the verticle axis it is possible to construct a direct graphical relation between dv/dx and x

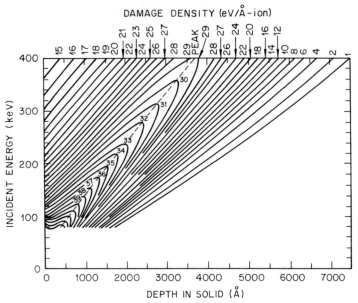

Fig. 43. Contours of constant damage density for phosphorus ions incident on an amorphous silicon target. Incident energies in the range of 100–400 keV (Brice, 1970).

and examples for phosphorus and boron implants into silicon are shown in Fig. 44. The damage distributions have a similar form to the atom implantation profiles but are generally shifted towards the surface by some 20–30%

A wide variety of experimental techniques have been used to measure the depth distribution of ion implantation damage. In the majority of cases the sample is sectioned, by anodization and HF etching (see, for example, Bicknell, 1972) and the changes observed in a damage related property are used to obtain the profile. The experimental techniques have included optical measurements, both refractive index changes (Crowder *et al.*, 1970), changes in reflectivity (Nelson and Mazey, 1967, 1968; Crowder and Title, 1970), and ir absorption spectra (Stein *et al.*, 1970); by transmission and scanning electron microscopy (Bicknell, 1969; Mazey *et al*; 1968; Davidson and Booker, 1970); by x-ray anomalous transmission topography (Schwuttke *et al.*, 1968) or x-ray fluorescence (Nelson *et al.*, 1970), by radioactive copper decoration techniques (Masters *et al.*, 1970); by electron spin resonance (Brower *et al.*, 1970; Crowder and Title, 1970); by inference from electrical measurements (Crowder and Title, 1970; Bicknell, 1972); and from channelling measurements using proton or helium ion backscattering (Westmorland et al., 1970; Akasaka *et al.*, 1973). These various experimental studies view different aspects of crystal microstructure but all tend to give similar, if

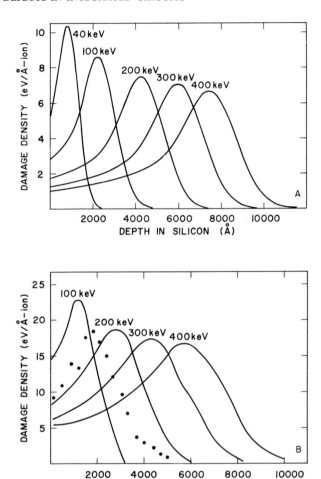

Fig. 44. Depth distribution of energy deposited into atomic processes as calculated from the energy contours of Brice (1970). The dotted points are experimental points from Crowder and Title (1970). A, boron; B, phosphorus.

often qualitative, distributions which are close to the theoretical curves of Brice (1970b).

A comprehensive experimental study of damage distribution in silicon by Group III and V dopant atoms has been published by Crowder and Title (1970). Using electrical measurements (resistivity and Hall coefficient) for low dose implants and ESR and refractive index changes for implants causing amorphous silicon, they studied a variety of dopants, implant voltages, and doses. The results for the most probable damage depth (X_D) are

TABLE VI

MOST PROBABLE DAMAGE DEPTH IN Si IMPLANTED
WITH LOW ION DOSES[a]

Ion	Energy (keV)	Dose (cm^{-2})	$\langle X_D \rangle^a$ (μm)	$\langle X \rangle^b$ (μm)	$(\langle X_D \rangle / \langle X \rangle)^c$
P^{31}	70	1×10^{13}	0.06_5	0.11	0.5_9
	140	3×10^{12}	0.11	0.21	0.5_3
		1×10^{13}	0.12	0.21	0.5_7
		2×10^{13}	0.12	0.21	0.5_7
	200	3×10^{12}	0.18	0.29	0.6_2
Si^{29}	150	2×10^{13}	0.14	0.22	0.6_4
B^{11}	60	1×10^{14}	0.16	0.23	0.7_9
		3×10^{13}	0.17	0.23	0.7_4
As^{75}	280	3×10^{11}	0.14	0.19	0.7_4
		1×10^{12}	0.12	0.19	0.6_3

[a] Crowder and Title (1970).

[b] The most probable damage depth $\langle X_D \rangle$ is defined as the distance from the surface of the sample to the location of the peak in the damage distribution. The accuracy of the measurements is approximately $\pm 10\%$.

[c] The most probable projected range $\langle X \rangle$ (frequently designated as R_p) refers to the peak in the distribution produced by ions impinging upon the sample in a random direction.

[d] This ratio is probably accurate to within $\pm 10\% - 20\%$.

[e] This wafer was not misoriented 7 degrees from $\langle 100 \rangle$ to minimize channelling.

shown in Table VI in conjunction with corresponding range data for the dopant distribution, the damage peak occurring about 50–70% of the depth for the most probable projected range. The dose had little effect on X_D. When the dose is sufficient to cause an amorphous layer, the width of this layer is a function of dose and the data are shown in Table VII. Crowder and Title also show damage profiles, obtained from electrical measurements and anodic stripping, for phosphorus implanted silicon and the result for 200 keV implants is shown in Fig. 44.

2. AMORPHOUS LAYERS PRODUCED BY ION IMPLANTATION

Above a critical dose the ion implantation damage is so extensive that the substrate is termed "amorphous." The critical dose is dependent on the mass and voltage of the implanted ion and on the temperature of the target as shown in Fig. 45. The structure of the amorphous phase is not certain but it is thought to consist of a modified form of silicon in which there is no long-range order but the close order atomic arrangement of tetrahedrally bonded atoms is preserved (Bicknell, 1969). Its presence can be detected by

TABLE VII

WIDTH OF AMORPHOUS LAYER PRODUCED BY 280 keV P^{31} IONS IN SILICON IMPLANTED AT ROOM TEMPERATURE[a]

ESR studies			Visual observation and stripping width[c]	Interference Fringes in 1–5 spectral region width[d]
Spins[a]		Width[b]		
(cm^{-2})	(cm^{-2})	(μm)	(μm)	(μm)
1×10^{16}	1.1×10^{16}	0.53 ± 0.05	—	0.56 ± 0.03
3×10^{15}	0.89×10^{16}	0.46 ± 0.05	0.50 ± 0.03	0.53 ± 0.03
1×10^{15}	0.72×10^{16}	0.37 ± 0.04	0.48 ± 0.05	0.50 ± 0.03
6×10^{14}	0.65×10^{16}	0.34 ± 0.04	—	$(<0.48)^e$
3×10^{14}	0.34×10^{16}	0.18 ± 0.02	0.25 ± 0.05	$(<0.42)^e$
2×10^{14}	0.20×10^{16}	0.10 ± 0.01	0.10 ± 0.05	—

[a] Crowder and Title (1970).

[b] This experimental data was taken from Crowder *et al.* (1970). The ESR signal at $g = 2.005_g \pm 0.000_5$ characteristic of amorphous Si prepared at room temperature is the line referred to in this column.

[c] The strength of the ESR signal was used to compute the corresponding width of a completely amorphous region using the spin density of 2×10^{20} cm^{-3} observed in amorphous Si films produced by rf sputtering (Brodsky and Title, 1969).

[d] The sample was stripped in 500-Å steps using anodic oxidation and HF etching of the oxide. The visual appearance of the adjacent implanted/unimplanted regions of the sample was compared to an amorphous "standard" as described in the text.

[e] The location of the amorphous Si–crystalline Si interface was determined from interference fringes in the optical transmission spectra of the samples as described in Crowder and Title (1970). The errors indicated refer to the spread observed in the width calculated from different fringes and do not include any experimental uncertainties in the refractive index of amorphous Si. When the Si surface is amorphous, the location of this boundary corresponds to the width of the amorphous layer.

[f] The amorphouse layer was buried. The surface of the sample was not amorphous as determined by visual observation.

changes in the physical properties of the substrate using electron diffraction (Bicknell, 1969; Mazey *et al.*, 1967; Davidson and Booker, 1970), lateral stress (Eer Nisse, 1971), ESR (Brower *et al.*, 1970; Crowder and Title, 1970; Morehead *et al.*, 1972), lattice expansion (Beezhold, 1971), and optical measurements (Nelson and Mazey, 1967, 1968; Crowder and Title, 1970; Crowder *et al.*, 1970, Stein *et al.*, 1970). By purely visual observation the amorphous silicon takes on a "milky" hue due to a change in the reflectivity (Kurtin *et al.*, 1969) if the amorphous layer reaches the surface.

The mechanism for the transformation from crystalline to amorphous silicon has been the subject of several theories—homogeneous nucleation by creating a critical defect density or alternatively heterogeneous nucleation

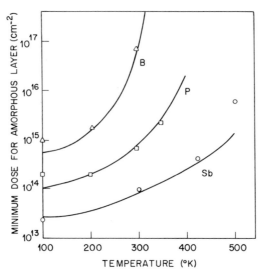

Fig. 45. Minimum dose required to produce an amorphous layer in silicon by ion implantation of boron, phosphorus and antimony. The solid lines are from calculations by Morehead and Crowder (1970), the data points for EPR measurements of Crowder *et al.* (1970).

by the growth of small amorphous regions until they overlap. Swanson *et al.* (1971) calculated that a spontaneous, homogeneous amorphization would occur in *germanium* if the crystal contained a defect concentration of 2–4%. At this point the free energy of the defective crystal becomes equal to that of the amorphous phase, and Swanson *et al.* suggest that a similar result would apply to silicon.

Recent experimental data, however, tend to support a heterogeneous nucleation process for heavy ion bombardment. Baranova *et al.* (1973) followed the change in refractive index as a function of dose for various dopant atoms. The variation was monotonic up to saturation for heavy ions such as antimony, suggesting the formation of isolated clusters of substantially amorphous content at low doses which on further bombardment coalesced to form amorphous zones. Similar data was obtained by Mayer *et al.* (1970), Stein *et al.* (1970), and Masuda *et al.* (1971) using backscattering, ir absorption, and ESR measurement, respectively.

The data for light ion implantation, such as boron, are less simple. The damage dependence on dose is itself dependent on the target temperature and dose rate (Eisen and Welch, 1970; Chadderton and Eisen, 1970; Hirvonen *et al.*, 1971) as shown in Fig. 46. For room temperature implantations (Fig. 46) the amount of disorder is proportional to the square root of the dose up to a specific dose (e.g., 4×10^{15} cm^2 at 5 μA/cm^2), where upon the dose dependence becomes linear. In addition, the amount of disorder for a con-

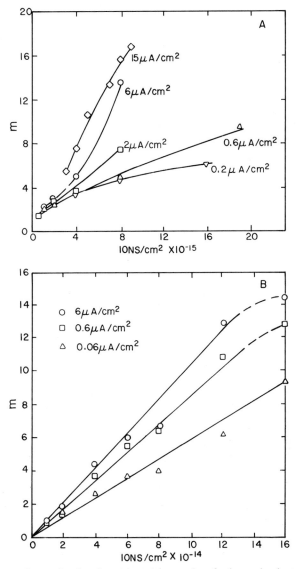

Fig. 46. Dependence of a disorder parameter m on dose for boron implanted into silicon at (A) room temperature and (B) −50°C (from Eisen and Welch, 1970).

stant dose rises with higher dose rates. Implants at −50°C (Fig. 46) show this latter dependence but for a given dose rate the damage is always linear with dose (until saturation sets in at the transition to an amorphous structure). At −120°C similar results are observed, although the damage is much less dose-rate dependent. These results have been rationalized as follows. At low doses and room temperature, where the damage is (dose)$^{1/2}$ dependent, the damage clusters are homogeneously nucleated because under these conditions the point defects caused by the ion bombardment are relatively mobile. However, as the dose rises and the crystal becomes more heavily damaged the defects are trapped by the preexisting damage clusters (heterogeneous nucleation) before they can migrate and create more nuclei. Under these conditions the observed linear dose dependence would be expected. At lower temperature the defect mobility is sufficiently restricted so that heterogeneous nucleation occurs at all dose levels.

3. BORON IMPLANTATION DAMAGE

The electrical behavior of boron implanted silicon has been well reviewed by Gibbons (1972). The effect of annealing is very dependent on the implant dose. If the dose ($> 2 \times 10^{16}$ cm^{-2}) is sufficient to drive the silicon amorphous, then a 650°C anneal for 30 min is sufficient to epitaxially regrow the implanted layer (although significantly defective) incorporating most of the boron into substitutionally active sites. For lower doses the annealing results are somewhat more complicated.

Seidel and MacRae (1970) showed that the free carrier concentration dependence on temperature showed three regions, as shown in Fig. 47. Initially, annealing to 500°C causes the concentration to rise, followed by a decrease between 500 and 600°C before rising to saturation at around 900°C. The decrease is generally referred to a negative annealing and has been explained (North and Gibson, 1970; Fladda *et al.*, 1970) in terms of a Watkins replacement mechanism in which silicon atoms, released by the annealing of damage clusters, replace boron atoms from interstitial sites. At lower doses there is less lattice damage and in consequence negative annealing is less important. The effect of negative annealing is less apparent if the resistivity is monitored as a function of temperature, as the mobility shows a rapid rise at these temperatures due to lattice damage annealing. However, it should be noted that for low doses of boron ($< 5 \times 10^{12}$ cm^{-2}) the low-temperature anneals (500°C) which are sufficient to restore most of the electrical activity are inadequate to restore the carrier mobility of the implanted samples (Darwish and Luginbühl, 1974).

In the intermediate dose range 2×10^{15}–2×10^{16} cm^2 (sufficient to produce an amorphous substrate) an anneal at 900°C, comparable to that used by Seidel and MacRae for lower doses, was found to be insufficient to produce 100% electrical activity (Schwettmann, 1974). This reduced activity was explained in terms of boron atom pairing. Long anneal times (10 hr at

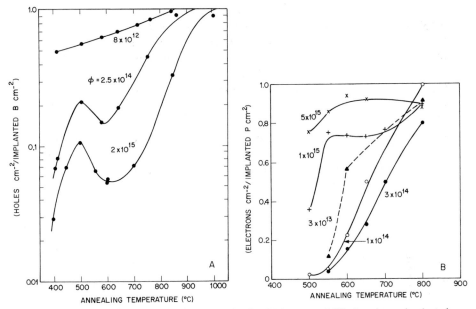

Fig. 47. Isochronal annealing data (30 min) for (a) boron and (b) phosphorus implanted silicon—the ratio of free carrier content from Hall measurements to the dose is plotted against annealing temperature for three doses (Seidel and MacRae, 1970; Crowder and Morehead, 1969).

900°C) were typically required to allow the boron to diffuse away from the implanted region and become active.

The crystal reordering of boron implanted silicon has been monitored by backscattering, transmission electron microscopy, ir absorption, and photoconductivity measurements. The latter two techniques are sensitive to annealing of vacancy related disorder (Chadderton and Eisen, 1970; Netange et al., 1972) which takes place below 250°C. The first evidence of recrystallization occurs on annealing at over 500°C, the temperature at which the mobility is seen to increase. This reordering of the lattice has been closely followed by several workers using transmission electron microscopy (Large and Bicknell, 1967; Bicknell, 1969; Mazey et al., 1968; Davidson, 1970; Davidson and Booker, 1970; Chadderton and Eisen, 1970, Bicknell and Allen, 1970). The results are in general agreement and can be summarized as follows.

Between 600 and 800°C a large number of rod-shaped defects lying along ⟨110⟩ directions appear. These defects, which do not occur if the implant is performed with a target temperature of 280°C, end abruptly inside the specimen. In consequence, they cannot be conventional dislocations which would terminate at the surface or on another line. The most likely interpretation is that the defects are rows or cylinders of interstitials, vacancies, or impurities. Above 800°C these linear defects begin to disappear, being replaced by a large number of dislocation loops. Chadderton and Eisen (1970), using (110) starting material, observed mostly pure edge dislocations on (110) planes with Burgers vector $(a/2) \langle 110 \rangle$, with additional loops on $\{111\}$ planes; Davidson and Booker, using (111) starting material observed the reverse. In both cases the dislocation loops are extrinsic with the extra half plane of atoms probably silicon (Bicknell, 1969). Annealing above 900°C causes a steady decrease in the number of dislocation loops although residual damage has been observed even after 1000°C annealing (Chadderton and Eisen, 1970).

4. PHOSPHORUS IMPLANTATION DAMAGE

The electrical annealing data for phosphorus is shown in Fig. 47 (Crowder and Morehead, 1969). At low doses $(< 10^{13} \text{ cm}^{-2})$ an annealing treatment at 600°C is sufficient to cause over 50% dopant activity (electrons cm^{-2} per dopant cm^{-2}) as compared with only 10%–20% activity at this temperature if higher doses in the low 10^{14} cm^{-2} range are used. For these dopant levels anneals of 600°C or above are required for 50% activity. For the highest doses $(> 10^{15} \text{ cm}^{-2})$, where amorphization of the substrate has occurred, very substantial electrical activity can be obtained at 500°C. This is in marked contrast to the degree of activity in boron implants of the same dose (which cause less crystal damage), where barely one tenth of the activity is achieved unless annealing above 800°C is used. Negative annealing is less marked in phosphorus implantation although it is apparent to some extent at doses of $1 \times 10^{15} \text{ cm}^{-2}$ shown in Fig. 47. It is more obvious, however, if implants are made at low temperature (North and Gibson, 1970) or at low voltage (Bicknell, 1972). The cause of carrier removal in phosphorus, however, is postulated to be different from the boron replacement reaction. Bicknell (1972) has suggested that the formation of E centers in which the substitutional phosphorus is coupled to a vacancy in a more probable mechanism to account for the loss of conductivity. Transmission electron microscopy showed that vacancy loops were generated at 650°C in sufficient numbers to account for the loss of phosphorus activity. The E centers are subsequently destroyed at higher temperatures with the release of the vacancies which subsequently migrate to form intrinsic loops (also observed by electron microscopy).

Additional electron microscopy studies of silicon damage caused by phosphorus implantations have been made by Davidson and Booker (1970),

Davidson (1970), Tamura (1973), and Wu and Washburn (1974). For low doses, only small (50–100 Å) interstitial dislocation loops are observed regardless of dose and annealing temperature. The loops lie in the $\langle 111 \rangle$ plane parallel to the surface with an $a/2 \langle 110 \rangle$ Burgers vector. The observation of contrast changes when the greater g vector is reversed suggest that precipitation occurs near or around the loop. Higher doses ($> 1 \times 10^{15}$ cm^{-2}), rendering the substrate surface amorphous, produce a highly irregular structure on annealing. Even if annealed at 1000°C these films do not recrystallize epitaxially and give complicated electron diffraction patterns showing the presence of single crystal, twin and polycrystalline spots, and a large number of double diffraction spots.

5.   RECOIL DAMAGE FROM "OXIDE MASKING"

Pattern-delineated silicon oxide layers are often used to mask ion implants during device fabrication. However this can lead to additional "anomalous" damage effects caused by the penetration of recoil oxygen atoms from the oxide layer into the underlying silicon substrate as shown schematically in Fig. 48. This problem is particularly severe for heavy ion implantation when nuclear energy transfer to substrate atoms is most effective, and the resulting light oxygen recoil has a much greater range than the primary ion, resulting in substrate damage deep into the substrate (Bogardus and Poponiak, 1973).

Recoil damage from oxide masking was first observed by Cass and Reddi (1973). Using a tapered window structure similar to that shown in Fig. 48, 150-ke V arsenic implants (10^{16} cm^{-2}) and 1000°C post implant anneals, they found high residual defect densities anywhere on the substrate which was covered by 1500 Å of oxide or less. Thicker oxides provided adequate

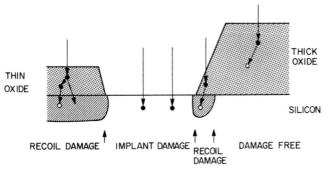

Fig. 48. Schematic diagram of oxygen recoil damage caused by implantation through a silicon dioxide masking layer; ●, implanted atom; ○, oxygen recoil atom.

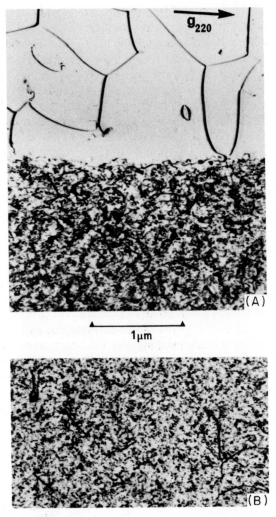

Fig. 49. Bright field transmission electron micrographs of (A) arsenic implanted (150 keV, 10^{15} cm^{-2}) silicon after 30-min anneal at 1000°C in dry nitrogen. Lower half of specimen was masked with 430 Å of thermal SiO$_2$. (B) Silicon specimen sequentially implanted with both arsenic (150 keV, 10^{15} cm^{-2}) and oxygen (10 keV, 10^{16} cm^{-2}) after 30-min anneal at 1000°C in dry nitrogen. (From Moline and Cullis, 1974.)

protection and unmasked areas recrystallized epitaxially during the annealing stage. The difference in postanneal crystal damage between direct arsenic implantation and oxygen recoil effects is shown in Fig. 49A (Moline and Cullis, 1975). The arsenic implantation led to extensive dislocation networks permeating the implanted volume while the arsenic implants through a thin oxide resulted in a dense defect structure which was essentially polycrystalline. For comparison Fig. 49B shows a region of the substrate subjected to combined arsenic and oxygen implants (not through an oxide) and a very similar defect structure to that under thin oxide was observed.

If thick oxides (1500 Å) are used the substrate is largely protected from recoil atoms, but any taper in the oxide at a window will result in localized damage at the window edge as shown in Fig. 48. Any metallic precipitation at this location is liable to cause particularly severe electrical problems due to the proximity of the implanted junction. The crystal damage itself, however, is insufficient to degrade reverse junction characteristics (Bogardus and Popniak, 1973) if substrates are adequately gettered.

Moline et al. (1973) have investigated the effect of krypton ($m = 84$, cf. As $m = 75$) implanation on the yield of oxygen atoms recoiled from SiO_2 layers. The recoil yield is at a maximum approaching the expected range of the krypton; for 24-keV krypton this thickness is approximately 150 Å and the recoil yield about 2.5. Calculations based on this work were used to predict the oxygen profile which would be expected in the experimental studies of Cass and Reddi (1973) and Moline and Cullis (1975) for 10^{15} cm^{-2}, 150-keV As$^+$ through 430 Å of SiO_2. The results are shown in Fig. 50 along with the arsenic implantation profile onto the exposed silicon. Similar results were obtained by Chu et al. (1974) using backscattering measurements. The oxygen concentrations are very high (cf background oxygen in float zone silicon rarely exceeds 10^{16} cm^{-3}) with the resulting possibilities of precipitation as discussed in Section II,C.

6. Junction Leakage in Implanted Devices

The effect of implanation on junction leakage has been investigated by Pickar and Dalton (1970) for electrically inert ions (He, C, Si) and by Michel et al. (1974) and Prussin (1974) for boron, phosphorus, and arsenic. The leakage current is very dependent on the dose and the implanted species as shown in Fig. 51A. Under the experimental conditions used, which were to implant the ions into a prediffused p^+n junction of depth 0.55 μm, even a dose of 10^{11} cm^{-2} of silicon ions increased the leakage current by three orders of magnitude. Isochronal (30 mins) annealing of these junctions caused a progressive reduction of leakage current until at 900°C it was similar to the original unimplanted diode. Similar annealing results were obtained by Michel et al. (1974) for electrically active ions.

The diodes are most degraded when the implantation damage peak is located near to the junction. This is shown in Fig. 51B where junctions of different depth were implanted with carbon ions of various energies. For the deeper junction ($X_j = 0.75$ μm) lower energy implants, less than 200 keV, had relatively little effect and the junction leakage increased with ion energy up to 600 keV. In contrast the shallow junction ($X_i = 0.3$ μm) was more susceptible to low-energy implants and degraded rapidly in the 250–400-keV range and then saturated above 400 keV. The LSS range figures shown in Fig. 51B are somewhat larger than would be expected from the experimental data on junction leakage even assuming that the region of maximum damage is approximately 30% shallower than the LSS range (Crowder and Title, 1970). Michel *et al.* (1974) have shown that if the implant damage and the junction depth coincide then somewhat higher annealing temperatures (up to 1100°C) are required to reduce the leakage current to its preimplant value.

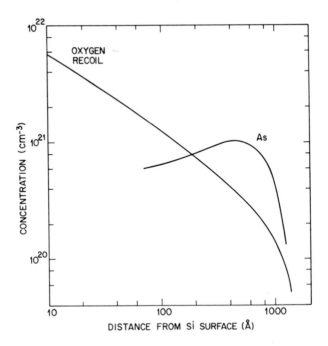

Fig. 50. Concentration profiles of oxygen recoils in a silicon substrate from 150 keV arsenic implant (10^{16} cm^{-2}) through 430 Å of SiO$_2$. The arsenic profile for the implant directly into silicon is shown for comparison. (From Moline and Cullis, 1975.)

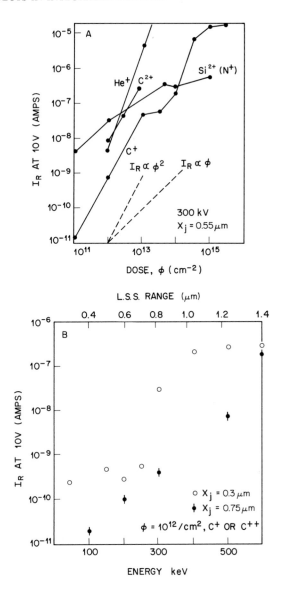

Fig. 51. (A) Reverse leakage, for 20 mil diodes at 10-V bias, as a function of implanted dose for various ions (from Pickar and Dalton, 1970). (B) Reverse leakage for two different junction depths, as a function of implant energy for carbon ions (from Pickar and Dalton, 1970).

ACKNOWLEDGMENTS

The author would like to acknowledge the many contributions he received from his colleagues in preparing this manuscript and especially to A. G. Cullis, K. A. Jackson, L. C. Kimerling, D. M. Maher, J. R. Patel, P. Petroff, McD. Robinson, G. A. Rozgonyi, and T. E. Steidel for their help in preparing and revising the manuscript. In addition, thanks are due to Mrs. S. Wert for her efforts in typing and retyping the manuscript and to Professor R. K. MacCrone for his patience in waiting for it.

References

Abe, T., Abe, Y., and Chikawa, J. (1973). *In* "Semiconductor Silicon" (H. R. Huff and R. R. Burgess, eds.), pp. 95–106. Electrochem. Soc. Softbound Symp. Ser., Chicago, Illinois.

Adams, A. C., and Kaiser, R. H. (1970). Ext. Abstr., Electrochem. Soc. Meeting, Los Angeles, May.

Aeschlimann, R., Gassmann, F., and Woodmann, T. P. (1970). *Mater. Res. Bull.* **5**, 167–171.

Akasaka, Y., Horie, K., Yoneda, K., Sakurai, T., Nishi, H., Kawabe, S., and Tohi, A. (1973). *J. Appl. Phys.* **44**, 220–224.

Akiyama, N., Yatsurugi, Y., Endo, Y., Imayoshi, Z., and Nozaki, T. (1973). *Appl. Phys. Lett.* **22**, 630–631.

Allen, J. W. (1957). *Phil. Mag.* **2**, 1475–1481.

Ang, C. Y., and Manasevit, H. M. (1965). *Solid State Electron.* **8**, 994–996.

Balluffi, R. W., and Ruoff, A. L. (1963). *J. Appl. Phys.* **34**, 1634–1647.

Baranova, E. C., Gusev, J. M., Martynenko, Yu. V., Starinin, C. V., and Hailbullin, I. B. (1973). *In* "Ion Implantation in Semiconductors" (W. L. Crowder, ed), pp. 59–71. Plenum Press, New York.

Barson, F., Hess, M. S., and Roy, M. M. (1969). *J. Electrochem. Soc.* **116**, 304–307.

Batavin, V. V. (1970). *Sov. Phys.–Semiconduct.* **4**, 641–644.

Batavin, V. V., Popova, G. V., and Batavina, L. A. (1967). *Sov. Phys.–Solid State* **8**, 2005–2006.

Batsford, K. O., and Thomas, D. J. D. (1962). *Solid State Electron.* **5**, 353–360.

Batsford, K. O., and Thomas, D. J. D. (1962). *Solid State Electron.* **5**, 353–360.

Batsford, K. O., and Thomas, D. J. D. (1963). *Elec. Commun.* **38**, 354–362.

Bean, A. R., and Newman, R. C. (1971). *J. Phys. Chem. Solids* **32**, 1211–1219.

Beezhold, W. (1971). *In* "Ion Implantation in Semiconductors" (I. Ruge and J. Gaul, eds.), pp. 267–273. Springer-Verlag, New York and Berlin.

Benson, K. E. (1969). *In* "Semiconductor Silicon" (R. R. Haberecht and E. L. Kern, eds), pp. 97–123. Electrochem. Soc. Softbound Symp. Ser., New York.

Benson, K. E., and Stewart, W. W. (1968). *Chem. Eng. Progr.* **64**, 93–101.

Bernewitz, L. I., and Mayer, K. R. (1973). *Phys. Status Solidi (a)* **16**, 579–583.

Bernewitz, L. I., Kolbesen, B. O., Mayer, K. R., and Schuh, G. E. (1974). *Appl. Phys. Lett.* **25**, 277–279.

Bicknell, R. W. (1969). *Proc. Roy. Soc.* **A311**, 75–78.

Bicknell, R. W. (1972). *Phil Mag.* **26**, 273–286, 911–927.

Bicknell, R. W., and Allen, R. M. (1970). *In Proc. Int. Conf. Ion Implantation, 1st* (F. H. Eisen and L. T. Chadderton, eds.), pp. 63–68. Gordon and Breach, New York.

Bicknell, R. W., Joyce, B. A., Neave, J. H., and Stirland, D. J. (1966). *Phil. Mag.* **14**, 31–46.

Billig, E. (1956). *Proc. Roy. Soc.* **A235**, 37–55.

Blech, I. A., Meieran, E. S., and Sello, H. (1965). *Appl. Phys. Lett.* **7**, 176–178.
Bogardus, E. H., and Poponiak, M. R. (1973). *Appl. Phys. Lett.* **23**, 553–555.
Booker, G. R. (1964). *Discuss. Faraday Soc.* **38**, 298–304.
Booker, G. R. (1965). *Phil. Mag.* **11**, 1007–1019.
Booker, G. R., and Howie, A. (1963). *Appl. Phys. Lett.* **3**, 156–157.
Booker, G. R., and Joyce, B. A. (1966). *Phil. Mag.* **14**, 301–315.
Booker, G. R., and Stickler, R. (1962). *J. Appl. Phys.* **33**, 3281–3290.
Booker, G. R., and Stickler, R. (1963). *Appl. Phys. Lett.* **3**, 158–160.
Booker, G. R., and Stickler, R. (1965). *Phil. Mag.* **11**, 1303–1308.
Booker, G. R., and Tunstall, W. J. (1966). *Phil. Mag.* **13**, 71–83.
Booker, G. R., Joyce, B. A., and Bradley, R. R. (1964). *Phil. Mag.* **10**, 1087–1091.
Brice, D. K. (1970a). *Appl. Phys. Lett.* **16**, 103–106.
Brice, D. K. (1970b). *Radiat. Effects* **6**, 77–87.
Brodsky, M. H., and Title, R. S. (1969). *Phys. Rev. Lett.* **23**, 581.
Brower, K. L., Vooker, F. L., and Borders, J. A. (1970). *Appl. Phys. Lett.* **16**, 108–110.
Buck, T. M. (1960). *In* "The Surface Chemistry of Metals and Semiconductors" (H. C. Gatos, ed), p. 107. Wiley, New York.
Buck, T. M., and Meek, R. L. (1970). *In* "Silicon Device Processing" (C. P. Marsdes, ed), pp. 419–430. NBS Spec. Publi. N. 337, Gaithersburg, Maryland.
Buck, T. M., Caisey, H. C., Dalton, J. V., and Yamin, M. (1968). *Bell Syst. Tech. J.* **47**, 1827–1854.
Buck, T. M., Pickar, K. A., Poate, J. M., and Hsieh, C-M. (1972). *Appl. Phys. Lett.* **21**, 485–487.
Buck, T. M., Poate, J. M., and Pickar, K. A. (1973). *Surface Sci.* **35**, 362–379.
Cagnina, S. F. (1969). *J. Electrochem. Soc.* **116**, 498–502.
Cass, T. R., and Reddi, V. G. K. (1973). *Appl. Phys. Lett.* **23**, 268–270.
Chadderton, L. T., and Eisen, F. H. (1970). *In Proc. Int. Conf. Ion Implantation, 1st* (L. T. Chadderton and F. H. Eiser, eds.), pp. 445–454. Gordon and Breach, New York.
Charig, J. M., Joyce, B. A., Stirland, D. J., and Bicknell, R. W. (1962). *Phil. Mag.* **7**, 1847–1860.
Chiang, Y. S., and Looney, G. W. (1973). *J. Electrochem. Soc.* **120**, 550–553.
Chu, T. L., and Gavaler, J. R. (1963). *J. Electrochem. Soc.* **110**, 388–393.
Chu, T. L., and Gavaler, J. R. (1964). *Phil. Mag.* **9**, 993.
Chu, W. K., Müller, H., Mayer, J. W., and Sigmon, T. W. (1974). *Appl. Phys. Lett.* **25**, 297–299.
Chynoweth, A. G., and McKay, K. G. (1956). *Phys. Rev.* **102**, 369–376.
Chynoweth, A. G., and Pearson, G. L. (1958). *J. Appl. Phys.* **29**, 1103–1110.
Collins, C. B., and Carlson, R. O. (1957). *Phys. Rev.* **108**, 1409–1414.
Corbett, J. W. (1970). *In Proc. Int. Conf. Ion Implantation, 1st* (L. T. Chadderton and F. H. Eisen, eds.), pp. 1–8. Gordon and Breach, New York.
Cottrell, A. H. (1953). "Dislocations and Plastic Flow in Crystals." Oxford. Univ. Press (Clarendon), London and New York.
Crowder, B. L., and Morehead, F. F. (1969). *Appl. Phys. Lett.* **14**, 313–315.
Crowder, B. L., and Title, R. S. (1970). *Radiat. Effects* **6**, 63–75.
Crowder, B. L., Title, R. S., Brodsky, M. H., and Pettit, G. D. (1970). *Appl. Phys. Lett.* **16**, 205–208.
Cullen, G. W. (1971). *J. Cryst. Growth* **9**, 107–125.
Cullis, A. G., and Katz, L. E. (1974). *Phil. Mag.* **30**, 1419–1443.
Czaja, W. (1966). *J. Appl. Phys.* **37**, 3441–3446.
d'Aragona, F. S. (1971). *Phys. Status Solidi (a)* **7**, 577–582.
Darwish, M. Y., and Luginbühl, H. W. (1974). *Appl. Phys. Lett.* **25**, 390–391.
Das, G. (1973). *J. Appl. Phys.* **44**, 4459–4467.

Dash, H. C. (1959). *J. Appl. Phys.* **30**, 459–474.

Dash, H. C. (1960). *J. Appl. Phys.* **31**, 763–737.

Dash, S., and Joshi, M. L. (1970). *In* "Silicon Device Processing" (C. P. Marsden, ed.), pp. 202–222. NBS Spec. Publ. No. 337, Gaithersburg, Maryland.

Davidson, S. M. (1970). *In Proc. Eur. Conf. Ion Implantation 1 st* pp. 238–241.

Davidson, S. M., and Booker, G. R. (1970). *In Proc. Int. Conf. Ion Implantation, 1st* (L. Chadderton and F. Eisen, eds.), pp. 51–61. Gordon and Breach, New York.

de Kock, A. J. R. (1970). *Appl. Phys. Lett.* **16**, 100–102.

de Kock, A. J. R. (1971). *J. Electrochem. Soc.* **118**, 1851–1856.

de Kock, A. J. R. (1974). Private communication.

de Kock, A. R., Roksnoer, P. J., and Boonen, P. G. T. (1973). *In* "Semiconductor Silicon" (H. R. Huff and R. R. Burgess, eds.), pp. 83–94. Electrochem. Soc. Softbound Symp. Ser., Chicago, Illinois.

de Kock, A. J. R., Roksnoer, P. J., and Boonen, P. G. T. (1974). *J. Cryst. Growth* **22**, 311–320.

DeLuca, R. D. (1969). *In* "Semiconductor Silicon" (R. R. Haberecht and E. L. Kern, eds.), pp. 299–315. Electrochem. Soc. Softbound Symp. Ser. New York.

Dermatis, S. N., and Faust, J. W. (1963). *IEEE Trans. Commun. Electron.* **65**, 194.

Dobson, P. S., and Filby, J. D. (1968). *J. Cryst. Growth*, **3**, 209–213.

Dumin, D. J. (1965). *J. Appl. Phys.* **36**, 2700–2703.

Dumin, D. J., and Robinson, P. H. (1968a). *J. Cryst. Growth* **3**, 214–218.

Dumin, D. J., and Robinson, P. H. (1968b). *J. Appl. Phys.* **39**, 2759–2765.

Dumin, D. J., and Silver, R. S. (1968). *Solid State Electron.* **11**, 353–363.

Drum, C. M., and van Gelder, W. (1972). *J. Appl. Phys.* **43**, 4465–4468.

Duffy, M. C., Barson, F., Fairfield, J. M., and Schwuttke, G. H. (1968). *J. Electrochem. Soc.* **115**, 84–88.

Eer Nisse, E. P. (1971). *In* "Ion Implantation in Semiconductors" (I. Ruge and J. Gaul, eds.), pp. 17–22. Springer-Verlag, New York.

Eisen, F. H., and Welch, B. (1970). *In Proc. Int. Conf. Ion Implantation, 1st* (F. H. Eisen and L. T. Chadderton, eds.), pp. 459–464. Gordon and Breach, New York.

Fairfield, J. M., and Schwuttke, G. H. (1966). *J. Electrochem. Soc.* **113**, 1229–1231.

Fairfield, J. M., and Schwuttke, G. H. (1968). *J. Electrochem. Soc.* **115**, 415–422.

Faust, J. W., John, H. F., and Stickler, R. (1966). *In* "Physics of Failure in Electronics," pp. 376–378. AD-637529.

Feist, W. M., Steele, S. R., and Readey, D.W. (1969). *Phys. Thin Films* **5**, 237–322.

Fiermans, L., and Vennik, J. (1965). *Phys Status Solidi* **12**, 277–289.

Fiermans, L., and Vennik, J. (1967a). *Phys. Status Solidi* **21**, 627–634.

Fiermans, L. and Vennik, J. (1967b). *Phys. Status Solidi* **22**, 463.

Filby, J. D., and Nielson, S. (1967). *Brit. J. Appl. Phys.* **18**, 1357–1382.

Finch, R. H., Queisser, H. J., Thomas, G., and Washburn, J. (1963). *J. Appl. Phys.* **34**, 406–415.

Fisher, A. W., and Amick, J. A. (1966). *J. Electrochem. Soc.* **113**, 1054–1060.

Fladda, G., Bjorkqvist, K., Eriksson, L., and Sigurd, D. (1970). *Appl. Phys. Lett.* **16**, 313–315.

Föll, H., and Kolbesen, B. O. (1975). *Appl. Phys.* **8**, 319–331.

Frank, F. C., and Turnbull, D. (1956). *Phys. Rev.* **104**, 617–618.

Fuller, C. S., and Logan, R. A. (1957). *J. Appl. Phys.* **28**, 1427–1436.

Gibbons, C. F. (1970). *In* "Silicon Device Processing" (C. P. Marsden, ed.), pp. 21–35. NBS Spec. Publi. No. 337, Gaithersuburg, Maryland.

Gibbons, J. F. (1972). *Proc. IEEE* **60**, 1062–1096.

Girifalco, L. A., and Grimes, H. H. (1958). NASA Tech. Note 4408.

Glaenzer, R. H., and Jordan, A. G. (1968). *Bull. Am. Phys. Soc. Ser. II* **13**, 497.

Glaenzer, R. H., and Jordan, A. G. (1969a). *Solid State Electron.* **12**, 247–258.

Glaenzer, R. H., and Jordan, A. G. (1969b). *Solid State Electron.* **12**, 259–266.
Glinchuk, K. D., Litovchenko, N. M., and Novikova, V. A. (1966). *Sov. Phys.-Solid State* **8**, 777–778.
Goetzberger, A., and Shockley, W. (1959). *Bull. Am. Phys. Soc. Ser. II*, **4**, 409.
Goetzberger, A., and Shockley, W. (1960). *J. Appl. Phys.* **31**, 1821–1824.
Goetzberger, A., and Stephens, C. (1962). *J. Electrochem. Soc.* **109**, 604–607.
Green, D., Glaenzer, R. H., Jordan, A. G., and Noreika, A. J. (1968). *J. Appl. Phys.* **39**, 2937–2939.
Grienauer, H. S., Kolbesen, B. O., and Mayer, K. R. (1974). *In* "Lattice Detects is Semiconductors," pp. 531–537. Inst. Phys. Conf. Ser., London.
Gupta, D. C. (1971). *Solid State Tech.* **14**, 33–40.
Grove, A. S. (1967). *In* "Physics and Technology of Semiconductor Devices," p. 40. Wiley, New York.
Groundy, P. C., and Boatman, J. (1966). Govt. Rep. AF33, 615–2881.
Gupta, D. C. (1971). *Solid State Tech.* **14**, 33–40.
Haitz, R. H., Goetzberger, A., Scarlett, R. M., and Shockley, W. (1963). *J. Appl. Phys.* **34**, 1581–1590.
Hall, R. N., and Racette, J. H. (1964). *J. Appl. Phys.* **35**, 379–397.
Ham, F. S. (1959). *J. Appl. Phys.* **30**, 915–926.
Handelman, E. T., and Povilonis, E. I. (1964). *J. Electrochem. Soc.* **111**, 201–206.
Haneta, Y. (1964). *Jpn. J. Appl. Phys.* **4**, 69–70.
Hart, P. B., Etter, P. J., Jervis, B. W., and Flanders, J. M. (1967). *Brit. J. Appl. Phys.* **18**, 1389–1398.
Heiman, F. P. (1967). *IEEE Trans. Electron Devices* **ED-14**, 781–784.
Heine, V. (1966). *Phys. Rev.* **146**, 568–570.
Henderson, J. C. (1964). *Discuss. Faraday Soc.* **38**, 318.
Hirvonen, J. K., Brown, W. L., and Glotin, P. M. (1971). *In Proc. Int. Conf. Ion Implantation, 2nd* (I. Ruge and J. Graul, eds.), pp. 8–16. Springer-Verlag, New York.
Holland, L. (1964). *In* "The Properties of Glass Surfaces." Chapman and Hall, London.
Holt, D. B. (1966). *J. Phys. Chem. Solids* **27**, 1053–1067.
Horn, F. H. (1955). *Phys. Rev.* **97**, 1521–1525.
Hornstra, J. (1958). *Phys. Chem. Solids* **5**, 129–141.
Hrostowski, H. J., and Kaiser, R. H. (1959). *J. Phys. Chem. Solids* **9**, 214–216.
Hsieh, C-M., and Maher, D. M. (1973). *J. Appl. Phys.* **44**, 1302–1306.
Hsieh, C-M., Mathews, J. R., Seidel, H. D., Pickar, K. A., and Drum, C. M. (1973). *Appl. Phys. Lett.* **22**, 238–240.
Hu, S. M. (1973a). *Appl. Phys. Lett.* **22**, 261–264.
Hu, S. M. (1973b). *In* "Atomic Diffusion in Semiconductors" (D. Shaw, ed.), pp. 217–350. Plenum Press, New York.
Hu, S. M. (1975a). *J. Appl. Phys.* **46**, 1465–1469.
Hu. S. M. (1975b). *Appl. Phys. Lett.* **27**, 165–167.
Hu, S. M., and Poponiak, M. R. (1972). *J. Appl. Phys.* **43**, 2067–2073.
Hu, S. M., and Schmidt, S. (1968). *J. Appl. Phys.* **39**, 4272–4283.
Hu, S. M., and Yeh, T. H. (1969). *J. Appl. Phys.* **40**, 4615–4620.
Hu, S. M., and Yeh, T. H. (1970). *J. Appl. Phys.* **41**, 2153–2155.
Ing. S. W., Morrison, R. E., Alt. L. L., and Aldrich, R. W. (1963). *J. Electrochem. Soc.* **110**, 533–537.
Ino, H., Kawamura, T., and Yasufuku, M. (1964). *Jpn. J. Appl. Phys.* **3**, 692–697.
Irving, J. C. (1962). *Bell Syst. Tech. J.* **41**, 387–410.
Jaccodine, R. J. (1963). *Appl. Phys. Lett.* **2**, 201–202.

Jaccodine, R. J. (1964). *Appl. Phys. Lett.* **4**, 114–115.
Jaccodine, R. J., and Drum, C. M. (1966). *Appl. Phys. Lett.* **8**, 29–30.
Jaccodine, R. J. (1968). *J. Appl. Phys.* **39**, 3105–3108.
Jesser, W. A., and Kuhlmann-Wilsdorf, D. (1967). *Phys. Status Solidi* **19**, 95–105.
Joshi, M. L. (1965). *J. Electrochem. Soc.* **112**, 912–916.
Joshi, M. L. (1966). *Acta Metall.* **14**, 1157–1172.
Joshi. M. L., and Dash, S. (1966). *IBM J. Res. Develop.* **10**, 446–454.
Joshi. M. L., and Dash, S. (1967). *IBM J. Res. Develop.* **11**, 271–283.
Joshi, M. L., and Howard, J. K. (1970). *In* "Silicon Device Processing" (C. P. Marsden, ed.), pp. 313–364. NBS Spec. Publ. No. 337, Gaithersburg, Maryland.
Joshi, M. L., and Wilhelm, F. (1965). *J. Electrochem. Soc.* **112**, 185–188.
Joshi, M. L., Ma, C. H., and Makris, J. (1967). *J. Appl. Phys.* **38**, 725–734.
Joyce, B. A. (1974). *Rep. Progr. Phys.* **37**, 363–420.
Joyce, B. A., Bennett, R. J., Bicknell, R. W., and Etter, P. J. (1965). *Trans. Metall. Soc. AIME* **233**, 556–562.
Jungbluth, E. D., and Wang, P. (1965). *J. Appl. Phys.* **36**, 1967–1973.
Kaiser, W. (1957). *Phys. Rev.* **105**, 1751–1756.
Kaiser, W., and Breslin, J. (1958). *J. Appl. Phys.* **29**, 1292–1294.
Kaiser, W., and Keck, P. H. (1957). *J. Appl. Phys.* **28**, 882–887.
Kaiser, W., Frisch, H. L., and Reiss, H. (1958). *Phys Rev.* **112**, 1546–1554.
Kaller, A. (1959). *Jenaer Jahrb.* 181.
Katz, L. E. (1974). *J. Electrochem. Soc.* **121**, 969–972.
Kendall, D. L., and DeVries, D. B. (1969). *In* "Semiconductor Silicon" (R. R. Haberecht and E. L. Kern, eds.), pp. 358–403. Electrochem Soc. Softbound Symp. Ser.
Kikuchi, M., and Tachikawa, K. (1960). *J. Phys. Soc. Jpn.* **15**, 835–852.
Kinchin, G. H., and Pease, R. S. (1955). *Rep. Progr. Phys.* **18**, 1–51.
Kimerling, L. C., and Poate, J. M. (1975). *In* "Lattice Defects in Semiconductors, 1974," pp. 126–148. Inst. Phys., London.
Kohn, J. A. (1956). *Am. Mineral.* **41**, 778–784.
Kohn, J. A. (1958). *Am. Mineral.* **43**, 263–284.
Kooi, E. (1964). *J. Electrochem. Soc.* **111**, 1383–1387.
Kranzer, D. (1974). *Appl. Phys. Lett.* **25**, 103–105.
Kurtin, S., Shifrin, G. A., and McGill, T. C. (1969). *Appl. Phys. Lett.* **14**, 223–225.
Kurtz, A. D., Kulin, S. A., and Averbach, B. L. (1956). *Phys. Rev.* **101**, 1285–1291.
Lambert, J. L., and Reese, M. (1968). *Solid State Electron.* **11**, 1055–1061.
Lander, J. J., Schreiber, H., Buck, T. M., and Matthews, J. R. (1963). *Appl. Phys. Lett.* **3**, 206–207.
Large, L. N., and Bicknell, R. W. (1967). *J. Mater. Sci.* **2**, 589–609.
Lawrence, J. E. (1965). *J. Electrochem. Soc.* **112**, 796–800.
Lawrence, J. E. (1966a). *J. Electrochem. Soc.* **113**, 819–824.
Lawrence, J. E. (1966b). *J. Appl. Phys.* **37**, 4106–4112.
Lawrence, J. E. (1967). *Brit. J. Appl. Phys.* **18**, 405–409.
Lawrence, J. E. (1968a). *J. Electrochem. Soc.* **115**, 860–865.
Lawrence, J. E. (1968b). *Trans. Metall. Soc. AIME* **242**, 484–489.
Lawrence, J. E. (1969). *J. Appl. Phys.* **40**, 360–365.
Lawrence, J. E. (1973). *In* "Semiconductor Silicon" (H. R. Huff and R. R. Burgess, eds.), pp. 17–34. Electrochem. Soc. Softbound Symp. Ser. Chicago, Illinois.
Leamy, H. J., de Kock, A. J. R., Kimerling, L. C., and Ferris, S. D. (1975). *Appl. Phys. Lett.* **27**, 313–315.
Lederhandler, S., and Patel, J. R. (1957). *Phys. Rev.* **108**, 239–242.
Lemke, H. (1965). *Phys. Status Solidi* **12**, 125–131.

Levine, E., Washburn, J., and Thomas, G. (1967a). *J. Appl. Phys.* **38**, 81–87.

Levine, E., Washburn, J., and Thomas, G. (1967b). *J. Appl. Phys.* **38**, 87–95.

Li, C. H. (1966). *Phys. Status Solidi* **15**, 3–56, 419–450.

Light, T. B. (1962). *In AIME Metall. Soc. Conf. Proc.* (J. B. Schroeder, ed), Vol. 15, p. 137. Wiley (Interscience), New York.

Livshits, A. L. (1960). *In* "Electro-Erosion Machining of Metals," Butterworths, London.

Logan, R. A., and Peters, A. J. (1957). *J. Appl. Phys.* **28**, 819–820.

Madin, R., and Asher, W. (1950). *Rev. Sci. Instrum.* **21**, 881–883.

Maher, D. M., and Eyre, B. L. (1971). *Phil. Mag.* **17**, 1–6.

Manasevit, H. M., and Simpson, W. I. (1964). *J. Appl. Phys.* **35**, 1349–1451.

Manasevit, H. M., and Forbes, D. H. (1966). *J. Appl. Phys.* **37**, 734–739.

Manasevit. H. M. Miller, A., Morritz, F. L., and Nolder, R. (1965). *Trans. Metall. Soc. AIME* **233**, 540–549.

Martin, J. A. (1969). *In* "Semiconductor Silicon" (R. R. Haberecht and E. L. Kerns, eds.), pp. 547–557. Electrochem. Soc. Softbound Symp. Ser., New York.

Masters, B. J., Fairfield, J. M., and Crowder, B. L. (1970). In "Proc. 1st Int. Conf. Ion Implantation" (L. Chadderton and F. Eisen, eds.), pp. 81–85. Gordon and Breach, New York.

Masuda, K., Namba, S., Gamo, K., and Murakami, K. (1971). *In Proc. U.S.-Jpn. Semin. Ion Implantation in Semicond.* (S. Namba, ed.), pp. 19–26. Jpn. Soc. Prom. Sci., Tokyo.

Matare, H. F. (1969). *In* "Semiconductor Silicon" (R. R. Haberecht and E. L. Kern, eds.), 249–290. Electrochem. Soc. Softbound Symp. Ser., New York.

Matukura, Y., and Miura, Y. (1963). *Jpn. J. Appl. Phys.* **2**, 518–519.

Mayer, A. (1970). *RCA Rev.* **31**, 414–430.

Mayer, J. W., Ericksson, L., and Davies, J. A. (1970). *In* "Ion Implantation in Semiconductors," Academic Press, New York.

Mazey, D. J., Nelson, R. S., and Barnes, R. S. (1968). *Phil. Mag.* **17**, 1145–1161.

McDonald, R. A., Ehlenberger, G. G., and Huffman, T. R. (1966). *Solid State Electron.* **9**, 807–812.

McGreivy, D. J., and Viswanathan, C. R. (1974). *Appl. Phys. Lett.* **25**, 505–506.

Meek, R. L., and Gibbon, C. F. (1974). *J. Electrochem. Soc.* **121**, 444–447.

Meek, R. L., and Seidel, T. E. (1975). *J. Phys. Chem. Solids.* **36**, 731–740.

Meek, R. L., Seidel, T. E., and Cullis, A. G. (1975). *J. Electrochem. Soc.* **122**, 786–796.

Mendelson, S. (1964). *J. Appl. Phys.* **35**, 1570–1581.

Mendelson, S. (1965). *Bull. Am. Phys. Soc.* **10**, 365.

Mendelson, S. (1966). *Mater. Sci. Eng.* **1**, 42–64.

Mendelson, S. (1967). *J. Appl. Phys.* **38**, 1573–1578.

Mercier, J. (1968). *Rev. Phys. Appl.* **3**, 127–130.

Mercier, J. (1970). *J. Electrochem. Soc.* **117**, 812–814.

Mercier, J. (1971). *J. Electrochem. Soc.* **118**, 962–966.

Mets, E. J. (1965). *J. Electrochem. Soc.* **112**, 420–425.

Michel, A. E., Fang, F. F., and Pan, E. S. (1974). *J. Appl. Phys.* **45**, 2991–2996.

Miller, L. E. (1959). *In* "Properties of Elemental and Compound Semiconductors" (H. C. Gatos, ed.), Vol. 5, pp. 303–321. Metall. Soc. Conf. Ser., Boston, Massachusetts.

Miller, D. P., Moore, J. E., and Moore, C. R. (1962). *J. Appl. Phys.* **33**, 2648–2652.

Miller, D. P., Watelski, S. B., and Moore, C. R. (1963). *J. Appl. Phys.* **34**, 2813–2821.

Moline, R. A., and Cullis, A. G. (1975). Private communication.

Moline, R. A., Reutlinger, G. W., and North, J. C. (1973). *In Proc. Int. Conf. on At. Collisions in Solids, 5th*, pp. 159–173 Plenum Press, New York.

Morehead, F. F., Crowder, B. L., and Title, R. S. (1972). *J. Appl. Phys.* **43**, 112–1118.

Morizane, K., and Gleim, P. S. (1969). *J. Appl. Phys.* **40**, 4104–4107.

Morizane, K., Witt, A., and Gatos, H. C. (1967). *J. Electrochem. Soc.* **114**, 738–742.

Mroczkowski, R. S., Witt, A. F., and Gatos, H. C. (1968). *J. Electrochem. Soc.* **115**, 750–752.

Murray, L. A., and Kressel, H. (1967). *Electrochem. Tech.* **5**, 406–407.

Nakamura, M., Kato, T., and Oi, N. (1968). *Jpn. J. Appl. Phys.* **7**, 512–519.

Nelson, R. S., and Mazey, D. J. (1967). *J. Mater. Sci.* **2**, 211–216.

Nelson, R. S., and Mazey, D. J. (1968). *Can. J. Phys.* **46**, 689–694.

Nelson, R. S., Cairns, J. A., and Blanires, N. (1970). *In Proc. Int. Conf. Ion Implantation, 1st* (F. H. Eiser and L. T. Chadderton, eds.), pp. 305–308. Gordon and Breach, New York.

Nes, E., and Washburn, J. (1972). *J. Appl. Phys.* **43**, 2005–2006.

Netange, B., Cherki, M., and Baruch, P. (1972). *Appl. Phys. Lett.* **20**, 349–351.

Newman, R. C., and Wakefield, J. (1962). *In* "Metallurgy of Semiconductors" (J. Shroeder, ed.), pp. 201–207.

Newman, R. C., and Willis, J. B. (1965). *J. Phys. Chem. Solids* **26**, 373–379.

Nishizawa, J., Terasaki, T., Yagi, K., and Miyamoto, N. (1975). *J. Electrochem. Soc.* **122**, 664–669.

Noack, J. (1969). *Phys. Status Solidi* **32**, K17–K19.

Nolder, R. L., and Cadoff, I. (1965). *Trans. Metall. Soc. AIME* **233**, 549–556.

Nolder, R. L., Klein, D. J., and Forbes, D. H. (1965). *J. Appl. Phys,* **36**, 3444–3450.

Norris, C. B. (1972). *Appl. Phys. Lett.* **20**, 187–190.

North, J. C., and Gibson, W. M. (1970). *In Proc. Int. Conf. Ion Implantation 1st* (E. H. Eiser and L. T. Chadderton, eds.), pp. 143–148. Gordon and Breach, New York.

Notis, M. R., and Conard, G. P. (1964). *J. Appl. Phys.* **35**, 695–697.

Nozaki, T., Yatsurugi, Y., and Akiyama, N. (1970). *J. Electrochem. Soc.* **117**, 1566–1568.

Patel, J. R. (1964). *Discuss. Faraday Soc.* **38**, 1–10.

Patel, J. R. (1968). *In* "Semiconductor Silicon" (R. R. Haberecht and E. L. Kern, eds.), pp. 632–637. Electrochem. Soc. Softbound Symp. Ser., New York.

Patel, J. R. (1973). *J. Appl. Phys.* **44**, 3903–3906.

Patel, J. R., and Batterman, B. W. (1963). *J. Appl. Phys.* **34**, 2716–2721.

Patel, J. R., and Chaudhuri, A. R. (1963). *J. Appl. Phys.* **34**, 2788–2799.

Patel, J. R., and Freeland, P. E. (1967). *Phys. Rev. Lett.* **18**, 833–835.

Patrick, W. J. (1970). *In* "Silicon Device Processing" (C. P. Marsden, ed.), pp. 442–449. NBS Spec. Publ. No. 337, Gaithersburg, Maryland.

Pavlov, P. V., Tetel'baum, D. I., Zor n, E. I., and Alekseer, V. I. (1967). *Sov. Phys.-Solid State* **8**, 2141–2146.

Pearce, (1974). Private communication.

Penning, P. (1958). *Philips Res. Rep.* **13**, 17–36

Penning, P. (1959). *Philips Res. Rep.* **14**, 337–345.

Petroff, P. M., and de Kock, A. J. R. (1975). *J. Cryst. Growth* **30**, 117–124 and also *Ext. Abstr. Am. Conf. Cryst. Growth, 3rd, Palo Alto, 1975*.

Petroff, P. M., Rozgonyi, G. A., Sheng, T. T. (1976). *J. Electrochem. Soc.* **123**, 565–570.

Pickar, K. A., and Dalton, J. V. (1970). *Radiat. Effects* **6**, 89–94.

Picraux, S. T. (1972). *Appl. Phys. Lett.* **20**, 91–93.

Plantinga, G. H. (1969). *IEEE Trans. Electron. Develop.* **ED-16**, 394–400.

Plaskett, T. S. (1965). *Trans AIME* **233**, 809–812.

Pomerantz, D. (1967). *J. Appl. Phys.* **38**, 5020–5026.

Pomerantz, D. I. (1972). *J. Electrochem. Soc.* **119**, 255–265.

Porter, W. A. (1970). Texas A&M Univ. Rep. No. TR3.

Porter, W. A., Drew, D. D., and Linder, J. S. (1972). *J. Appl. Phys.* **43**, 1477–1479.

Prussin, S. (1961). *J. Appl. Phys.* **32**, 1876–1881.

Prussin, S. (1974). *J. Appl. Phys.* **45**, 1635–1642.

Pugh, E. N. and Samuels, L. E. (1964). *J. Electrochem. Soc.* **111**, 1429–1431.

Queisser, H. J. (1961). *J. Appl. Phys.* **32**, 1776–1780.

Queisser, H. J., and van Loon, P. G. G. (1964). *J. Appl. Phys.* **35**, 3066–3067.
Queisser, H. J., Finch, R. H., and Washburn, J. (1962). *J. Appl. Phys.* **33**, 1536–1537.
Ravi, K. V. (1972). *J. Appl. Phys.* **43**, 1785–1792.
Ravi, K. V. (1973). *Metall. Trans.* **4**, 681–689.
Ravi, K. V., and Varker, C. J. (1973). *In* "Semiconductor Silicon" (H. R. Huff and R. R. Burgess, eds.), pp. 136–149. Electrochem. Soc. Softbound Symp. Ser., Chicago, Illinois.
Ravi, K. V., and Varker, C. J. (1974a). *J. Appl. Phys.* **45**, 263–271.
Ravi, K. V., and Varker, C. J. (1974b). *Appl. Phys. Lett.* **25**, 69–71.
Ravi, K. V., Varker, C. J., and Volk, C. E. (1973). *J. Electrochem. Soc.* **120**, 533–541.
Read, W. T. (1954). *Phil. Mag.* **45**, 775–796.
Rieger, H. (1964). *Phys. Status Solidi* **7**, 685–699.
Rinder, W. (1962). *J. Appl. Phys.* **33**, 2479–2480.
Robinson, P. H., and Dumin, D. J. (1968). *J. Electrochem. Soc.* **115**, 75–78.
Robinson, P. H., and Heiman, F. P. (1971). *J. Electrochem. Soc.* **118**, 141–143.
Robinson, P. H., and Wance, R. O. (1973). *RCA Rev.* **34**, 616–629.
Ronen, R. S., and Micheletti, F. B. (1975). *Solid State Technol.* 39–46.
Ronen, R. S., and Robinson, P. H. (1972). *J. Electrochem. Soc.* **119**, 747–752.
Rozgonyi, G. A., Petroff, P., and Read, M. H. (1975). *J. Electrochem. Soc.* **121**, 1725–1729.
Rupprecht, H., and Schwuttke, G. H. (1966). *J. Appl. Phys.* **37**, 2862–2866.
Sanders, I. R., and Dobson, P. S. (1969). *Phil. Mag,* **20**, 881–893.
Sandmann, H. (1969). *In* "Semiconductor Silicon" (R. R. Haberecht and E. L. Kern, eds.), pp. 124–131. Electrochem. Soc. Softbound Symp. Ser., New York.
Sato, Y. and Arata, H. (1964). *Jpn. J. Appl. Phys.* **3**, 511–515.
Schlotterer, H. (1968). *Solid-State Electron.* **11**, 947–956.
Schmidt, P. F., and Stickler, R. (1964). *J. Electrochem. Soc.* **111**, 1188–1189.
Schonherr, E. (1968). *J. Cryst. Growth* **2**, 313–321.
Schroder, D. K., and Rai-Choudhury, P. (1973). *Appl. Phys. Lett.* **22**, 455–457.
Schwettman, F. N. (1974). *J. Appl. Phys.* **45**, 1918–1920.
Schwuttke, G. H. (1961). *J. Electrochem. Soc.* **108**, 163–167.
Schwuttke, G. H. (1962). *J. Appl. Phys.* **33**, 1538–1540.
Schwuttke, G. H. (1963). *J. Appl. Phys.* **34**, 1662–1664.
Schwuttke, G. H. (1974). Tech. Rep. 1–5 from Contract DAHC15-72-C-0274.
Schwuttke, G. H., and Fairfield, J. M. (1966). *J. Appl. Phys.* **37**, 4394–4396.
Schwuttke, G. H., and Queisser, H. J. (1962). *J. Appl. Phys.* **33**, 1540–1542.
Schwuttke, G. H., Brack, K., Gardner, E. E., and DeAngelis (1968). *In* "Radiation Effects in Semiconductors" (F. L. Vook, ed.), p. 406. Plenum Press, New York.
Seeger, A., and Chik, K. P. (1968). *Phys. Status Solidi* **20**, 455–542.
Seidel, T. E., and MacRae, A. U. (1970). *In Proc. Int. Conf. Ion Beams, 1st* (F. H. Eisen and L. T. Chadderton, eds.), pp. 149–154. Gordon and Breach, New York.
Seidel, T. E., and Meek, R. L. (1973). *In* "Ion Implantation in Semiconductors and Other Materials" (W. L. Crowder, ed.), pp. 305–315. Plenum, New York.
Seidel, T. E., Meek, R. L., and Cullis, A. G. (1975a). *J. Appl. Phys.* **46**, 600–609.
Seidel, T. E., Payne, R. S., Moline, R. A., Costello, W. R., Tsai, R. C. C., and Gardner, K. R. (1975b). Ext. Abstr. IEDM, Washington, D.C.
Sell, D. D., and MacRae, A. U. (1970). *J. Appl. Phys.* **41**, 4922–4932.
Senitzky, B., and Moll, J. L. (1958). *Phys. Rev.* **110**, 612–620.
Sequin, C. H. Zimany, E. J., Tompsett, M. F., and Fuls, E. N. (1976). *IEEE J. Solid State Circuits* **Sc-11**, 1–8.
Sheff, S. (1967). *Electrochem. Tech.* **5**, 47–52.
Shiraki, H. (1975). *Jpn. J. Appl. Phys.* **14**, 747–752.
Shockley, W. (1953). *Phys. Rev.* **91**, 228.

Shockley, W. (1961). *Solid-State Electron.* **2**, 35–67.

Shul'pina, I. L., Lainer, L. V., Mil'vidskii, M. G., and Rashevskaya, E. P. (1967). *Sov. Phys. Solid State* **9**, 1291–1294.

Sigmund, P. (1969). *Appl. Phys. Lett.* **14**, 114–117.

Sigmund, P., and Sanders, J.-B. (1967). *In Proc. Int. Conf. Appl. Ion Beams Semicond. Tech.* (P. Glotin, ed.), pp. 215–233. Centre d'Etudes Nucleaires, Grenoble.

Silcock, J. M., and Tunstall, W. J. (1964). *Phil. Mag.* **10**, 361–389.

Sirtl, E., and Adler, A. (1961). *Z. Metall.* **52**, 529–531.

Soper, R. B. (1970). *In* "Silicon Device Processing" (C. P. Marsden, ed.), pp. 412–418. NBS Spec. Publ. No. 337.

Stein, H. J., Vook, F. L., Brice, D. K., Borders, J. A., and Picraux, S. T. (1970). *In Proc. Int. Conf. Ion Implantation, 1st* (L. Chadderton and F. Eisen, eds.), pp. 17–24. Gordon and Breach, New York.

Stickler, R., and Booker, G. R. (1962). *J. Electrochem. Soc.* **109**, 743–744.

Stickler, R., and Booker, G. R. (1963). *Phil. Mag.* **8**, 859–876.

Stickler, R. and Faust, J. W. (1966). *Electrochem. Tech.* **4**, 339–401.

Struthers, J. D. (1956). *J. Appl. Phys.* **27**, 1560.

Sugita, Y., Tamura, M., and Sugawara, K. (1969). *J. Appl. Phys.* **40**, 3089–3094.

Sukhodreva, I. M. (1964). *Sov. Phys-Solid State* **6**, 311–313.

Swanson, M. L., Parsons, J. R., and Hoelke, C. W. (1971). *In* "Radiation Effects in Semiconductors" (J. W. Corbett and G.D. Watkins, eds.), pp. 359–367. Gordon and Breach, New York.

Sylwestrowicz, W. (1962). *Phil. Mag.* **7**, 1825–1845.

Taloni, A., and Rogers, W. J. (1970). *Surface Sci.* **19**, 371–379.

Tamura, M. (1973). *Appl. Phys. Lett.* **23**, 651–653.

Tamura, M., and Sugita, Y. (1970). *Jpn. J. Appl. Phys.* **9**, 368–375.

Tamura, M., and Sugita, Y. (1973). *J. Appl. Phys,* **44**, 3442–3444.

Tannenbaum, E. (1961). *Solid State Electron.* **2**, 123–132.

Teal, G. K., and Buehler, E. (1952). *Phys. Rev.* **87**, 190.

Theuerer, H. C. (1956). *Trans. AIME* **206**, 1316.

Thomas, D. J. D. (1963). *Phys. Status Solidi* **3**, 2261–2273.

Thomas, R. N., and Francombe, M. H. (1967). *Appl. Phys. Lett.* **11**, 134–136.

Thomas, R. N., and Francombe, M. H. (1971). *Surface Sci.* **25**, 357–378.

Trumbore, F. A. (1960). *Bell Syst. Tech. J.* **39**, 205–233.

Tung, S. K. (1962). *In* "Metallurgy of Semiconductor Materials" (*AIME Metall. Soc. Conf., Los Angeles, 1961*), p. 87. Wiley (Interscience), New York.

Turner, A. P. L., Vreeland, T., and Pope, D. P. (1968). *Acta Crystallogr.* **A24**, 452–458.

Unvala, B. A., and Booker, G. R. (1964). *Phil. Mag.* **9**, 691–701.

van der Merwe, J. H. (1963). *J. Appl. Phys.* **34**, 117–127.

van der Merwe, J. H. (1964). *In* "Single Crystal Films," pp. 139–163. Mcmillian, New York.

van der Merwe, J. H. (1970). *J. Appl. Phys.* **41**, 4725–4731.

Varker, C. J., and Ravi, K. V. (1974). *J. Appl. Phys.* **45**, 272.

Voltmer, F. W., and Padovani, F. A. (1973). *In* "Semiconductor Silicon" (H. R. Huff and R. R. Burgess, eds.), pp. 75–82. Electrochem. Soc. Softbound Symp. Ser., Chicago, Illinois.

Wagner, R. S., and Doherty, C. J. (1966). *J. Electrochem. Soc.* **113**, 1300–1305.

Wang, C. C., Gottlieb, G. E., Cullen, G. W., McFarlane, S. H. III, and Zaininger, K. H. (1969). *Trans. Metall. Soc. AIME* **245**, 441–454.

Wang, P., Pink, F. X., and Cupta, D. C. (1970). *In* "Silicon Device Processing" (C. P. Marsden, eds.), pp. 285–291. Nat. Bur. Std. Spec. Publ. No. 337, Gaithersburg, Maryland.

Washburn, J., Thomas, G., and Queisser, H. J. (1964). *J. Appl. Phys.* **35**, 1909–1914.

Webb, W. W. (1962). *J. Appl. Phys.* 33, 1961–1971.

Westmoreland, J. E., Mayer, J. W., Eisen, F. H., and Welch, B. (1970). *In Proc. Int. Conf. Ion Implantation, 1st* (L. Chadderton and F. Eisen, eds.), Gordon and Breach, New York.

Whitten, W. M., Heitz, A. J., and McNamara, J. R. (1964). Paper presented at Electrochem. Soc. Fall Meeting, Washington, D.C.

Willoughby, A. F. W. (1968). *J. Mater. Sci.* 3, 89–98.

Witt, A., Lichtensteiger, M., and Gatos, H. C. (1973). *J. Electrochem. Soc.* **120**, 1119–1123.

Wu, W-K., and Washburn, J. (1974). *J. Appl. Phys.* **45**, 1085–1090.

Yoshida, M., Arata, H., and Terunuma, Y. (1968). *Jpn. J. Appl. Phys.* **7**, 209–219.

Yu, K. K., Jordan, A. G., and Longini, R. L. (1967). *J. Appl. Phys.* **38**, 572–583.

Microstructure of Some Noncrystalline Ceramics: Origin and Meaning

L. D. PYE

New York State College of Ceramics
Alfred University
Alfred, New York

I. Introduction and Overview

During the past several decades the extensive growth of the field of materials science encompassing glass, polymers, ceramics and metallurgy, can generally be associated with advances in one of the following areas: processing, characterization, and properties of solids. Throughout much of this effort the structure of these materials, e.g., atomistic, microstructure, and phenomenalogical, has served as a focal point for not only developing new materials but increasing our understanding and exploitation of older ones as well. This has been especially true in ceramics where elementary microstructure analysis in terms of phase composition, geometry, and distribution was absolutely essential, if not critical, for the creation of several new classses of highly useful ceramics. This would include transparent sintered oxides, glass ceramics, and bioceramics, to name but a few.

None too surprising then, entire conferences have been assembled to deal with the subject of microstructures alone; their production, their analysis, and their significance (Fulrath and Pask, 1968).

Historically, the observation and interpretation of microstructures in ceramics have only dealt with crystalline materials. Indeed, until the late 1950s noncrystalline ceramics, e.g., oxide glasses, were regarded as sensibly free of all microstructure, at least in the sense of that encountered in crystalline ceramics. This viewpoint, while not held universally, was fostered in part by earlier triumphs of the random network theory of glass structure (Zachariassen, 1932; Warren, 1937) and was strengthened by the implicit glass engineering goal of this era: produce a homogeneous, single-phased product through fusion. As will be discussed below this concept of a microstructure-free vitreous state has been seriously questioned in the last decade (Roy, 1972), and while falling short of the extreme position of the earlier crystallite theory (Lebedev, 1921; Randall et al., 1930), this challenge has progressed sufficiently to occasion a rethinking of nearly all aspects of the fabrication and characterization of noncrystalline ceramics.

To a very large extent this more modern concept of vitreous ceramics evolved from the discovery of an ability to induce, by appropriate thermal treatment, a clearly discernible and well-controlled microstructure in ostensibly homogeneous glasses. This microstructure proved essential for achieving the catalyzed crystallization of glass, and is now recognized to be either crystalline (Armistead and Stookey, 1964) or noncrystalline (Nordberg, 1944; Hammel, 1965).[†] Moreover, the occurrence of the noncrystalline variety is so ubiquitous that it has been rightfully inquired if its occurrence is not an intrinsic characteristic of all glass-forming melts (Roy, 1972). At present the more extreme position, especially when dealing with multicomponent glasses[‡], is to maintain that one truly has a homogeneous single-phased glass. Such claims can only be made after careful characterization, and even then some questions can nearly always be raised regarding techniques used, sensitivity, and thoroughness of the characterization undertaken. Extension of these ideas has led some to hold that an ultrafine-grained microstructure in glass is to be expected rather than automatically assumed to be absent. This would include single-component glasses as well as multicomponent ones. Thus, a decreased emphasis on a structural description

[†] The phenomenon of noncrystalline microphase separation is sometimes referred to as metastable liquid–liquid immiscibility, glass-in-glass separation, amorphous phase separation, or more recently as amorphous phase decomposition. Often, it is described simply as "phase separation." In classical phase equilibria concepts, however, the term "phase" includes both crystalline and glassy phases. Hence, "phase separation" is not as definitive as the other terms listed above and whenever possible, its usage should be avoided.

[‡] This would include glasses from several different chemical families, e.g., fluorides, metals, nonoxide chalcogenides, as well as more conventional oxide glasses.

based on a random atomic network has been offered (Porai-Koshits, 1960; Vogel, 1966; Roy, 1972). Closely related to this question, Ernsberger (1972) has suggested that at the very least a reconsideration of all properties of glass in light of a heterogeneous, nonuniform atomic structure is now in order.

The rather extensive research effort that led to this heterogeneous concept and much of the subsequent work performed in recognition of its importance, can be roughly categorized as follows.

1. Clarification of the energetics, kinetics, and modes of microstructure formation.

2. Establishment of the relationships between bulk properties and well-defined microstructure, especially in terms of modes of formation, growth or ripening processes, and extent of occurrence.

3. Application of this concept to explain a variety of experimental observations, including glass formation itself.

An underlying technological impetus for these efforts, quite apart from the obvious need to increase our understanding of glass structure, is the implicit recognition that microstructure development in viscous liquids (e.g., glasses) is a process which is highly dependent on composition, time, and temperature. Because of the ease with which these parameters can be varied, the process lends itself a fine degree of control—one that might be unique in materials science. It is this control, acting in concert with an inherent tendency for the process to begin at the atomic level, that is so attractive to the materials scientist. This effect has given rise to the concept of derivative materials technology (Britton, 1974). In other words, microstructure development in glass is widespread, easily controlled, and is a tool which holds promise as a precursor for producing new materials by altering the structure of glass and, hence, its properties. In the field of glass science, recognition of this overall concept ranks second only in importance to the idea that these same viscous liquids convert to a vitreous state in the first place.

The present survey was undertaken to provide a partial review of recent activity in this area. No attempt will be made to include all work reported thus far on this subject, much of which is first rate. Rather, the main purpose here is to first review the formation of microstructure in glasses, real and potential, and then summarize its influence on the properties of glass. The latter will be accomplished by identifying the major parameters of microstructures unique to glass-forming melts and then establishing their influence on selected properties of glass. It is believed that this dual approach is essential, since the properties of many glasses having a well-defined microstructure (e.g., heterogeneous glasses) are highly dependent on the manner in which this heterogeneity is produced. Moreover, it is useful to know those conditions which may induce heterogeneity so that in preparing new glasses, or examining older ones, the possibility of such glasses possessing a microstructure can be assessed. In other words, a full characterization of glasses through property measurements can be inferred only in the context of noting its method of preparation and attendant thermal history which might have given rise to or excluded microstructure development (this point is stressed below). For similar reasons, establishing some generalized microstructure–property relationships should be of equal value.

To keep this review within manageable bounds emphasis will be placed on bulk oxide glasses prepared through fusion. Thus glazes, thin films, and other nonoxide glasses will not be included. Also, there is a need to define at the outset a working definition of microstructure in glasses. Ideally, this definition would not be made unduly restrictive by incorporating arbitrary restraints related to size, composition, extent, or structure, e.g., crystalline versus noncrystalline. However, a material prepared by crystallizing a glass might be better described as a crystalline material with a glassy phase if the amount of crystalline microstructure is greatly in excess of the residual parent glass. Thus these materials will not be included. At the same time, however, other glasses can be thermally decomposed into two thoroughly interdispersed liquids which, when cooled to room temperature, convert to the vitreous state. Since both phases remain noncrystalline, this derivative solid will be regarded as appropriate for the present review even though the volume fractions of each can be equal. A more difficult problem arises when considering a potential ordering (in a crystalline sense) of a group of ions or atoms over an adjacent area of the glass network. This potential ordering follows, e.g., from an ability to distinguish between respective coordination numbers of the cations which might compose the group —B, Ti, Co, Si, etc., and also from our knowledge of bond distributions, interatomic distances, and identification of ionic configurations comprising the group. Although one might construe this ordered region as a "microstructural component" in an otherwise completely random network, it is suggested here that ordering within such groups might be better discussed in the context of an atomic theory of glass structure as opposed to a microstructural or heterogeneous one. Yet it must also be recognized that such a "microstructural component," when existing in its highest ordered state, constitutes the foundation of the crystallite theory of glass structure. Since no sharp delineations can be made between these extremes, the differences between the classical random network and crystallite theories of glass structure are sometimes diffuse at the local atomic level. For our purposes here then, we shall consider the microstructure of glass as "any component consisting of ions or atoms, or groups of ions and atoms, which represent a variation in composition or structure from a completely random network beyond or at the limit of short range order." This definition is similar to one proposed by Rindone (1974). We shall now turn our attention to the production of microstructure in glass and examine its significance.

II. Origin of Microstructure

A. *General Thermodynamics Basis*

The subject of the origin and occurrence of microstructure in noncrystalline solids is best introduced by discussing the thermodynamics and kinetics of phase transformation. The latter topics are treated extensively in Gibbs (1931), Slater (1939), Turnbull (1956), Swalin (1962), and Burke (1965). The main emphasis here will be to identify the conditions which give rise to microstructures in glass, and similarly, those which favor its exclusion. To accomplish this˙ a brief review of the thermodynamics necessary to describe equilibrium of phases within a system will be given. By extending these concepts to glass-forming systems, it will be seen that microstructure of several types are easily developed through suitable control of composition, melting techniques, and thermal history. Proceeding in this manner, their effects on some of the properties of glass will follow at once.

For our purposes here it is first necessary to note that the Gibbs free energy G, the heat content H, and the Helmholtz free energy F, are defined by the relations

$$G = H - TS \tag{1}$$

$$H = E + PV \tag{2}$$

$$F = E - TS \tag{3}$$

in which V, E, T, P, and S are volume, internal energy, temperature, pressure, and entropy, respectively. At constant temperature and pressure, the differential forms of these equations are

$$dG_{T,P} = dH - T\,ds \tag{4}$$

$$dH = dE + P\,dV \tag{5}$$

$$dF_{T,P} = dE - T\,dS \tag{6}$$

In dealing with solid and liquid ceramics at atmospheric pressure, the $P\,dV$ term in Eq. (5) is quite small. Thus,

$$dH \cong dE \tag{7}$$

and by comparing Eqs. (4) and (6) in light of Eq. (7), we conclude that

$$dG_{T,P} \cong dF_{T,P} \tag{8}$$

These equations are quite general; they may be applied to the system as a whole or to any phase within the system. Because of Eq. (8), we may also discuss the equilibrium of this system in terms of either G or F.

For true equilibrium, we require F to have the lowest possible numerical value and that

$$dF_{T,P} = 0 \tag{9}$$

A phase within a system fulfilling these requirements is said to be stable. Other phases having numerically greater values of F but still satisfying Eq. (9) are said to be metastable. Thus it is possible for a metastable phase to transform into a more stable one, e.g., one with a lower value of F. This energy decrease constitutes a driving force for such a transformation to take place. It follows that in order for a transformation to occur, a stable phase must necessarily be brought to a condition where it becomes metastable. This is usually accomplished by altering temperature or pressure, or both.

It is important to note that the existence of an energy difference between two phases is no guarantee that one will transfer into the other (Burke, 1965). In spite of such differences, there are countless examples in materials science where many systems can exist indefinitely in a metastable state, e.g., diamond at room temperature and pressure, or for that matter, noncrystalline solids. The reason for this is that the velocity of a transformation does not depend entirely on the driving force giving rise to the transformation. In other words, the kinetics of a transformation can be governed by factors other than this driving force. This follows from a requirement that a phase undergoing transformation must first pass through a transition state which momentarily increases its energy. This concept is more clearly illustrated in Fig. 1. Here, the Gibbs free energy G is shown for an atom or group of atoms during reaction at an initial, intermediate (transition), and final state. The reaction coordinate is any variable that defines the progress of the reaction. ΔG is the driving force for the reaction. The energy difference between the initial

Fig. 1. Schematic relationship between thermal activation energy and driving force of a reaction.

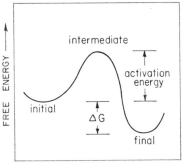

and intermediate state is referred to as an activation free energy. Because of the role of temperature in providing sufficient energy for the atoms to overcome this energy barrier and thus undergo reaction, it is sometimes called a thermal activation energy. In phase transformation phenomena, this activation energy is sometimes associated with diffusion across an interface. The field of catalysis is based on a proven ability to reduce this activation energy, e.g., increase the reaction velocity, independently of the driving force. As will be explained more fully, similar remarks are applicable to the catalyzed crystallization of glass.

We shall now attempt to apply the above concepts to both one- and multicomponent systems which form glasses when cooled from their liquid state, and also show the potential formation of microstructures in such glasses. To this end Fig. 2 describes the free energy–temperature relationships for a one-component system at atmospheric pressure. In Fig. 2a these relationships are given for two solid phases S_1 and S_2 which undergo a

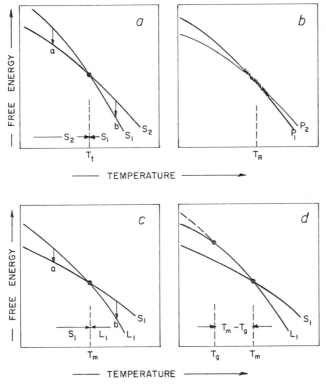

Fig. 2. Schematic free energy–temperature relationships for a solid–solid transformation (a); second-order transformation (b); solid–liquid transformation (c); and glass transformation (d).

reversible transformation at T_t. The overall decrease in free energy for each phase with increasing temperature follows directly from Eq. (3) in which the term $(-TS)$ becomes increasingly dominate at higher temperatures and, hence, a drop in F. In Fig. 2a, S_1 is stable above T_t; below T_t, S_2 is stable. This corresponds to our previous concept that the phase with the lowest value of F at a given temperature is the stable phase. By supercooling S_1, below T_t or by superheating S_2 above T_t, both phases become metastable and it is seen that a driving force now exists for their respective transformations at (a) and (b). As mentioned earlier, the existence of this driving force is not a guarantee that the transformation will take place, only that it can. At T_t the free energies of both phases are sensibly equal and each can coexist indefinitely at this temperature.

It is often useful to classify transformations in terms of the continuity of the first-, second-, and third-order derivatives of the Gibbs free-energy function. This classification scheme was first suggested by Eherenfest (1933), and while subject to some ambiguity (Ubbelohde, 1957), nonetheless provides some insight for the present discussion. In this scheme, transformations showing discontinuities in the first derivatives of G are termed first order and may be associated with sharp intersections of the respective functions for each phase at the transformation point, e.g., Fig. 2a. Since

$$\left(\frac{\partial G}{\partial T}\right)_P = -S, \qquad \left(\frac{\partial G}{\partial P}\right)_T = V \tag{10}$$

and

$$\left[\frac{\partial(G/T)}{\partial(1/T)}\right]_P = H \tag{11}$$

we may expect discontinuities in entropy, volume, and enthalpy at the transformation point. Fusion, vaporization, and many other solid–solid transformations are often described as being first order. The author is unaware of any liquid–liquid transformation classified as first order. Second-order transformations are defined as those having a continuous Gibbs function and its first derivatives, but having discontinuities in its second derivatives. In this case, the intersection of the Gibbs function for each phase is thought to be smeared as is indicated in Fig. 2b (Ubbelohde, 1952). Evidently, in this case by Eqs. (10) and (11), we anticipate continuous changes of S, V, and H and discontinuous changes in heat capacity C_P, compressibility β, and thermal expansion α, since

$$\left(\frac{\partial^2 G}{\partial T^2}\right)_P = \frac{-C_P}{T} \tag{12}$$

$$\left(\frac{\partial^2 G}{\partial P^2}\right)_T = -\beta V \tag{13}$$

and

$$\frac{\partial^2 G}{\partial P \partial T} = \left(\frac{\partial V}{\partial T}\right)_P = \alpha V \tag{14}$$

Examples of second-order transformations include order–disorder phenomena, onset of ferromagnetism, ferroelectricity, and superconductivity. Ostensibly, second-order transformations take place without an accompanying change of phase, at least in the sense described above. Although third-order transformations can be described by extension of these ideas, no such transformations have yet been reported (Swalin, 1962).

B. Application to Glass-Forming Systems

1. THE GLASS TRANSITION

We shall now attempt to apply these concepts to glass-forming systems. For clarity, an overview of glass melting (couched in the above context) will be presented first. This will be followed by an illustration of the role of these concepts in the potential formation of microstructure in such systems. To this end, we should first note that there are two distinct steps required to form a glass by cooling a melt. The first involves heating a crystalline phase S (or other disordered states, e.g., gels) above their melting point T_m so as to induce the solid–liquid transformation, $S_1 \rightarrow L_1$, at b (see Fig. 2c). The second, and by far the more critical, is to quench the liquid phase thus formed to a lower temperature where it solidifies in the absence of reverse transformation $L_1 \rightarrow S_1$ at some temperature a. Inherent in this process of solidifying without crystallization is a continuous and rapid increase of melt viscosity with falling temperature, e.g., $10^{-3} \rightarrow 10^{14}$ N sec m^{-2}.[†] Achieving this higher viscosity represents a quasiconversion to the solid state, albeit a noncrystalline one, and is indicative of a rigidity which allows a molecular rearrangment of the melt to take place over a period of 10^5 sec (Turnbull, 1969). This rigidity is sufficient to allow transmission of a shear ware, i.e., a solid.

The temperature at which the melt attains this viscosity is commonly referred to as the glass transformation temperature T_g.[‡] It is closely associated with the temperature at which the liquid phase gains an excess of free energy. This latter situation can be depicted as shown in Fig. 2d. Designation of a glass transformation temperature in a thermodynamic sense is regarded here as the most fundamental view of this effect since it represents that temperature at which the melt becomes metastable with respect to both the liquid and crystalline phases of this system. In short, the melt is no longer strictly a supercooled liquid; it has transformed into a new state, one

---

[†] Poise = 10^{-1} N sec m^{-2}

[‡] T_g is also referred to as the glass transition temperature.

which is described as vitreous. This transformation is sometimes classified as second order (Gee, 1947; Gibbs, 1960).[†] Coincident with this phenomenon, and in keeping with our previous definitions of second-order transformation, there are measurable changes in heat capacity, specific volume, and expansion coefficient of the melt which are induced by this transformation. By measuring these properties it is possible to estimate the value of T_g rather precisely.

Since the transformation temperature is moderately dependent on quenching rate (higher rates increase T_g), this has led to the concept of a glass transformation range as opposed to a single and wholly unique transformation temperature. It is for this reason that the properties of glass are slightly dependent on the thermal history of the melt range. Alternatively, the properties of glass change at measurable rates when annealed in the transformation range. Measurement of T_g is useful in establishing the temperature interval through which a melt must be quenched $(T_m - T_g)$ if a vitreous state is to be achieved. It is during this interval (or time span) that the two time-dependent events associated with crystallization—nucleation and growth—normally take place. Both events (quenching and crystallization) are time dependent and, by inference, competitive (Pye, 1972). At the same time, $T_m - T_g$ represents the interval where a glass can normally be reheated to induce crystallization. Since there is a continuous decrease of viscosity with increasing temperature, and in view of the dependence of crystallization on viscosity, this latter process lends itself to a high degree of control by correct choice of temperature and time of reheating.

2. MELTING PHENOMENA

a. Silica glass. In this section an attempt will be made to apply some of the above thermodynamic concepts to silica. In order to correlate much of the data to be presented and also to provide a basis for discussion, a schematic free-energy diagram has been constructed for this system (Fig. 3). This diagram is based on the work of several authors as is indicated in the caption of this diagram. Its approximate nature is emphasized.

In principle, the free-energy–temperature relationship of each phase can be calculated through Eq. (3) providing adequate thermodynamic data are available. In the absence of such data, dF/dT is assumed here to be constant at all temperatures and for each phase and while not completely rigorous, is sufficient for present purposes. Also the ordinate in Fig. 3 has been greatly expanded for clarity. Accordingly, the relative free energy of quartz, cristobalite, and liquid silica are depicted in Fig. 3. Thus, using the minimum free energy criterion for establishing the relative stability of these phases, it is

† However, see Rehage and Bonchard (1973) and Davies and Jones (1953) for a discussion of this matter.

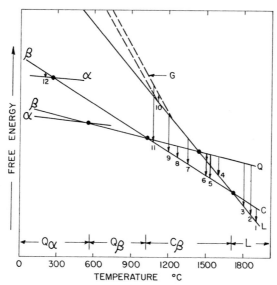

Fig. 3. Schematic free energy–temperature relationship for pure silica based on reported transformations in this system. Transformations 1, 2, 3, 4, and 5 after Mackenzie (1960); 6 (Chaklader and Roberts, 1958); 7 (Wagstaff, 1968); 8 (Grimshaw *et al.*, 1956); 9 (Gokularathnam *et al.*, 1972); 11 (Joly, 1901); 12 (Gorlich *et al.*, 1974). Q—quartz; C—cristobalite; L—liquid.

seen that this construction allows for the experimental observations that α-quartz is more stable from 0–573°C, β-quartz from 573–1025°C, cristobalite from 1025–1710°C, and above 1710°C, liquid silica.[†] The construction also accounts for the metastable melting of quartz at 1440°C, the α–β cristobalite inversion at 259°C, and a glass transition range spanning 1060–1200°C. This range, as opposed to a fixed temperature, is derived from the variation of the glass transition temperature of silica with water content (Bruckner, 1971). Depending on water content, melt viscosity within this range is approximately 10^{14} N sec m^{-2}, e.g., the Turnbull criterion was applied in estimating this range. Thus, a range of temperature is more appropriate in the absence of specifying water content.

The vertical lines indicate those temperatures where respective metastable and stable phase transformations have been reported. The transformation of one metastable phase into another before final conversion into a stable phase, such as indicated in transition (6), is sometimes referred to as Ostwald's law. Yet this law is not always observed, as is shown in transition (2). In

---

[†] Tridymite is not indicated here as a stable polymorph of *pure* silica (see Holmquist, 1961). Much controversy surrounds the existence of this phase; this question will be dealt with in a future publication.

keeping with our previous discussion, Fig. 3 provides no information regarding the amount of time required for the noted transitions to take place, only that they can.

Consider now the superheating of either cristobalite or quartz so as to induce transitions (1), (2), or (3), and then examining the quenched melt at room temperature in order to determine the extent to which melting has taken place. Appropriate tools for making this determination might include x-ray diffraction, electron microscopy, infrared absorption, and hardness–temperature relationships of the cooled samples. Such an experiment was conducted by MacKenzie (1960). His results indicated a gradual conversion to liquid phase, one which could be quenched to a noncrystalline solid at room temperature, e.g., vitreous silica. For melting times of one-half hour at $1760-1800°C$, residual microcrystals $(0.1 \rightarrow 0.5 \ \mu m)$ of either quartz or cristobalite were observed, depending, of course, on the starting material. At temperatures greater than $1800°C$ no unmelted crystals were detected. However, hardness–temperature relationships for glasses melted at $1800°C$ for one-half hour, were very similar to quartz. It was, therefore, inferred the latter glass might also contain microcrystals, albeit nondetectable by electron microscopy and/or x-ray diffraction. Glasses melted at $1900°C$ did not show this behavior.

At a slightly earlier date, Westbrook (1960) conducted similar hardness–temperature studies on both pure and multicomponent silica glasses. Like MacKenzie, his results indicated discontinuities at the quartz inversion temperature, but he also observed discontinuities around $70°C$ which were associated with a potential $\alpha-\beta$ cristobalite inversion. Westbrook's data are summarized in Fig. 4. Here, Fig. 4a was obtained on vitreous silica produced by melting quartz (no melting temperatures or times were given), Fig. 4b on vitreous silica from melting cristobalite (note the absence of a discontinuity at $573°C$). Perhaps most important of all, Fig. 4c represents data on vitreous silica produced by vapor-phase hydrolysis of $SiCl_4$. In Westbrook's view, the data of Fig. 4a and b imply a structural vestige which is reminiscent of the parent crystal structure from which the vitreous silica is derived. Much of MacKenzie's work also correlates with this concept. However, in the case of vitreous silica produced by hydrolysis of $SiCl_4$, there is no polymorphic parent; molecular SiO_2 created by reaction of $SiCl_4$ with H_2O is allowed to condense on a substrate whose temperature is well below the melting point of cristobablite. Groupings of such silica molecules, once deposited, are probably closest in atomic structure to liquid silica, e.g., maximum disorder. To account for regions in the cooled glass which might be structurally similar to the vestiges suggested by Westbrook, and thus give rise to the data of Fig. 4c, one might assume that the transitions (7) or (9) have taken place to a limited extent during the condensation pro-

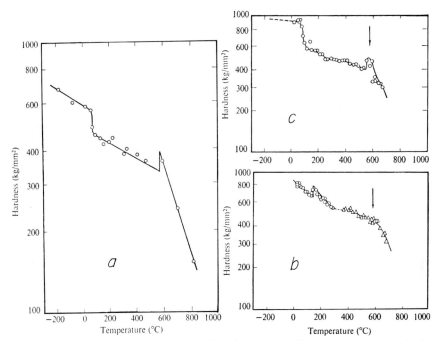

Fig. 4. Hardness–temperature relationships for vitreous silica; (a) quartz melted, (b) cristobalite melted, (c) hydrolysis of $SiCl_4$, Corning 7943 type (after Westbrook, 1960).

cess. A *limited* conversion of a liquid phase below its freezing point to a more stable crystalline modification which might extend over a few inter-atomic distances, is not new in materials science; it is often referred to as nucleation. Whether this mechanism is the only one contributing to West-brook's observation is by no means certain, at least on the basis of data now available. It is at least clear that the properties of vitreous silica are related to its method of fabrication and the starting materials. This relation-ship will be raised again in the context of a general discussion on the structure of glass.

Other property measurements on vitreous silica are also suggestive of structural transitions taking place within this material when heated. For example, Görlich *et al.* (1974) compared the infrared spectra of α and β cristobalite to that observed for vitreous silica after prolonged heat treat-ment at 210°C. They concluded that transition (12) in Fig. 3 of $\beta \rightarrow \alpha$ cris-tobalite was operative in vitreous silica. In this case, their material was prepared by melting quartz. Similarly, Stozharov (1958) discussed the cor-relation of refractive index with thermal history below the glass transition temperature. Again the temperature dependence of refractive index was thought to relate to possible phase transformation within several glasses. Thermal expansion measurements are also discussed in this latter work in the context of the above. These same property measurements were important in Lebedev's original concept of glass structure (as discussed by Volenkov and Porai-Koshits, 1936).

b.   Multicomponent glasses

(1)   *General considerations.*   In discussing the structure of multicomponent glasses after a series of measurements, it is often assumed that the glasses used in such studies were originally homogeneous in both a chemical and physical sense. To obtain such homogeneity, experimental glass melts are often carried out in inert crucibles (e.g., platinum), stirred for extended periods of time, or sometimes even quenched, crushed, and remelted repeatedly. Occasionally, the use of a gel technique is employed (Roy, 1969). In the absence of these precautions, regions of obvious inhomogeneity in the cooled glass, e.g., bubbles, cords, unmixed and undissolved oxides, are often visible to the naked eye. For small laboratory melts, the origin of chemical inhomogeneity, aside from corrosion effects, is found in both a tendency for the denser oxide components to sink to the bottom of the crucible during melting, and from insufficient melting times for complete interdiffusion of all components (Cooper, 1972). Upon casting, the regions of nonuniform composition are intermixed. The resulting nonhomogeneous areas are sometimes referred to as straie or cords. In commercial glasses, straie is often caused by batching errors or extended refractory corrosion. Physical inhomogeneities can be caused by improper casting techniques, nonuniform quenching, and also from poor annealing.

The extent to which these difficulties are overcome is always relative; one simply controls his melting procedures until the homogeneity of the resulting glass falls within acceptable limits. The key phrase here is "acceptable limits." Obviously, it is strongly influenced by one's ability to detect inhomogeneity and also by the intended use of the resulting glass. Similarly, as we have seen in previous sections, property measurements are often used to infer the presences of inhomogeneity on both an atomic and microstructural level. The danger of claiming a completely homogeneous glass is obvious; one's method of detection may simply lack sensitivity. With these considerations in mind, a careful explanation of melting history and procedures is highly desirable when reporting new results which relate to the homogeneity and structure of glass.

(2)   *Laminar microstructure.*   The occurrence of a laminar microstructure in flat and container glasses has been recognized for some time (Englert and Fuhrman, 1968). A typical example is shown in Fig. 5. This microstruc-

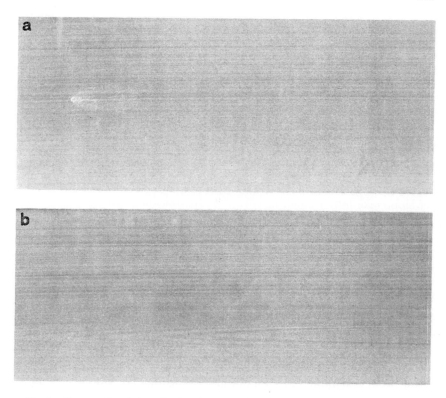

Fig. 5. Cross sectional view of striated structure in window glass (ream); (a) normal structure, parallel layers, (b) abnormal structure, intertwined layers. For clarity photographs do not include surfaces. Magnification $\sim 40 \times$ (courtesy of E. L. Swartz, PPG Industries).

ture, sometimes referred to as ream in deference to its similarity to a ream of paper, is composed of heterogeneous layers (~ 10-μm thick) which run parallel to the direction in which the glass is drawn from a manufacturing tank. The chemical differences between these layers are estimated to be less than 0.010%. Yet they are readily observed in a microscope with oblique illumination. Their origin lies in incomplete melting times, insufficient mixing, and the overall melting process used to produce these types of glasses, e.g., continuous melting, surface charging, and drawing techniques. This stratification is also observed in rolled optical glass (microscope slides).

For plate or sheet glass, the parallelism of these layers is crucial. With high parallelism there is little optical distortion of objects viewed flatwise through a pane of glass. When these layers are bent, twisted or somehow intertwined, however, distortion is quite evident. As a result, process control in plate or sheet glass engineering is acutely concerned with the melting and forming conditions which aid or impede this parallelism. Thus, we have here a prima facie example of the need to control microstructure of a common commercial glass product which is ordinarily construed to be free of any microstructure in the classical materials science concept. In container glass, ream is of less importance since optical distortion of the contents of a container by this mechanism is not especially evident, even when the parallelism of these layers is at a minimum. This is probably due to the proximity of the container walls to its contents as opposed to viewing a distant object through a pane of glass.

The literature abounds, of course, with scores of property studies on microscope slides which fail to even mention their inherent microstructure let alone take into account its effect on experimental results.

(3) Melting history. As indicated in the above, it has been recognized for some time that the quality of glass obtainable for commercial and experimental use is strongly influenced by the melting process. Thus, it is observed that producing a homogeneous product void of unmelted batch, entrapped gaseous inclusions, and unmixed glass requires control over melting temperatures, times, and melt atmospheres. The concept that beyond these obvious factors there exists a relationship between the properties of glass and a melt structure sensitive to the melting process itself, was probably first suggested by McKinnis and Sutton (1959). In advancing this concept, these authors sought to distinguish between the properties of glass which arise from structural units established in melting as opposed to those depending on the physical arrangement of these units with respect to one another. In their view, the latter are controlled by thermal history, e.g., annealing, whereas the former depend on melt history. Hence, they coined the term "melting history" as a means of more clearly delineating its influence on properties. As an example of this influence, a correlation between the strength and viscosity of E-glass with melting times and temperatures was offered in this same work.

In a later series of investigations, Rindone and coworkers (Rindone, 1969, 1974; Sproull and Rindone, 1973, 1974) attempted to verify and expand this concept. Besides viscosity and strength, other property measurements such as internal friction, infrared absorption, density, and light scattering were studied in addition to electron microscopy analysis. Glasses from the $Li_2O–K_2O–SiO_2$ system were chosen for this work. Initially all melting was done in air in platinum for varying times and temperatures; later efforts concentrated on varying partial pressure of oxygen over the melt by melting in vacuum, N_2, O_2, CO_2, and Ar. Results obtained were not construed to be an effect of base-glass volitilization although some initial volitilization was noted. Rod samples of glasses were drawn directly from the melt. Like McKinnis *et al.* the latter investigators found significant

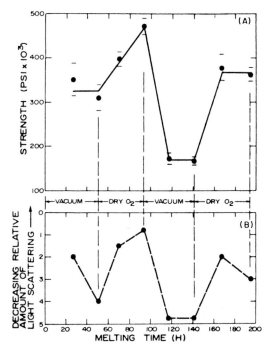

Fig. 6. Relationship of light scattering to measured strength of pristine glass when melted alternately in vacuum and dry O_2 for various times (after Sproull and Rindone, 1974).

changes in pristine strength by varying melting times. The effect was particularly influenced by melt atmosphere and could be reversed by alternate melting in dry O_2 and vacuum (Fig. 6).

A melt homogeneity argument was invoked to explain these observations; more homogeneous melts giving higher strengths. Direct transmission micrographs indicated an elongated potassium- or lithium-rich second phase about 500–1000-Å long and 200-Å wide. The amount of this phase was found to be proportional to the amount of light scattered by these glasses; hence, the correlation of Fig. 6.

The extent to which melt homogeneity manifested itself was viewed as an interplay of three distinct effects:

1. the degree of initial mixing and diffusion of the batch components (tending to increase homogeneity);

2. amount of platinum dissolved in the glass which would facilitate precipitation of a second phase;

3. the amount of oxygen available to preserve a homogeneous melt after initial mixing.

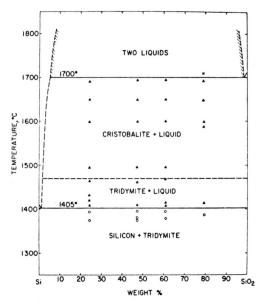

Fig. 7. Proposed liquid immiscibility in the system Si–SiO$_2$ (after Johnson and Maun, 1968).

The latter two points are of special interest here since a good deal of experimental glass melting is carried out in platinum. The relationship of both phase dissolution and development to melt atmosphere is also generally ignored. Johnston (1965), however, noted that soda–silica glasses containing SnO$_2$, TiO$_2$, or Sb$_2$O$_3$ developed a microstructure when melted in various atmospheres. With the exception of a melt containing Sb$_2$O$_3$, most samples gave no x-ray patterns suggesting there was sufficient time available for crystallization or that the microstructure was of a noncrystalline nature. At a somewhat later date, Johnson and Muan (1968) reported a large range of compositions in the Si–SiO$_2$ system where two liquid phases coexist (Fig. 7). This diagram, while of a preliminary nature, lends support for the need to control the oxygen content of the melt if maximum melt homogeneity is to be obtained and subsequently preserved. Thus, the inference by Sproull and Rindone (1974) that, at least in this system, melt homogeneity is related to phase equilibria in the Li$_2$O–K$_2$O–SiO$_2$–O$_2$ system is taken here as basically correct. Extension of these concepts to other systems will be followed with much interest.

3. LIQUID–LIQUID TRANSFORMATIONS

a. General aspects. It has long been recognized that binary and ternary oxide systems have areas where a homogeneous single-phase melt will decompose or transform into two liquid phases, e.g., exhibit liquid immiscibility.

An appreciation that this transformation could also take place metastably and that it is describable in terms of solution thermodynamics is a more modern emphasis. Several recent reviews treat this subject extensively (Cahn and Charles, 1965; Cahn, 1969; Seward, 1970). Hence, no effort will be made here to summarize all aspects of this phenomenon, rather several important features germane to the present discussion will be reviewed.

Accordingly, both metastable and stable liquid immiscibility for two hypothetical glass forming systems A–B and C–D are illustrated in Fig. 8. For composition C in Fig. 8a, above T_1, a single-phase liquid is stable. Reducing the melt temperature to T_2 will cause the melt to transform into two stable liquid phases. The compositions of these liquids are governed by the Lever rule as applied to the boundary limits of the two-liquid region, $X-Y-Z$.

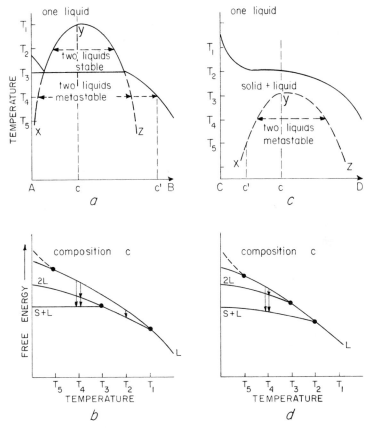

Fig. 8. Stable and metastable liquid immiscibility in hypothetical binary systems. Free energy–temperature relationships of (b) correspond to (a), those in (d) correspond to (c), the Ostwald transformations at T_4, as shown here, yield a solid and new liquid; in many glass–ceramic systems, the end product is simply a new stable crystalline phase (or phases).

This limit is sometimes referred to as a coexistence curve, whereas the area it encloses is often called a miscibility gap or an immiscibility dome. The process by which a viscous melt separates into two liquid phases is described as amorphous phase decomposition, or more simply, amorphous phase separation, since both phases are noncrystalline. At T_4 this same process is again possible, but in this case, the resulting liquid phases are metastable since ultimately they can in turn transform into a stable liquid of composition C′ and a polymorph of A (crystalline). Thus, the coexistence curve can be extended below T_3 (dashed lines). If, however, a melt undergoes amorphous phase separation only at either T_2 or T_4, and is then quenched below the glass transformation temperature T_5, the resulting glass will retain both liquid phases in a vitreous state. Hence, the occasional description, glass-in-glass separation.

This sequence can also be described in a schematic free-energy–temperature curve such as that shown in Fig. 8b. Observe that the metastable transformation L → 2L at T_4 is but another example of Ostwald's law. Also, here the free energy difference between the single and two-liquid melt curves represents the *overall* driving force for this transformation. It is not necessarily equivalent to the driving force for the nucleation of a new liquid phase (Christian, 1965).

In the system C–D, Fig. 8c, amorphous phase separation always takes place metastably, for example at T_4, since the immiscibility dome falls below the stable liquidus. Alternatively, a single-phase melt of composition C at T_1 must be supercooled below T_3 in order to observe the metastable transformation at T_4. As before, this situation can be depicted as shown in Fig. 8d. It is interesting to note that there may be a range of temperatures and compositions within the immiscibility dome where the *direct* transformation of a single phase liquid into solid and new liquid, e.g., L → S + L at T_4 in Fig. 8d is thermodynamically forbidden (Cahn, 1969). In other words, Ostwald's law must hold.

Two remarks regarding the glass transformation temperature are appropriate. First, it is possible to find glass-forming melts in many systems whose transformation temperature lies above the coexistence curve. For example, composition C′ in C–D might have a glass transformation temperature T such that $T_5 < T < T_4$. In this case, one would be less concerned about the development of a two-liquid microstructure in a melt of this composition since above the transformation temperature, an absence of driving force for amorphous phase separation is evident. Even though this driving force may develop below T_5, the process is not expected for kinetic reasons. Hence, the melt should remain homogeneous until it converts to the vitreous state. It is possible that the coexistence curve always lies below the glass transformation temperature. This may be the reason for the absence of liquid immiscibility in such systems as K_2O–SiO_2, Cs_2O–SiO_2, and Rb_2O–SiO_2 (Charles, 1967). An obvious corollary here is that if one wishes to avoid a two-liquid-phase glass, or conversely, enhance its development, the coexistence curve must be known in relation to the glass transformation temperature for all compositions of interest.

Second, if a melt undergoes amorphous phase separations during quenching or subsequent annealing, then the resulting glass should exhibit two glass transformation temperatures when heated, one for each phase developed. This has been observed in several systems utilizing either thermal expansion or heat capacity measurements; for example, Na_2O–SiO_2 (Maklad and Kreidl, 1971—thermal expansion), Pb–Se–Ge–As (Moynihan et al., 1971—heat capacity), and PbO–GeO_2–SiO_2 (Topping et al., 1974—thermal expansion). Some of the measurements taken by the latter authors are presented in Fig. 9.

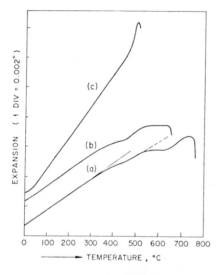

Fig. 9. Expansion curves showing effects of immiscibility; (a) double transition (PbO–GeO_2–SiO_2 glass), (b) plateau (PbO–GeO_2–SiO_2 glass), (c) standard PbO–SiO_2 glass. (After Topping et al., 1974).

Finally, it is of interest to note that the Ostwald transformations at T_4 in Fig. 8b and 8d are among the most important in all of materials science for they represent one way in which the catalyzed crystallization of glass has been achieved. The importance of this transformation was apparently recognized independently by both Stookey (1959) and Roy (1960).

b. Microstructure of amorphously phase-separated glass. There are two principle ways of producing amorphously phase-separated glass. First, one may simply prepare a glass-forming melt, quench it to room temperature and by appropriate physical and chemical methods, assess the extent to which the process has taken place. Control over this method is weak. For a given composition the parameters important to this process are: (1) quench rate, (2) type of quench, (3) quench temperature, (4) melting crucible, and (5) the sensitivity of one's technique of assessing the presences of two phases. These same parameters are also important to the question of glass formation (Pye, 1972). Obviously there are times when one hopes that amorphous phase separation is minimized, e.g., property–composition studies, or conversely, when a highly developed microstructure is necessary such as for certain opal glasses. Proving its absence or presence is, of course, crucial to the question of the intrinsic structure of glass.

A second and more controllable method involves the isothermal annealing of glass above its transformation temperature. By this approach one in effect extends the amount of time available for the process to proceed. By intermittent examination of samples for telltale opalescence of various times and temperatures, it is possible to construct a time–temperature–transformation diagram similar to one prepared by Burnett and Douglass (1970) (Fig. 10a). Figure 10b shows the onset of opalescence for a soda–silica glass heat treated at 640°C for the indicated times.

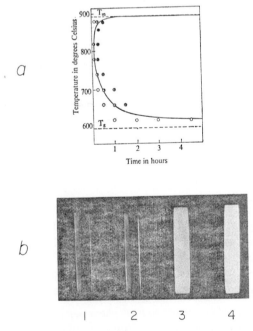

Fig. 10. Temperature–time relationships for opalescence produced by amorphous phase separation; (a) Na_2O–CaO–SiO_2 glass; ○—clear glass, ●—opalescence, ◐—partial opalescence, (after Burnett and Douglas, 1970), (b) Na_2O–SiO_2 glass; 1—zero hours at 640°C, 2—three hours at 640°C, 3—ten hours at 640°C, 4—thirty hours at 640°C (courtesy of L. Ploetz). Size of samples in (b) ∼0.5 × 4.0 × 1.0 cm.

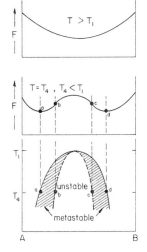

Fig. 11. Free-energy–temperature relationships associated with amorphous spinodal decomposition.

The microstructure developed in this process is of intense interest in glass science. In order to more thoroughly discuss the morphology of this microstructure, recourse will be made to the thermodynamic character of a liquid phase giving rise to the immiscibility domes in Fig. 8. For clarity, only the liquid phase of a binary system will be considered. Accordingly, a free-energy–composition curve for a homogeneous single-phase liquid at some temperature T is shown in Fig. 11. For this condition a smooth, continuous curve with a single minimum is required. When reducing the melt temperature to T_4, however, the free-energy curve develops a region of negative curvature and two minima, providing, of course, that the melt is prone towards amorphous phase separation at this temperature (Cottrell, 1948). The equilibrium compositions for a melt undergoing transformation at T_4 are a and d. Outside these extremes, only a single liquid phase is stable; between them a two-phase melt has a lower free energy and is thus the preferred structure. The loci of equilibrium compositions trace out the immiscibility dome. The loci of the inflection points delineate the so-called spinodal. Between a–b and c–d (hashed area), single-phase melts are said to be metastable and transform to a two-phase melt by classical nucleation and growth. Between b–c all melts are regarded as unstable; they transform to a two-phase melt by "spinodal decomposition" (Cahn, 1961).[†]

The anticipated microstructures for initial stages of transformation in the metastable and unstable regions have been summarized by Cahn and Charles (1965), Table I. We note that a microstructure characterized by discrete

---

[†] For a summary discussion of alternative concepts related to this transformation, see James (1975).

TABLE I

Potential Microstructure Associated with Amorphous
Phase Separation[a]

Nucleation and growth	Spinodal decomposition
Invariance of second-phase composition to time at constant temperature	Continuous variation of both extremes in composition with time until equilibrium compositions are reached
Interface between phases is always same degree of sharpness during growth	Interface between phases initially is very diffuse, eventually sharpens
Tendency for random distributions of particle sizes and positions in matrix	Regularity of second-phase distribution in size and position characterized by a geometric spacing
Tendency for separation of second-phase spherical particles with low connectivity	Tendency for separation of second-phase nonspherical particles with high connectivity

[a] After Cahn and Charles (1965).

particles dispersed through a continuous matrix forms one extreme; the other consists of irregular diffuse boundaries and high interconnectivity of both phases. Both extremes are sensitive to time and temperature. In the former, the particles will increase in number and size until the compositional limits are reached. In the latter, the interconnectivity apparently decreases with time, tending to become more like the former, e.g., coalescence of the smaller volume phase due to surface tension forces and a gradual sharpening of interfaces. Examples of each type, as produced by heat treating soda-silica glasses are given in Fig. 12.

There is dramatic proof of the interconnectivity associated with the initial stages of spinodal decomposition. After decomposition, one phase may be chemically leached from the bulk biphase glass leaving a residual porous glass whose composition is that of the more chemically durable second phase. This process would not be possible if both phases were not continuous (Nordberg, 1944). The resulting porous glass can now be reheated to form a nonporous article whose composition is generally much higher in silica than the parent glass. Such a process is important for manufacturing low-cost high-silica glasses suitable for a variety of industrial applications, e.g., Vycor brand glass products. Alternatively, the porous glass can be impregnated with various metals to form new composites with decidedly different electrical properties (Watson, 1966). Other variations of this latter procedure have been reported (Elmer and Nordberg, 1966). The point here

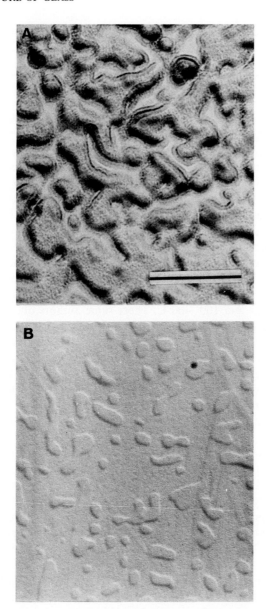

Fig. 12. Microstructures associated with amorphous phase separation in a soda–silica glass; (A) two interconnected phases, (B) one continuous phase containing a dispersed phase. Coarsening by prolonged heat treatment may cause (A) to convert to (B). White bar is approximately 0.25 μm.

is that once again we find a demonstrated commercial need for controlling the microstructure of glass.

c. Other phenomena. Liquid–liquid phase transformations in glass-forming melts are not limited to immiscibility effects. For example, it is known that the viscosity of liquid sulfur changes by a factor of two thousand when heated over the narrow temperature range 158–166°C (Bacon and Fanelli, 1943). Maximum viscosity is achieved at 187°C and is followed by a gradual decrease with increasing temperature up to several hundred degrees. The effect is completely reversible. The initial increase in viscosity is in sharp contrast to the normal viscosity–temperature relationship of many glass-forming melts. Obviously, it is easier to convert the more viscous melt to the vitreous state. The low-temperature low-viscosity melt is thought to be constructed of 8-member rings, whereas the high-temperature high-viscosity melt is regarded as made up of long sulfur chains. It has been suggested that both liquids are in thermodynamic equilibrium (Powell and Eyring, 1943). Later work by Vezzoli *et al.* (1969) identified what was believed to be four different liquid fields for this element over the pressure and temperature range 0–35 kbar and 100–400°C. These same authors offered evidence for a second-order phase transformation between two of these liquid fields.

Although the above viscosity effects apparently relate more to changes of atomic structure (e.g., polymerization) rather than microstructure differences, they serve to point out the rather subtle influences that many melts are subjected to prior to the time they are converted to a vitreous state. None too surprising, partial second-order transformations, with attendant effects on properties, have been suggested for liquid silicates as well (Lacey, 1968).

In addition to these considerations there is some evidence for clustering effects in silicate glasses. Clusters have been defined as cooperative compositional fluctuations surrounded by melt of less organized structure (Ubbelohde, 1965). These clusters are not necessarily equivalent to the crystalline nuclei previously discussed, although both structures are governed by the same small particle thermodynamics. It has been suggested that such clusters could be "frozen in" during quenching of a glass-forming melt (Maurer, 1956). This was established by determining the influence of various thermal histories on observed light scattering of small glass chips. The scattered intensity was ten to one hundred times larger than that calculated for simple room temperature agitation. Moreover, it was determined that the scattering intensities were not related to melting times or incipient crystallization, and that the scattering level was characteristic of the annealing treatment employed. The scattering changes could not be explained by changes in bulk refractive index alone since glasses with the same refractive index gave different scattering intensities. On this basis, Maurer concluded that the origin of the high scattering intensities derives from physical inhomogeneities, e.g., clusters which are principally thermal fluctuations in the melt and are frozen into the cooled glass.

The practical implications of this proposal are great since in fiber transmission optics, the lower limit of optical losses may ultimately be dictated by scattering losses. Thus, despite recourse to ultrapure materials, highly sophisticated fabrication techniques, and various cladding schemes, the upper limit of transmission may be limited by the one crucial step required of all fiber manufacturing, namely, the melting of glass.

4. CATALYZED CRYSTALLIZATION OF GLASS

Much of the recent work concerned with the catalyzed crystallization of glass, e.g., glass ceramics, has dealt with the relationship of the microstructure of the resulting crystalline material to its bulk properties. Since these materials contain only a small volume fraction of residual glass, they fall outside the scope of the present review even though they are derived from a homogeneous glass. Yet it is recognized that this conversion to a crystalline material nearly always involves an intermediate microstructure in the parent glass. Moreover, one is not always concerned with achieving a full slate of crystallization; as will be discussed below, many partially crystallized glasses have important commercial value. In this latter case, one is actually dealing with a glass–crystal composite in which the crystalline phase may either be (1) low volume (e.g., dispersed) or (2) comparable in volume to the glassy phase. For these reasons a brief review of the various ways by which one may achieve either total or partial crystallization of glass is appropriate.

Stewart (1972) has summarized the major reaction paths for achieving the crystallization of glass (Fig. 13). Path (1) represents the direct transformation of a pure single-phase glass into a more stable crystalline phase. This transformation involves both the creation of stable nuclei and the subsequent growth of the crystalline phase around these nuclei. Both processes, nucleation and growth, have their own characteristic temperature and time dependence, and while this direct transformation is relatively rare in oxide glasses

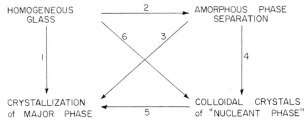

Fig. 13. Various paths associated with potential crystallization of glass forming melts (after Stewart, 1972).

it is important to note that even here the transformation begins with a development of microstructure in the parent glass, e.g., the nuclei. Very little is known about the structure of these nuclei.

In contrast to path (1) the development of an intermediate amorphous phase as in path (2, 3) and also for path (2, 4, 5) is well documented for a number of glass-forming systems. A good representative of the former is the crystallization of $Al_2O_3-SiO_2$ glasses (MacDowell and Beall, 1969). The latter path has been observed in classical glass ceramic compositions, e.g., $Li_2O-SiO_2-Al_2O_3-TiO_2$ (Doherty et al., 1967). On cooling, glasses in this system show amorphous phase separation on a scale of about 50 Å. On reheating, this microstructure promotes formation of a nucleant phase, $Al_2Ti_2O_7$, which in turn crystallizes a major crystalline phase of this system, β-eucryptite.

Path (6, 5) represents the case where small amounts of metals such as Ag, Au, Pt, Cu, Rh, Pd, etc., are incorporated into a base glass and by suitable control of initial concentration, melting conditions, thermal history, and in some cases (Au, Ag) exposure to actinic radiation, colloidal particles of these metals are precipitated in the base glass. Occasionally, this precipitation is spontaneous, as on cooling; other times, extended isothermal annealing is required (striking). Since the process involves both a reduction to a metallic state and diffusion of the reduced species to form a particle, their mean size may vary over a wide range depending on the interplay of the above factors. For Pt in Grahm's glass ($NaPO_3$), Gutzow et al. (1968) have shown that a maximum number of particles (5×10^9 particles/cm^3) is achieved when the Pt concentration is about 0.01%. For this condition the mean particle size is about 2000 Å. Maurer (1959) has shown that the minimum size of a gold particle capable of catalyzing lithium metasilicate is about 80 Å (10,000 atoms). This is in sharp contrast to the smallest, stable gold particle which may contain only three or four atoms (Maurer, 1958). Thus the need for microstructure control to achieve catalyzed crystallization is apparent.

Of special interest to the present work is the case where final crystallization in path (6, 5) does not occur, leaving instead a homogeneous glass containing a dispersion of metallic particles. It has been recognized for some time that the presence of these metal particles has a pronounced effect on the optical properties of glass. That is, a highly dispersed small-volume microstructure induces a charactersitic color in the bulk glass. Thus, gold and copper particles give rise to a beautiful magneta color, silver to a decided yellow cast, and platinum to a somewhat dull grey appearance. Similarly, selenium gives a characterstic pink color in soda–lime silica glasses (average particle size 50–200 Å). Since this pink color is complimentary to the bluish green arising from ionic iron in these same glasses, this element is used extensively in the glass container industry as a decolorizer. It is here especially that the control of microstructure in glass has real meaning for it has been determined that the precise shade of pink required to achieve satisfactory decolorization depends on (in a very complex way) initial composition, furnace atmosphere, and of course, thermal history (Paul, 1975). Evidently, then, we have another outstanding example of the need to understand the origin and significance of microstructure in glass to render a product more commercially acceptable.

Still another group of technologically important glasses which are partially crystalline is the photochromic glasses reported by Armisted and Stookey (1964), Smith (1966), and Araujo and Stookey (1967). Their photochromic behavior arises from minute silver halide crystals which may be regarded as being suspended in an inert glass. Again, initial composition, melting history, and thermal history play an important role in determining the volume fraction of these crystals and their mean size. The latter parameters strongly influence the resulting photochromic behavior. In general, glasses with particle diameter less than 50 Å are not photochromic. Above 300 Å these glasses become translucent and are less desirable. Satisfactory results are obtained when the average particle size is about 50–100 Å at a concentration of $\sim 10^{15}$ particles/cm^3. This corresponds to an average particle separation distance of about 600 Å. This separation distance is crucial since the inert glass host prevents diffusion of the halogens freed by absorption of light by the halide crystal. Thus, recombination with free silver within the halide crystal is enhanced. It is this alternate decomposition and recombination that gives rise to the variable optical transmission of these glasses. This sequence of events should be contrasted with that found in conventional photographic materials. In the latter case, because of their composition and microstructure, the halogens diffuse away from the free silver created by absorption of light. Hence, they are no longer available for recombination and upon cessation of irradiation, the decomposition is permanent (Megla, 1966). Evidently, then, the variable transmission character of the photochromic glass is derived from an understanding of the role of their very special microstructure and a knowledge of how to obtain that microstructure.

Finally, we note that there are many further examples of glass–crystal composites in which full crystallization of the parent glass is not achieved, purposely or otherwise. This is generally the situation when the crystalline phase is not a glass former, e.g., magnetite (Ballard, 1972), barium titanate (Borelli and Layton, 1969), and lead–metaniobate (Anderson and Friedberg, 1962). In these situations, one is generally concerned with changes in properties of the bulk glass created by their partial crystallization. Clearly, as was the case in photochromic glasses, the intrinsic properties of the crystallizing phase may influence or dominate the overall bulk properties. For these latter systems, the amount of crystallization (10–70%) is far in excess of that found for photochromic glasses (0.7%). In both situations, however, the extent of crystallization, the type of crystallizing phase, and phase geometry have marked influences on bulk properties. As such, these considerations underpin an overall understanding of microstructure–property relationships in oxide glasses, and, accordingly are discussed in some detail in the next section.

C. The Structure of Glass

From the above it is apparent that any serious discussion of the structure of glass must include reference to its microstructure, real and potential (Vogel, 1966). Thus, when assessing the chemical and physical homogeneity of glass one should consider its composition, mode of preparation, thermal history, methods of analysis, and propensity towards amorphous phase separation and/or partial crystallization. Clearly, several commercial processes and products rest on an understanding of how one can induce and control a desired microstructure, and ultimately alter the properties of the resulting biphase glass in a useful manner.

Quite aside from such considerations, several theories have been offered which hold that the intrinsic structure of glass is inherently nonrandom. That is, there are regions in a cooled glass where the atomic structure is assumed to be more ordered than a completely random assemblage of ions which comprise the glass. Included here are the original crystallite theory, the vitron concept of Tilton (1957), the micellar theory of Prebus and Michner, (1954) and Zarzycki and Mezard (1962), and the pentagonal dodecahedral model of Robinson (1965). These theories should be viewed from a perspective that they offer alternatives to a description of the vitreous state as a completely amorphous solid, e.g., the random network theory. In proposing these alternatives, the respective authors sought to provide a basis for understanding certain bulk properties which are not otherwise readily explainable from a random network approach. For example, it is occasionally observed that discontinuities appear in property–composition plots in certain glass-forming systems. It has been suggested that these discontinuities are related to composition areas of the primary crystallization fields (Babcock, 1968). According to the classical random network theory, such a correlation is not warranted. Likewise, the similarity of certain properties such as the hardness–temperature relationships, for both the crystalline and vitreous phases of the same composition, e.g., SiO_2, are beyond explanation by the network theory. Yet, at the same time, the network theory, with its corollary concepts of modifiers and intermediates, has served as a useful first approximation for the description of silicate and borate melts. Later modifications, as introduced by Smyth (1972), have also been valuable.

It is beyond the scope of this work to review these classically antagonistic approaches to an inferred structure of glass and attempt to decide which has the greater validity. It is noted, however, that many glass-forming systems fall outside the original Zachariassen concept, e.g., nitrate glasses. Moreover, it is possible that both viewpoints represent extreme positions and in the long term a more useful approach will be found in a general formulation which treats both extremes as limiting cases. Eckstein (1968) has reviewed such a possibility. For our purposes, it will be assumed that many glasses possess a demonstrable microstructure, and without speculating whether this microstructure is inherent in all oxide glass forming systems, the remaining portion of this review will seek to identify the parameters of such a biphase glass which influence its properties.

III. Microstructure–Property Relationships in Oxide Glasses

A. General Aspects

There are essentially two approaches for discussing the subject matter suggested by the title of this section. First, one might consider the major physical and chemical properties individually and, by referring to the various parameters of an assumed microstructure, illustrate how each property depends on the noted parameters. This approach offers an in-depth, state-of-the-art summary of each property and while inherently valuable, represents a rather extensive undertaking which falls outside the scope of the present work. Alternatively, one might first summarize the salient features of an assumed microstructure and then, where possible, show how each has been observed to affect selected properties. This latter approach is adopted here for several reasons. First of all, to the author's knowledge a summary discussion of the major features of microstructure in glasses has not yet been undertaken. For that matter, the same is true for polycrystalline oxides as well. Second, not all the properties respond with equal force to the various microstructure parameters. As a result, by considering only a single property with perhaps low sensitivity to some of these parameters, an important microstructure–property relationship might be obscured or diluted. In contrast, by being able to select the most sensitive property among several an intrinsic relationship can be more clearly demonstrated. Third, not all of the microstructure–property relationships have been studied to the same extent for all properties. Evidently then, a more thorough discussion of this entire subject should be possible by considering several properties as needed.

It is—perhaps—worthwhile to note here that the subject of microstructure–property relationships in oxide glasses is ultimately a subset of the more general theories of composite materials (Lewis, 1976). It is likely, therefore, that a complete understanding of this subject will be found in the context of these theories. Yet given the state of technology of oxide glasses, it is perceived here that an introductory approach used in this work is the more valuable one at this time. For all of the above reasons then, a summary of the salient features of microstructures in oxide glasses is presented first, followed by illustrations of their influence on selected properties.

B.   Salient Features of Microstructure in Oxide Glasses

Consider a general biphase material consisting of a dispersed phase α in a homogeneous matrix β (Fig. 14). In keeping with the scope of this review, β is regarded as noncrystalline; α may be crystalline or noncrystalline and β is regarded as having the larger volume fraction so long as α is crystalline. In other instances, such as amorphous phase separation, the equilibrium volume fractions (as predicated by the Lever rule) can be made nearly equal by correct choice of compositions within the miscibility domes. It is important to note that the development of a microstructure in homogeneous glasses via partial crystallization or amorphous phase separation always leads to changes in volume fractions of α and β. Compositional changes in α and β, however, may or may not occur depending on the nature of the process involved. For example, the crystallization of a parent glass to yield an α phase of the same composition is well established, e.g., cristobalite in vitreous silica. In other cases, such as when α is $\alpha-Fe_2O_3$ (formed by crystallizing melts in the $Na_2O-SiO_2-Fe_2O_3$ system), α is again crystalline but differs markedly from the overall parent glass composition. In this case, during the formation of α, the volume fraction of both α and β vary with time as will the composition of β, whereas the composition of α is sensibly fixed.[†] Other combinations are possible; equal volume fractions of α and β while their compositions and properties change as is, perhaps, the case in the initial stages of spinodal decomposition; or when the volume fractions, compositions, and properties all change with the appearance and growth of α. The latter case might arise when α is a solid solution with a wide range of chemical solubilities. Still further combinations can be achieved by preparing glass–composites through such techniques as hot pressing (Davidge and Green, 1968). Finally, α is regarded for this review as inherently nongaseous.

With these constraints it turns out that at least eleven different parameters can be identified as having a potential effect on the bulk properties of these biphase glasses. These parameters are conveniently divided into three groups; those associated with the β phase, those associated with α, and those related to the interaction of both phases. These parameters are summarized in Table II.

[†] Clearly, this applies mainly to the initial onset of crystallization.

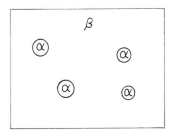

Fig. 14. Schematic representation of biphase glass.

To a very large extent, the fixation of these parameters is, of course, dependent on the system chosen for study and the subsequent thermal history used to produce biphase glasses. That is, the base glass composition, its propensity to partially crystallize or undergo amorphous phase separation, its mode of formation, and subsequent thermal history all combine to introduce α at some specified volume fraction, mean size, crystallinity, etc. Some of these parameters can be varied independently, e.g., mean separation distance, orientation, etc. As discussed above, compositional changes of α may or may not be concomitant with those of β while α is being formed. There is an important corollary to the situation where such compositional changes occur; namely, the formation of an interface between α and β. This interface can be quite sharp as is apparently the case when α is metallic and is formed under conditions of low viscosity, e.g., during melting. Alternatively, the interface may extend over several interatomic distances. The latter situation results directly from the role of diffusion in both the precipitation and growth of α, whether it is crystalline or noncrystalline. In such a process ion depletion or enhancement around α is inherent; hence, the formation of an interface between α and β is to be expected.

Another interaction parameter arises from the formation of α at one temperature and then measuring the properties of the resulting biphase

TABLE II

PARTIAL SUMMARY OF MICROSTRUCTURE PARAMETERS OF
BIPHASE OXIDE GLASSES

β phase	α phase	α–β interactions
Composition	Composition	Interfacial
Properties	Properties	Thermal
	Crystallinity	
	Volume fraction	
	Mean size	
	Mean separation distance	
	Orientation	

material at another. If the volume changes of each phase are unequal during heating or cooling to this second temperature, then one may expect the creation of tensile or compressive stresses around α depending, of course, on the respective temperature dependence of each phase. Hence, this interaction is denoted here as a thermal one. The compositions of α and β may or may not be the same; the more fundamental requirement is that the thermal expansion of each phase be different.

Ideally, one would like to be able to control all of the above parameters in a given system in attempting to unequivocally assess their respective influences on the properties of bulk glasses. Owing to the large number of parameters involved, this situation will probably never be achieved. It also is possible that these parameters are not truly independent over all limits. Yet as is shown below, considerable information is presently available which at least qualitatively demonstrates their respective relationships to the properties of biphase glasses.

C. *Properties of Biphase Oxide Glasses*

To the extent possible, the following will attempt to illustrate the various effects of the parameters listed in Table II on the bulk properties of biphase glasses. In some cases, it will be more convenient to discuss these parameters in pairs or groups. Also, some of the data available in the literature are conveniently discussed in the context of more than one parameter.

1. COMPOSITIONS, PROPERTIES, AND VOLUME FRACTIONS OF α AND β

As noted, the formation of α in a homogeneous glass may lead to a variety of volume fractions of α and β with corresponding similar or dissimilar compositions and properties. Fortunately, it is unlikely that the chemical durabilities of each phase are exactly equal. This follows from the overall dependence of this property on the structure and composition of any phase. Since, by definition, biphase glasses are nonhomogeneous, point-to-point differences in chemical durability of the bulk glass are to be expected. When such glasses are exposed to various liquid media such as H_2O, dilute HF, etc., their unequal corrosion rates may change an atomically smooth fracture surface into one which is uneven, undulating, and in some cases, pitted. It is this change in surface topology which has proved so valuable for revealing the presence of second phase through replica electron microscopy. This in turn has allowed much to be said about the energetics, kinetics, and mode of formation of the second phase by itself or in relationship to concurrent changes of β. It is probably unwise to unequivocally infer the absence of a second phase when these procedures give negative results without a thorough investigation of etchant type, strength, and etchant times. Even then some ambiguity may remain. The main point here, of course, is that much of the literature dealing with the characterization of biphase glasses is based on the difference of the chemical properties of the phases which constitute these glasses. Without such differences these techniques would be severely limited.

Additional studies of biphase glasses in the context of the volume fractions, composition, and properties of their constituent phases have included investigations on electrical properties (Hakim and Uhlmann 1967; Charles, 1965; Kinser and Hench, 1968); mechanical properties (Pye et al., 1974, Shaw and Uhlmann, 1969; Redwine and Field, 1968); and rheological properties (Simmons et al., 1974; Haller et al, 1971; Mazurin et al., 1970). There are many features common to these studies; the change in properties related to the growth and ripening of the various phases, analysis in terms of the phase continuity and/or dispersion, and the effect of compositional changes of each phase during growth. Of particular interest is the large change of viscosity (several orders of magnitude in some cases) affecting, among other things, the glass transformation temperature, the annealing temperature, and the softening temperature (both shear and dilatometric). These viscosity parameters are of considerable importance in commercial glass manufacturing; the value of extending these concepts into commercial glass compositions suggests itself.

Not all bulk properties are affected to this extent. For example, Pye et al. (1974) noted that when comparing densities of amorphously phase-separated soda–silica glasses with homogeneous ones of the same composition no differences were evident providing that both glasses were reannealed at a common temperature. It was also suggested that no change is expected for equilibrium amorphous phase separation as long as the molar volume changes linearly with composition across the miscibility gap as apparently is the case for glasses in the soda–silica system. Without annealing, slight density differences were observed (~ 0.0005 g/cm^3). This was attributed to a faster volume relaxation of a soda-rich phase over the homogeneous glass. Shaw and Uhmann (1969), however, demonstrated that density measurements of binary glasses could be used to indicate the presence of an immiscibility region within many binary systems. Density, of course, is an additive property and should therefore be independent of the morphology of both α and β depending instead on the volume fraction and density of each phase. Sharp interfaces are required for this concept to be valid.

2. COMPOSITION, VOLUME FRACTION OF β

It is often convenient to think of β as an inert suspension medium for a second phase (α) whose properties, e.g., dielectric, optical, etc., completely overshadow those of β. As remarked earlier, this appears to be the case for photochromic glasses. In other instances, e.g., viscosity measurements, the viscosity of β will dominate the bulk viscosity of biphase glasses as long as α is dispersed. Consequently, the contribution of the composition, volume fraction, and properties of β to the overall properties of biphase glasses should only be assessed in the context of the particular property under consideration.

3. COMPOSITION, VOLUME FRACTION, PROPERTIES OF α

As noted earlier, α need not always be noncrystalline; indeed the literature abounds with studies where α is strongly dielectric, magnetic, or optically active. Of particular value are those cases where the strength (magnitude) of a bulk property depends almost entirely on the amount of α present. This fact may be used to carry out kinetic studies on the formation of α itself (Ballard, 1972). An understanding of the effect of compositional changes of α on its properties is vital in such studies. For example, Ballard (1972) observed significant differences of the Curie temperature of a soda–iron–silica glass that contained Fe_3O_4 crystals (magnetite) depending on whether this phase was nucleated by platinum or by titania additions to the base glass (Fig. 15). The platinum nucleated biphase glass gave a Curie temperature very close to pure magnetite (570°C), whereas for titania induced nucleation the Curie temperature was substantially lower (420°C). This observation was easily explained by assuming that a titanomagnetite solid solution $[(Fe_3O_4)_{1-x}(TiFe_2O_4)_x]$ formed during crystallization. Stephensen (1972) had previously shown that the Curie temperature of this solid solution decreases linearly with x. The data of Fig. 15, then, are not all that surprising.

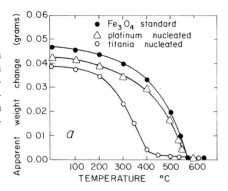

Fig. 15. Apparent weight change of an iron–soda–silica glass containing Fe_3O_4 in a nonhomogeneous magnetic field with increasing temperature. Curie points are estimated by noting temperature at which apparent weight change falls to zero (extrapolated) (after Ballard, 1972).

4. SEPARATION DISTANCE, MEAN SIZE OF α

We have previously noted that a large separation distance of a dispersed phase (hallide crystals) is necessary for proper photochromic behavior of glasses that contain these crystals. More generally, when the mean separation distance falls to zero and the volume fraction of α is sufficient, e.g., $>15\%$, it is possible for this phase to become continuous. Corresponding effects on the properties of the bulk glass, e.g., chemical durability, viscosity, etc., follow at once.

The mean size of α is generally the more significant parameter. For example, it has been observed that a partially crystallized glass containing a dispersed crystalline phase (Fe_3O_4 or $\gamma\text{-}Fe_2O_3$) is supermagnetic so long as the mean particle size is less than 100 Å (O'Horo and Steinitz, 1968). Increasing this particle size leads to a biphase glass with normal ferromagnetic behavior with higher coercive forces. Prior to crystallization, the glass shows paramagnetic behavior. Similar observations were made by Collins and Mulay (1971) in soda–iron–silica glasses containing $\beta\text{-}NaFeO_2$.

An even more dramatic illustration of the importance of the size of α has been provided by Sproull and Rindone (1973). In a series of experiments dealing with the pristine strength of glass rods of composition $0.5Li_2O\cdot 0.5K_2O\cdot 2SiO_2$, a correlation was offered between a measured phase (by TEM) and that predicted by the classical Griffith equation:

$$\sigma_f = (KE\gamma/C)^{1/2}$$

Here, σ_f is the measured fracture strength, γ is the fracture surface energy, E is Young's modulus, K is a constant related to flaw geometry, and C the size of the flaw leading to fracture. These data are summarized in Table III. It is seen that the measured strengths are highest when the flaw size is smallest, and that the agreement between measured and calculated values of flaw size is quite good. Until recent years, the interpretation of the strength of glass has been couched in terms of introducing flaws on the surface of glass by abrasion of other methods (Ernsberger, 1966). Yet, in this case, the origin of the flaws giving rise to fracture is regarded as internal. In other words, there may be an intrinsic relationship of the strength of glass to its microstructure. This concept was apparently first suggested by Watanabe and Moriya (1961),

TABLE III

CALCULATED GRIFFITH FLAW SIZE AND MEASURED
PHASE SIZE[a]

Strength (psi)	Calculated flaw size (Å)	Measured phase size (Å)
170,000	1240	1000–2000
220,000	740	600–1500
400,000	225	200–500
500,000	140	100–200

[a] After Sproull and Rindone (1973).

and as such, represents a milestone in the field of glass science. Rindone (1974) has provided an overview discussion of these considerations and has also indicated their potential application in commerical processes.

Another example of the importance of this parameter (mean size) is found in the crystallization studies carried out by Herczog (1964). In this work, $BaTiO_3$ was precipitated from a silicate or borate based glass and the dielectric properties of the resulting material were measured. Figure 16 shows the effect of both grain size and volume fraction of $BaTiO_3$ on dielectric loss and constant as a function of temperature. Note that as expected, larger volume fractions of $BaTiO_3$ increased the apparent dielectric constant of the biphase glass and that in decreasing the grain size to about 0.2μ the familiar tetragonal–cubic phase transformation of $BaTiO_3$ at low temperatures ($\sim 120°C$) is absent. This latter result was interpreted in terms of a Schottky-type depletion layer that becomes significant only when the surface-to-volume ratio is sufficiently large, e.g., for small particles.

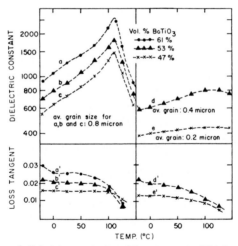

Fig. 16. Variation of dielectric constant and loss tangent of biphase glasses containing microcrystalline $BaTiO_3$ at 1 kHz (after Herczog, 1964).

Finally, biphase glasses may be completely transparent in the visible part of the spectrum providing that the dispersed phase is much smaller than the wavelength of light.[†] This condition is essential for minimizing losses due to scattering processes. An additional criterion is that the refractive index of α closely match that of β. Also, transparency is enhanced if the optical anisotropy (birefringence) of α is small. Thus, in this instance, the crystallinity of α

[†] This assumes, of course, that β is fully transparent.

is of major concern. Moreover, it appears that nearly all glasses scatter light to some extent (Smith, 1971). These considerations, then, are of paramount concern where optical transparency is highly desirable, e.g., optical wave-guides.

5. ORIENTATION OF α

It is possible to orient a dispersed phase in glass by appropriate drawing techniques (Seward, 1974; Stookey and Araujo, 1968). These latter authors attempted to produce a bulk polarizing material by stretching a lithia alu-minosilicate glass which contained a finely dispersed second phase consisting of metallic silver (Fig. 17). An asymmetry of a geometrical form of a metal particle can be expected to give rise to dichroism. Thus, a bulk biphase glass processed in this manner could be expected to act in much the same way as ordinary polarizing film. That this is indeed the case is shown in Fig. 18 where optical transmittance through a glass containing the silver particles is seen to be strongly dependent on orientation with respect to the longitudinal (pull) axis of the glass sample. Glasses containing optically active halide crystals also show this behavior when stretched (Araujo *et al.*, 1970).

If, however, α is nonspherical and has a different refractive index than β, and if there is a degree of alignment induced again by stretching, then the resulting glass is simply birefringent. This effect is sometimes referred to as birefringence of form and should not be confused with simple birefringence caused by transformation range behavior (strain). Indeed, it is possible to

Fig. 17. Elongated silver particles in a lithia–alumina–silicate glass. Bar is approximately 0.1 μm. (courtesy R. J. Araujo, Corning Glass Works).

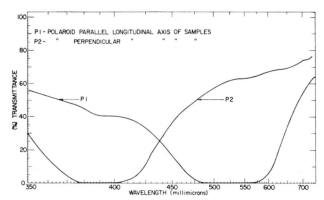

Fig. 18. Transmittance curves of a bulk polarizing glass. P1—light polarized parallel to longitudinal (pull) axis of sample; P2—light polarized perpendicular to pull axis of sample (after Stookey and Araujo, 1968).

produce stress-free glasses which are strongly birefringent. The effect is well documented (Takamori, 1974, Stirling, 1955; Botvinkin and Ananich, 1962) and has a strong theoretical basis (Coker and Filon, 1951). At this point, however, the number of glass systems studied is quite small; much additional work is required for its unambiguous interpretation. For example, certain stress-free glasses become birefringent when stretched even when a second phase is not discernible (Ernsberger, 1973). This somewhat startling result, by itself, suggests the need for further research. Is it possible that this simple procedure is much more sensitive to slight phase differences than traditional analytical tools such as TEM which, as we have seen, are really dependent upon chemical differences between phases? If so, the phenomenon of form birefringence may constitute a crucial frontier of glass science.

6. INTERFACIAL INTERACTIONS

As noted previously, an interface between α and β is closely associated with the formation of α. In this sense it is important to note that the diffusion zone (interface) around a minor glassy phase in Li_2O–SiO_2 glasses may, according to Tomozawa (1972), act as a site favorable to crystal nucleation. Evidently, then, this interface may be important not only to the properties of bulk glasses but to the catalyzed crystallization of glass as well. Regarding bulk properties, Hammel and Ohlberg (1965) were able to analyze light-scattering experiments with amorphously phase-separated glasses in the CaO–Al_2O_3–B_2O_3–SiO_2 system in terms of a diffusing species at some radial distance from a growing particle. Maxima in scattering at various angles could be equaled with interference effects between these particles and their diffusion fields. This is essentially Goldstein's diffusion zone model (Goldstein, 1963). Kerzwawycz and Tomozawo (1974), however, offered an alternative explanation to these findings based on thermal expansion differences between α and β. Both arguments are convincing.

A somewhat clearer example of the importance of this interface is found in the studies of Fahmy et al. (1972) on the magnetic properties of partially crystallized glasses in the B_2O_3–BaO–Fe_2O_3 system. In these studies, low values of saturation magnetization could be accounted for only by assuming that a ferrimagnetic particle (α-Fe_2O_3 or Fe_3O_4) was surrounded by an antiferrimagnetic surface. This conclusion was corroborated by both small angle x-ray scattering and electron microscopy. Thus it is apparent that interfaces can affect the optical, magnetic and, as we have seen before, the dielectric properties of biphase glasses.

7. THERMAL INTERACTIONS

As noted previously, by forming α at one temperature and then cooling (or heating) the biphase glass to another, a hydrostatic stress P may develop around α. The magnitude of this stress is given by the Weyl (1959)–Selsing (1961) equation

$$P = \frac{\Delta\alpha \, \Delta T}{(1 + v_\alpha)/2E_\alpha + (1 - 2v_\beta)/E_\beta}$$

where v_α, v_β, and E_α, E_β are Poisson's ratio and Young's modulus of each phase, $\Delta\alpha$ the difference in expansion coefficient, and ΔT the cooling or heating range. In cooling the biphase material of Fig. 14, ΔT may be taken as the annealing point of β to ambient. Over this range, the plasticity of β may be considered negligible. For $\alpha_\beta > \alpha_\alpha$, P is compressive (and positive); the tangential and radial stress of β are, respectively, $+PR^3/r^3$ and $-PR^3/r^3$, where R is the radius of α and r is distance from the center of α. For these latter equations, a positive sign denotes tension whereas a negative sign compression.

These equations have been tested extensively in hot-pressed glass–crystal composites (Binns, 1962; Davidge and Green, 1968; Frey and Mackenzie, 1967). By suitable choices of glass compositions such as a borosilicate, soda–lime, or potash lead for β, and crystalline inclusions of Al_2O_3, ThO_2, or ZrO_2, for α, values of P ranging from -26.7 to $+8.8 \times 10^8$ dyn/cm^2 have been achieved. These studies have yielded several interesting observations. For large negative P, circumferential cracks may occur around α during cooling from the fabrication temperature. When stressed, fracture paths tend to pass around α linking up with existing cracks. With smaller values of P, e.g., $\alpha_\beta \rightarrow \alpha_\alpha$, fracture paths tend to be relatively straight and may pass through α. With positive P, cracks tend to be generated which radiate from α and form a crazed network. When stressed to failure, the fracture path in this case is much more irregular and tortuous. Generally, the fracture strengths of these biphase glasses are observed to fall steadily with increasing size of α. With positive P, some increase in strengths have been observed (Frey and Mackenzie, 1966) but his appears to be the exception rather than the rule. Many of these effects are discussed in excellent reviews by Davidge (1973) and Lange (1973).

There are convincing data which illustrate that the thermal interactions which are so prominent in these model glass crystal composites are also operative in biphase glasses whose microstructure is thermodynamic in origin, e.g., amorphous phase separation, nucleation, and/or crystallization. For example, it has already been noted that Kerawawycz and Tomozawo (1974) used this concept in developing a light-scattering model for amorphously phase-separated glasses. Simiarly, these same tensile or compressive forces have been regarded as responsible for retarding or enhancing phase transformations of crystals included in a glassy matrix. An example of retardation is provided by Wagstaff (1969) who observed that β-cristobalite precipitated in vitreous silica can be quenched to a metastable state at room temperature. That is, tensile forces created by the differential contraction of this biphase glass prevent the conversion of beta to alpha cristobalite. This tensile force is thought to oppose the high volume contraction associated with this transformation and thereby depress the transformation temperature.

An inducement of phase transformation was observed by Lawless (1974) who studied a biphase glass consisiting of $SrTiO_3$ crystals (0.5 volume fraction, 5 μ in size) precipitated in an alumina–silica glass at 900°C. In this case, the forces that developed around α ($SrTiO_3$) were compressive and of sufficient magnitude to induce transformation to a lower symmetry phase around 33°K. This transformation caused the dielectric constant to increase from a room temperature value of 50 to a low temperature value of 520. By crystallizing α at 1100°C, a $SrAl_2Si_2O_8$ feldspar coprecipitates with $SrTiO_3$ and is thought to change the hydrostatic forces around α to a uniaxial stress. This shifts the transformation temperature from 33 to 70°K; this in turn allows this triphase glass–crystal material to be used as a low-temperature capacitance thermometer with a sensitivity of 350 pF/°K. Evidently, then, once more we have an outstanding example of the need to understand the origin and meaning of microstructure in such materials in terms of its effects on bulk properties.

IV.   Summary: Future Work

By the foregoing, it is overwhelmingly evident that microstructure in many oxide glasses is not only to be expected but can be induced, developed (ripened), and controlled in such a way as to render the resulting biphase glass technologically useful. It is also apparent that the subject of microstructure–property relationships in oxide glasses is both new and old—new in the sense that its role in affecting properties seems to be more and more appreciated with the issuance of each new journal dealing with noncrystalline solids, and old form a viewpoint that these relationships have been dealt with for several decades, albeit in many cases, unknowingly.

What of the future? It would not be too surprising if subsequent research and development in this area of glass science falls within the broad outlines suggested by this review—origin, meaning and exploitation. In this sense, the propensity to induce amorphous phase separation when melting in reducing atmospheres is of paramount importance. The implication here is that many glass-forming compositions, now thought to be outside regions of immiscibility, may be subjected to the thermodynamic conditions giving rise to this phenomenon unless proper control of this melting parameter is assured. At the same time, establishing the regions of immiscibility in common systems such as $Na_2O–Al_2O_3–SiO_2$ when melted in oxidizing conditions, or more clearly delineating the coexistence curves in many systems previously demonstrated to exhibit metastable immiscibility, will be of equal value. The mode, energetics, and kinetics of amorphous phase separation are obviously intimately associated with such studies.

Control of morphology, especially orientation of a dispersed phase in many glasses, will be of similar concern. It is not unlikely that anisotropic glass ceramics will be developed within the near future with attendant enhancement of physical properties. At the same time, the development of new research tools capable of detecting small volume fractions of a dispersed phase will be highly valuable. Given the ease in which birefringence measurements can be made in most laboratories, the field of form birefringence is a top candidate for extensive reserach to realize both its limitations and usefulness. It is possible that this method is far more sensitive to the presence of a small amount of a dispersed phase than more sophisticated techniques such as TEM.

Finally, it is expected that these overall concepts will be studied in common commercial glasses in hopes that products derived from these compositions can be made more valuable to society, e.g., stronger, lighter, less expensive. In any event, it is at least clear that the overall subject of microstructures in glasses is one of the most active frontiers in the field of glass science. Given its complexity and ubiquitous nature, it is possible that as of this writing, some of the best discoveries still lie ahead. To argue otherwise would be idle.

ACKNOWLEDGMENTS

The author gratefully acknowledges the invitation of Professor MacCrone to prepare this manuscript and thanks him for his patience and encouragement throughout this undertaking. The kind permission of the authors to quote various aspects of their previous work is also acknowledged. Thanks are also extended to Academic Press for the preparation of this volume.

References

Anderson, R. C., and Freidberg, A. L. (1962). *Symp. Nucel. Crystall. Melts* p. 29. Am. Ceram. Soc., Columbus, Ohio.

Araujo, R. J., and Stookey, S. D. (1967). *Glass Ind.* **48**, 687.

Araujo, R. J., Crammer, W. H., and Stockey, S. D. (1970). U.S. Patent No. 3,540,793.

Armistead, W., and Stookey, S. (1964). *Science* **144**, 150.

Babcock, C. L. (1968). *J. Am. Ceram. Soc.* **51**, 163.

Bacon, R., and Fanelli, R. (1943). *J. Am. Chem. Soc.* **65**, 639.

Ballard, C. P. (1972). Ph.D. thesis, Alfred Univ.

Binns, D. B. (1962). *Sci. Ceram.* **1**, 315.

Borelli, N. F., and Layton, M. (1969). *IEEE Trans. Electron. Develop.* **ED-16**, 511.

Botvinkin, O. K., and Ananich, N. I. (1963). "Advances in Glass Technology," p. 86, Plenum Press, New York.

Britton, M. (1974). *Mater. Sci. Res.* **8**, 165.

Brückner, R. (1971). *J. Non-Crystall. Solids* **5**, 177.

Burke, J. (1965). "The Kinetics of Phase Transformations in Metals." Pergamon, Oxford.

Burnett, D. G., and Douglas, R. (1970). *Phys. Chem. Glasses* **11**, 125.

Cahn, J. W. (1961). *Acta Metall.* **9**, 688.

Cahn, J. W. (1969). *J. Am. Ceram. Soc.* **52**, 118.

Cahn, J. W., and Charles, R. J. (1965). *Phys. Chem. Glasses* **6**, 181.

Chaklader, A. C., and Roberts, A. L. (1958). *Trans. Brit. Ceram. Soc.* **57**, 115.

Charles, R. J. (1965). *J. Am. Ceram. Soc.* **48**, 432.

Charles, R. J. (1967). *J. Am. Ceram. Soc.* **50**, 631.

Christian, J. W. (1965). "The Theory of Transformations in Metals and Alloys." Pergamon, Oxford.

Coker, E. G., and Filon, L. N. (1951). "Photoelasticity," 2nd ed., p. 276. Cambridge Univ. Press, London and New York.

Collins, D. W., and Mulay, L. N. (1971). *J. Am. Ceram. Soc.* **54**, 69.

Cooper, A. (1972). *In* "Introduction to Glass Science (L. D. Pye, H. J. Stevens, W. C. LaCourse, eds.), p. 563. Plenum Press, New York.

Cottrell, A. H. (1948). "Theoretical Structural Metallurgy." St. Martins Press, New York.

Davidge, R. W. (1973). *In* "Fracture Mechanics of Ceramics" (R. C. Bradt, D. P. Hasselman, F. Lange, eds.), p. 417. Plenum Press, New York.

Davidge, R. W., and Green, T. J. (1968). *J. Mater. Sci.* **3**, 629.

Davies, R., and Jones, G. (1953). *Proc. Roy. Soc.* **217A**.

Doherty, P. E., Lee, D. W., and Davis, R. S. (1967). *J. Am. Ceram. Soc.* **50**, 77.

Eckstein, B. (1968). *Mater. Res. Bull.* **3**, 199.

Eherenfest, P. (1933). *Proc. K. Akad. Wet. Amsterdam* **36**, 153.

Elmer, T. H., and Nordberg, M. E. (1966). *Proc. Int. Congr. Glass, 7th* Gorden and Breach, New York.

Englert, W., and Fuhrman, P. (1968). *Bull. Am. Ceram. Soc.* **47**, 562.

Ernsberger, F. (1966). *Glass Ind.* **47**, 481.

Ernsberger, F. (1972). *Ann. Rev. Mater. Sci.* **2**, 529.

Ernsberger, F. (1973). Personal communication.

Fahmy, M., Park, M. J., Tomozawa, M., and MacCrone, R. (1972). *Phys. Chem. Glasses* **13**, 21.

Frey, W. J., and Mackenzie, J. D. (1967). *J. Mater. Sci.* **2**, 124.

Fulrath, R. M., and Pask, J. A. (1968). "Ceramic Microstructures." Wiley, New York.

Gee, G. (1947). *Q. Rev. Chem. Soc. London* **1**, 265.

Gibbs, J. H. (1960). "Modern Aspects of the Vitreous State" (J. D. MacKenzie, ed.), Vol. 1, p. 152, Butterworths, London.

Gibbs, W. (1931). "Collected Works," Vol. 55. Longmans Green, New York.

Gokularathnam, C. V., Gould, R. W., and Hench, L. L. (1972). "Advances in Nucleation and Crystallization in Glasses," (L. Hench and S. Freiman, eds.). *Am. Ceram. Soc.*, Columbus, Ohio.

Goldstein, M. (1963). *J. Appl. Phys.* **34**, 1928.

Görlich, E., Blaszczak, K., and Sieminska, G. (1974). *J. Mater. Sci.* **9**, 1926.

Grimshaw, R. W., Hargreaves, J., and Roberts, A. L. (1956). *Trans. Brit. Ceram. Soc.* **55**, 36.

Gutzow, I., Toschev, S., Marinov, M., and Popov, E. (1968). *Proc. Conf. Silicate Ind., 9th.*

Hakim, P. M., and Uhlmann, D. R. (1971). *Phys. Chem. Glasses* **12**, 132.

Haller, W., Simmons, J. H., and Napolitano, A. (1971). *J. Am. Ceram. Soc.* **54**, 299.

Hammel, J. J. (1965). *Int. Congr. Glass, 7th, Brussels, June* Gorden and Breach, New York.

Hammel, J. J., and Ohlberg, S. (1965). *J. Appl. Phys.* **36**, 1442.

Herczog, A. (1964). *J. Am. Ceram. Soc.* **47**, 107.

Holmquist, S. B. (1961). *J. Am. Ceram. Soc.* **44**, 82.

Johnson, R. E., and Muan, A. (1968). *J. Am. Ceram. Soc.* **51**, 430.

James, P. F. (1975). *J. Mater. Sci.* **10**, 1802.

Johnston, W. D. (1965). *J. Am. Ceram. Soc.* **48**, 184.

Joly, J. (1901). *Nature (London)* **64**, 102.

Kerwawycz, J., and Tomozawo, M. (1974). *J. Am. Ceram. Soc.* **57**, 467.

Kinser, D. L., and Hench, L. L. (1968). *J. Am. Ceram. Soc.* **51**, 445.

Lacey, E. D. (1968). *J. Am. Ceram. Soc.* **51**, 150.

Lange, F. F. (1973). "Fracture and Fatigue of Composites." Academic Press, New York.

Lawless, W. N. (1974). *Ferroelectrics* **7**, 379.

Lebedev, A. (1921). *Trans. Opt. Inst. USSR.*

Lewis, D. (1976). Personal communication.

MacDowell, J. F., and Beall, G. H. (1969). *J. Am. Ceram. Soc.* **52**, 117.

Mackenzie, J. D. (1960). *J. Am. Ceram. Soc.* **43**, 615.

Maklad, M. S., and Kreidl, N. J. (1971). *Proc. Int. Congr. Glass, 9th* **1**, 75.

Maurer, R. D. (1956). *J. Chem. Phys.* **25**, 1206.

Maurer, R. D. (1958). *J. Appl. Phys.* **29**, 1.

Maurer, R. D. (1959). *J. Chem. Phys.* **31**, 244.

Mazurin, O. V., Kluyev, V. P., and Roskova, G. P. (1970). *Phys. Chem. Glasses* **11**, 192.

McKinnis, C. L., and Sutton, J. W. (1959). *J. Am. Ceram. Soc.* **42**, 250.

Megla, G. K. (1966). *Appl. Opt.* **5**, 57.

Moynihan, C. T., Macedo, P. B., Aggarwal, I. D., and Schnous, U. F. (1971). *J. Non-crystall. Solids* **2**, 232.

Nordberg, M. E. (1944). *J. Am. Ceram. Soc.* **27**, 299.

O'horo, M., and Steinitz, R. (1968). *Mater. Res. Bull.* **3**, 117.

Paul, A. (1975). *J. Mater. Sci* **10**, 415.

Porai-Koshits, E. A. (1960). "Structure of Glass," pp. 9–16. Acad. Sci. USSR, Moscow (Transl. 1960).

Powell, R. E., and Eyring, H. (1943). *J. Am. Ceram. Soc.* **65**, 648.

Prebus, A. F., and Michner, J. W. (1954). *Ind. Eng. Chem.* **46**, 147.

Pye, L. D. (1972). *In* "Introduction to Glass Science" (L. D. Pye, H. J. Stevens, W. C. LaCourse, eds.). Plenum Press, New York.

Pye, L. D., Ploetz, L., and Manfredo, L. (1974). *J. Non-Crystall. Solids* **14**, 310.

Randall, J. T., Rooksby, H. P., and Cooper, B. S. (1930). *J.Soc.Glass. Tech.* **14**, 219.

Redwine, R., and Field M. (1968). *J. Mater. Sci.* **3**, 380.

Rehage, G., and Bonchard, W. (1973). "The Physics of Glassy Polymers," p. 54. Wiley, New York.

Rindone, G. E. (1969). *Glass Ind.* **50**, 138.

Rindone, G. E. (1974). presented at *Int. Congr. Glass, 10th*, Kyoto, Japan.

Robinson, H. A. (1965). *J. Phys. Chem. Solids* **26**, 229.

Roy, R. (1960). *J. Am. Ceram. Soc.* **43**, 670.

Roy, R. (1969). *J. Am. Ceram. Soc.* **52**, 344.

Roy, R. (1972). "Advances in Nucleation and Crystallization in Glasses" (L. L. Hench and S. Freiman, eds.), p. 51. Am. Ceram. Soc., Columbus, Ohio.

Selsing, J. (1961). *J. Am. Ceram. Soc.* **44**, 419.

Seward, T. P. III (1970). "Phase Diagrams" (A. Alper, ed.), Vol. 1, p. 295. Academic Press, New York.

Seward, T. P. III (1974). *J. Non-Crystall. Solids* **15**, 487.

Shaw, R. R., and Uhlmann, D. R. (1969). *J. Non-Crystall. Solids* **1**, 474.

Shaw, R., and Uhlmann, D. R. (1971). *J. Non-Crystall. Solids* **5**, 237.

Simmons, J. H., Mills, S. A., and Napolitano, A. (1974). *J. Am. Ceram. Soc.* **57**, 109.

Slater, J. (1939). "Introduction to Chemical Physics," pp. 3–32, 166–182, 256–307. McGraw-Hill, New York.

Smith, G. P. (1966). *Proc. Int. Congr. Glass, 7th* Gordon and Breach, New York.

Smith, R. A. (1971). *Contemp. Phys.* **12**, 523.

Smyth, H. T. (1972). *In* "Introduction to Glass Science" (L. D. Pye, H. J. Stevens, and W. C. LaCourse, eds.), p. 61. Plenum Press, New York.

Sproull, J. F., and Rindone, G. E. (1973). *J. Am. Ceram. Soc.* **56**, 102.

Sproull, J. F., and Rindone, G. E. (1974). *J. Am. Ceram. Soc.* **57**, 160.

Stephensen, A. (1972). *Phil. Mag.* **25**, 1213.

Stewart, D. (1972). "Introduction to Glass Science" (L. D. Pye, H. J. Stevens, W. C. LaCourse, eds.), p. 237. Plenum Press, New York.

Stirling, J. F. (1955). *J. Soc. Glass Tech.* **39**, 134.

Stookey, S. D. (1959). *Proc. Int. Congr. Glass, 5th* 32K. Glass Tech. Ber.

Stookey, S. D., and Araujo, R. J. (1968). *Appl. Opt.* **7**, 777.

Stozharov, A. I. (1958). "The Structure of Glass," Vol. 1, p. 93. Consultants Bureau, New York.

Swalin, R. (1962). "Thermodynamics of Solids." Wiley, New York.

Takamori, T. (1974). *J. Am. Ceram. Soc.* **57**, 366.

Tilton, L. W. (1957). *J. Res. Nat. Bur. Std.* **59**, 139.

Tomozowa, M. (1968). *Phys. Chem. Glasses* **13**, 161.

Topping, J. A., Fuchs, P., and Murthy, M. K. (1974).´ *J. Am. Ceram. Soc.* **57**, 205.

Turnbull, D. (1956). *Solid State Phys.* **3**, 225.

Turnbull, D. (1969). *Contemp. Phys.* **10**, 473.

Ubbelohde, A. R. (1952). *Nature (London)* **169**, 832.

Ubbelohde, A. R. (1957). *Quart. Rev.* **11**, 246.

Ubbelohde, A. R. (1965). "Melting and Crystal Structure." Oxford Univ. Press (Clarendon), London and New York.

Vezzoli, G. C., Dachille, F., and Roy, R. (1969). *J. Polym. Sci. Part A-1* **7**, 1557.

Vogel, W. (1966). *Phys. Status Solidi* **14**, 255.

Volenkov, N., and Porai-Koshits, E. (1936). *Z. Kristallogr.* **95**, 195.

Wagstaff, F. E. (1968). *J. Am. Ceram. Soc.* **51**, 449.

Wagstaff, F. E. (1969). *Phys. Chem. Glasses* **10**, 50.

Warren, B. E. (1937). *J. Appl. Phys.* **8**, 645.

Watanabe, M., and Moriya, T. (1961). *Rev. Elec. Commun. Lab.* **9**, 50.

Watson, J. (1966). *Phys. Rev.* **148**, 223.

Westbrook, J. H. (1960). *Phys. Chem. Glasses* **1**, 32.
Weyl, D. (1959). *Ber. Deut. Keram. Ges.* **36**, 319.
Zachariassen, W. H. (1932). *J. Am. Chem. Soc.* **54**, 3841.
Zarzycki, J., and Mezard, R. (1962). *Phys. Chem. Glasses* **3**, 163.

Microstructure Dependence of Mechanical Behavior of Ceramics

ROY W. RICE

Naval Research Laboratory
Washington, D.C.

I. Introduction

A. Background and Scope

Mechanical behavior, especially strength and fracture, of ceramics generally exhibits the most important and complex dependence on microstructure. The importance of this dependency arises from the central role mechanical properties play in the use of ceramics. Regardless of whether ceramics are being used for their many important and often unique (electrical, magnetic, optical, mechanical, etc.), properities, prevention of mechanical failure is almost always an important, if not critical, requirement. Although almost all properties of ceramics generally depend on impurities and porosity, strength and fracture properties are generally unique in the scope and extent to which they also depend on grain size. Further, while most ceramic properties are more closely related to the average rather than the extremes of the microstructural character, strength and fracture very often depend critically on microstructural extremes rather than averages. The dependence of most mechanical behavior on grain size as well as other microstructural parameters, especially extremes of these parameters, makes study of such behavior especially complex.

There has been a substantial advancement of our understanding of mechanical properties, especially strength and fracture, in the past few years, making this review of such properties quite timely. Because new concepts have been displacing old ones, especially for behavior at moderate temperatures, these concepts and this temperature range will be emphasized. However, elevated temperature behavior will be discussed both for its own sake as well as for its contrast or continuity with lower temperature behavior. This will typically be done at the end of each microstructural, i.e., porosity, grain size, or impurity, section for a given property, or as a short high-temperature section at the end of the major section on a mechanical property. Elastic properties will be discussed first, crack propagation next, followed first by static tensile strength and fracture, then by briefer discussions of other forms of tensile (e.g., impact and thermal shock) failure. Next, other mechanical properties, especially hardness and compressive strength and related behavior will be discussed, the latter more briefly. Finally, a summary and brief discussion of the state of understanding of microstructural dependence of mechanical behavior and of needed research is presented.

In addition to the greater emphasis on lower, mostly room temperature properties, relative emphasis is also given to those areas whose microstructural dependence has not been reviewed, i.e., fracture energy, or not widely considered, e.g., hardness and compressive strength. Since further theoretical development is clearly needed and should take advantage of past work, considerable past theoretical work and equations are summarized and compiled, respectively. Similarly, where possible and consistent with the extent of the review of a topic, substantial data compilation is also given. Finally, it should be remembered that the theme of this chapter is the microstructural dependence of mechanical behavior, not mechanical behavior itself. Thus, for example, important mechanical behavior such as the strain rate and environmental dependence of strength, which have little known microstructural dependence, are only briefly discussed.

Two procedural notes are in order. First, notation generally occurring in more than one section is shown in Table I. Other notation, used only once or twice, is defined where it is introduced. Except as indicated, no duplicate use of symbols has been made. Second, it should be noted that all porosity dependance has been fit to the exponential relationship e^{-bP}, which is the most common relationship and generally fits data well. This has been done primarily in order to provide a common basis of comparing data and, secondarily, to provide a broader basis for choosing b values for data interpolation and extrapolation. Wherever sufficient data was given, e.g., as opposed to only a curve or an equation, the data were fitted by a least-squares technique. Since many authors only used visual fitting of data, the

TABLE I

GENERAL NOTATION[a]

Symbol	Meaning	Symbol	Meaning
a	Flaw size	S	Tensile (or flexure) strength
B	Constant in Petch equation	T	Temperature
b	Porosity factor in e^{-bP}	T_m	Absolute melting temperature
C	Compressive strength	v	Elastic wave velocity
e	Naperian or natural logarithm base	Y	Geometrical flaw factor in Griffith's equation
E	Young's modulus	γ	Fracture energy
G	Shear modulus[b]	ε	Strain
G	Grain size[b]	λ	Pore separation
H	Hardness[c]	v	Poisson's ratio
K	Bulk modulus	ξ	Pore shape factor
m	Empirical porosity exponent	ρ	Density
P	Volume fraction porosity	σ_c	Stress for activation of single crystal slip or twinning
P_c	Volume fraction porosity at which strength becomes 0	φ	Volume fraction second phase
R	Pore radius		

[a] Note use of subscript zero refers to property at $P = 0$.
[b] Use of the same symbol should not cause confusion.
[c] Subscripts V and K refer to Vickers hardness and Knoop hardness, respectively.

least-squares fitting can give somewhat different values than those obtained by the original author. Where only graphic presentation of data is available, errors in reading values can shift resultant b values, but generally by 10% or less. Because of such errors, and often a somewhat limited range of porosity studied, the standard deviations on the b values cannot be taken as true statistical measures of their accuracy but only as a good relative guide to the fit of the exponential relationship or the scatter of the data, with the latter being the predominant factor.

B. Overview of Microstructural Dependence of Mechanical Behavior

Before proceeding to a detailed review, a brief historical and technological perspective, primarily on the most important behavior, tensile, strength, and fracture, is in order. The ultimate goal or use of ceramic studies is to develop better ceramics. Originally, investigators relied exclusively on correlating processing variables and resultant mechanical behavior. However, the generally opposite trends of decreasing porosity and grain size on strength, generally leading to a maximum in strength as a function of firing temperature, was recognized fairly early. Empirical and analytical studies of porosity became quite extensive from the late 1950s into the 1960s. These generally treated pores as simply a means of reducing load-carrying capacity or as stress concentrations. More recently, pores have been viewed as an integral part of flaws themselves, based in part on fracture mechanics concepts, and pores have been shown, through fractography, to be common origins of failure.

Grain size dependence of strength (S) was generally based on adapting the Griffith theory for glass failure by assuming grain size control of flaw size, most commonly with the flaw size (a) being equal to the grain size (G), in any event giving $S \propto G^{-1/2}$. Based on empirical observations, Knudsen (1959) proposed that this be generalized to $S \propto G^{-a}$ and that this could be combined with porosity dependence: $S = S_0 G^{-a} e^{-bP}$. Carniglia (1965) made one of the earliest and more extensive applications of the Hall–Petch equation

$$S = \sigma_c + BG^{-1/2} \tag{1}$$

which Knudsen had rejected despite basing his original analysis on five studies of brittle metals and only one of a ceramic. Carniglia's application of the Hall–Petch equation came during a period of particular interest in the effects plastic processes, especially slip, have on mechanical behavior. Subsequently, intense interest developed in the application of fracture mechanics, especially in determining fracture energies for use in the Griffith equation. We now hopefully appear to be moving into a new phase, the first aspect of which is direct determination and characterization of fracture origins. The second aspect is integrating this with further development and refinement of each of the above phases: microstructure–strength correlation, and fracture mechanics studies.

II.   Microstructural Dependence of Elastic Properties

A.   Background and Theories of Porosity Effects

Elastic properties of ceramics, i.e., primarily Young's (E), bulk (K), and shear (G) moduli, as well as Poisson's ratio (v), are important since in the absence of nonelastic effects they determine the strain for a given stress, and strain is important in many applications. For example, strains must (1) often be limited, e.g., at a strain of 1.6%, the wing tips of an aircraft could bend so the tips would be vertical; (2) often be matched, e.g., in joining different bodies; and (3) sometimes maximized, e.g., to accommodate thermal strain. Elastic properties are also very important because of the broad dependence of other mechanical behavior on them, especially Young's modulus.

Elastic properties depend primarily on porosity and may often also have significant dependence on impurities. Hence, these topics are treated in the foregoing order and degree of emphasis. Elastic properties have no basic dependence on grain size. They do depend some on grain orientation, since materials of both cubic and non-cubic crystal structures are elastically anisotropic. Since pore size, shape, and location may often correlate with grain size, shape, or orientation, there can be an indirect or second-order correlation of elastic properties with such grain parameters. However, since effects of grain parameters on elastic properties are limited in most cases, they are not treated further.

Considerable development of empirical, semianalytical, and analytical equations relating porosity and elastic properties has occurred. The reader is referred to Wachtman's (1969) review for a summary of empirical equations. Table II lists, in approximate chronological order, the various semianalytical and analytical equations, their authors and bases, and their assumptions and related limitations. Several of these equations developed for two or multiphase bodies, e.g., Kerner's, Budiansky's, and all those associated with Hashin, are adapted for porosity effects by setting properties of all phases except the matrix equal to zero. Since some assumptions may depend on limited differences in the properties of the phases, setting values of one phase equal to zero can increase the error. Although all equations are based on the matrix being isotropic, essentially all ceramics are elastically anisotropic. This is a serious problem only in some highly anisotropic, noncubic materials.

TABLE II

THEORETICAL POROSITY–ELASTIC PROPERTY RELATIONS

Authors	Derivation[a] (basis)	Assumption–applicability	Equations
Dewey (1947)	A(S)	Limited concentration of spherical pores having uniform size	(1) $G = G_0\left[1 - \dfrac{15(1 - v_0)}{7 - 5v_0}P\right]$ (2) $K = K_0\left[1 - \dfrac{3(1 - v_0)}{2(1 - 2v_0)}P\right]$
Gatto (1950)	SA	Spherical pores	(3) $\dfrac{dE}{dP} = 2.36,\ E = E_0(1 - 2.36P)$ Uncertainties in the theory and translation question this commonly quoted numerical factor, e.g., it may be 2.636.
Mackenzie (1950)	A(S)	Limited concentration of spherical pores whose spatial and size distribution is statistically uniform	(4) $G = G_0\left[1 - \dfrac{5(3K_0 + 4G_0)P + \theta P^2}{9K_0 + 8G_0}\right]$ (5) $K = K_0\left[\dfrac{1}{1 - P} + \dfrac{3K_0}{4G_0}\dfrac{P}{1 - P} + \zeta P^3\right]^{-1}$
Kerner (1952)	A(S)	Randomly distributed pores that in the mean are spherical	(6) $K = \dfrac{4K_0G_0(1 - P)}{4G_0 + 3PK_0}$ (7) $G = G_0\dfrac{(1 - P)(7 - 5v_0)}{P(8 - 10v_0) + 7 - 5v_0}$

204

Reference		Description		Equation
Knudsen (1959); Duckworth (1953)	SA(LB)	Uniformly sized and spaced pores between sintering spheres (or cylinders) of uniform packing and size that are deforming to become cubes (or square prisms); $P \lesssim 50\%$	(8)	$E = E_0 e^{-bP}$
Rice (1976)	SA(LB)	Interchange of above solid and pore phases; i.e., spherical (or cylindrical) pores increasing their contact to become cubical (or square, prismatic) pores; $P \gtrsim 50\%$	(9)	$E = E_0[1 - e^{-b'(1-P)}]$
Hashin (1962)	A(S)	Low concentration of uniformly dispersed spherical pores that can be of different sizes	(10) (11)	$G = G_0\left[1 - \dfrac{15(1-v_0)}{7-5v_0}P\right]$ $K = K_0\left[1 - \dfrac{3(1-v_0)}{2(1-2v_0)}P\right]$
Hashin (1962)	A(S)	High concentration of uniformly dispersed spherical pores	(12) (13)	$G = G_0\dfrac{7-5v_0}{15(1-v_0)}(1-P)$ $K = K_0\dfrac{2(1-2v_0)}{3(1-v_0)}(1-P)$
Hashin (1962)	A(S)	Arbitrary concentration of pores of varying size such that the body is made up of spherical shells of the matrix that are sized so smaller and smaller ones fill all space between larger ones; the radii of each pore (a_n) and each shell (b_n) satisfy $(a_n/b_n) = P$	(14) (15) (16) (17)	$E = E_0\left(\dfrac{1-P}{1+k_1 P}\right) \quad k_1 = \dfrac{(1+v_0)(13-15v_0)}{2(7-5v_0)}$ $G = G_0\left(\dfrac{1-P}{1+k_2 P}\right) \quad k_2 = \dfrac{2(4-5v_0)}{7-5v_0}$ $K = K_0\dfrac{1-P}{1+k_3 P} \quad k_3 = \dfrac{1+v_0}{2(1-2v_0)}$ $v = \dfrac{E_0[(7-5v_0)+2(4-5v_0)P]}{G_0[2(7-5v_0)+(1+v_0)(13-15v_0)P]} - 1$

TABLE II (*Continued*)

Authors	Derivation[a] (basis)	Assumption–applicability	Equations
Hasselman (1962) (based on Hashin, 1962)	—	—	(18) $\dfrac{E}{E_0} = \dfrac{G}{G_0} = \dfrac{K}{K_0} = 1 - \dfrac{AP}{1 + (A - 1)P}$
Hashin and Rosen (1964)	A(S)	Limited concentration of uniformly disperse aligned cylindrical pores	*Stress parallel to pore axes*
			(19) $v = v_0$
			(20) $E = E_0(1 - P)$ $K = \dfrac{E_0 G_0(1 - P)}{9G_0 - 3E_0(1 + P)}$
			(21) $G = G_0\left(\dfrac{1 - P}{1 + P}\right)$
			Stress perpendicular to pore axes
			(22) $E = \dfrac{4G_0 K}{K + \psi G}$ $\psi = 1 + \dfrac{K v_0{}^2}{E_0(1 - P)}$
			(23) $K = \left(K_0 + \dfrac{G_0}{3}\right)\dfrac{2v_0(1 - P)}{P + 2v_0}$
			(24) $G = G_0\left[1 - \dfrac{4P(1 - v_0)}{1 + 2P(1 - 2v_0)}\right]$ to first power of P
Budiansky (1970)		Isotropic bodies	(25) $K = K_0\left[1 - \dfrac{3P(1 - v)}{2(1 - 2v)}\right]$
	A(S)		(26) $G = G_0\left(1 - 15P\dfrac{1 - v}{7 - 5v}\right.$

Chung (1972) (Birch)   (SA)   —

$$(27)\quad v = J(\bar{m})\rho^u \sim j(\bar{m}) + \tau\rho = j(\bar{m}) + \frac{\tau\rho_0}{1-P}$$

$\bar{m}$ = mean atomic weight; J, u, j, τ are constants;
u = Gruneisin constant minus $\frac{1}{3}$

Paul (1960) (see Table VII); Brown *et al.* (1964) (see Table XI)   A(LB)   Simple cubic stacking of nonintersecting spherical pores

$$(28)\quad \frac{E}{E_0} = 1 - \pi\left(\frac{3P}{4\pi}\right)^{2/3}$$

Simple cubic stacking of nonintersecting cubic pores

$$(29)\quad \frac{E}{E_0} = 1 - P^{2/3}$$

---

^{a} A = analytical, SA = semianalytical, (S) = based on stress concentration, (LB) = based on load-bearing area.

While the reader must be referred to the original authors' work for details of the derivation of each equation, some general and specific comments are given. The first general comment concerns the derivation and bases (Table II) that are directly related. Analytical derivations use mechanics to determine the average stress, which may or may not involve considering specific local values of stress. Such approaches, if fully valid, provide self-consistent elastic properties, i.e., E, K, G, and v can all be calculated from any two of the four (see Table III) as may some other properties (e.g., sound velocity v). Semianalytical derivations, commonly used to obtain E because of its broad importance, are generally based on load-bearing area, i.e., using the fact that the elastic property is directly proportional to the average cross-sectional area of solid material in planes normal to the stress, as determined through use of simple geometry or more sophisticated stereo-logical techniques. Semianalytical models are not self-consistent; i.e., if one uses the same functional form for E, K, G, and v as generally implied by the derivation, they will not be consistent with calculations from the equations of Table III.

TABLE III

STANDARD ELASTICITY RELATIONS

(1) $E = \dfrac{9KG}{3K + G}$

(2) $v = \dfrac{E}{2G} - 1 = \dfrac{9}{6 + 2G/K} - 1$

(3) $K = \dfrac{EG}{9G - 3E}$

(4) $G = \rho v_s^2$ $v_s = $ shear wave velocity

Despite this important deficiency, the semianalytical models are useful because they provide simple equations for: (1) extrapolations for elastic properties (e.g., E), which are widely needed for engineering and scientific purposes, and (2) interrelation of porosity effects (e.g., of S, E, and γ in the Griffith equation as noted later). Further, as shown below, none of the analytical models are both fully self-consistent and in good agreement with experimental results, increasing the utility of semianalytical models.

The second general comment is that several of the equations in Table II are nearly or exactly identical, falling into two groups. The first group is based on Hashin's equations (Eqs. 10 and 11) (Table II) for low concentration of pores and Dewey's equations, which are equal for both K and G and hence E and v. As discussed later and seen in Fig. 4, v for Budiansky's equations has a limited dependence on P, so his equations for K and especially G and hence E are also close to those of this first group. Using identities of Table III shows that Hashin's equations (Eqs. 14–17) (Table II) for any concentration of pores are identical to those for K and G of Kerner and hence also for E and v forming the second group. If the second-order P terms in Mackenzie's equations are neglected, i.e., since his equations are valid only for small P, his equations split between the above two groups. Thus, use of identities in Table III shows his equation for G is identical to that of group 1 above, while his equation for K is identical to that of group 2 above. As seen in Fig. 5, his E–P dependence falls very close to that of group 1 above. Finally, Hashin and Rosen's equation for G when stress is applied parallel to aligned cylindrical pores is nearly identical to that of group 2 above for any typical value of v (e.g., 0.2–0.4).

The third general comment is that completely valid theories must extrapolate to proper values at $P = 0$ and $P = 1$. Proper extrapolation at $P = 0$ is most important, since the low porosity region is usually of greatest interest, and it is often desired to obtain E_0, G_0, etc., by extrapolation. Most theories do properly extrapolate at $P = 0$. Hashin's theory for high concentrations of pores fails to extrapolate to E_0, G_0, and K_0 (see Figs. 5–8), but the theory is not valid at low porosity. Also Hashin and Rosen's theory for stress normal to aligned cylindrical pores fails to give K_0 at $P = 0$ (Fig. 7) due to anisotropy assumptions nor does Eq. 27 (Table II), and hence, neither gives E_0 as $P \rightarrow 0$. Extrapolation to zero as $P \rightarrow 1$ is important theoretically, but concern for this has often been overemphasized or unjustified, since many theories are based on limited porosity (Table II) and hence are not valid as $P \rightarrow 1$. Thus, for example, Eqs. (1), (2), (4—neglecting second-order term), (10), (11), (24), (26), and (27) of Table II all go to zero at $P \sim \frac{1}{2}$. Equations (5—neglecting second-order term), (6), (7), (9), (12–17), (20), (21), and (23) go to zero at $P = 1$, which is important mainly for Eqs. (9), (12), and (13), since they are valid for large P. Obtaining zero elastic properties at $P = 1$ is valid only for homogeneous distributions of most pore structures. Clearly, in practice, elastic properties will often go to zero before $P = 1$ because of inhomogeneous pore distribution or unusual pore structures (e.g., laminar pores normal to the stress axis). Different approaches, such as precolation theory (e.g., see Shante and Kirpatric, 1971) may be necessary to handle such situations.

Within the context of the above general comments, the following comments on the models of Table II are made, considering analytical models first. Dewey's (1947) model was derived to include the effects of gas pressure within pores of materials such as elastomers. Both algebraic errors and the assumptions for such pressure effects preclude the use of her simplified equations; however, her basic derivation of the general equations leading to those of Table II appear correct. Mackenzie's theory (1950), the only one derived directly for porosity effects, is most often used only to the first power of P, because it is simpler and over the region for which its assumptions are valid, higher-order terms should not be too significant. The coefficients (θ and ζ) of these terms are not given analytically in Mackenzie's treatment, and hence one cannot determine the extrapolation of his model as $P \to 1$ if these are included. However, θ can be determined semiempirically by letting G and $K \to 0$ as $P \to 0$, but similar treatment does not work for ζ. Kerner's (1952) and Budiansky's (1970) theories, which were derived by averaging techniques for multicomponent composites, have the least restrictive sets of assumptions, and give some good agreement with data.

Hashin's (1962, 1968) theory was derived for composite materials containing spherical second-phase particles. Hasselman (1962) was apparently the first to utilize these equations for porosity by setting the properties of the second phase equal to zero. He treated the coefficients of the porosity terms as empirical rather than analytical expressions derived from Hashin's theory. This gives the same form for E, G, and K and can allow empirical extrapolation to variations from limited pore concentration or spherical pores. (Note in his original derivation, Hasselman used a different expression from that of Table II but then changed to that shown in Table II by changing the sign of A, a negative coefficient in the original expression.) Weil (1964) subsequently gave the complete analytical expressions for porosity derived from Hashin's composite theory. Hasselman and Weil's use of Hashin's expression are identical, with the A parameter in Hasselman's equations being equal to the respective $k_i + 1$ in Weil's equations. Hashin (1962, 1968) applied his theory, simplified for low concentrations of the second phase, to a low concentration of matrix (i.e., high second-phase concentration). This gives the high-porosity relations in Table II, balancing his low-concentration approximation (which does not extrapolate to 0 at $P - 1$), by extrapolating to zero at $P = 1$. (Note his limited concentration theory, though not valid at $P = 1$, does extrapolate to zero.)

Rossi (1968) used Hashin's low-concentration approximation to extend the theory to nonspherical pores by essentially a perturbation technique. This was done by recognizing that the coefficients of the second phase (i.e., porosity terms in Table II) were directly related to the stress concentrations (q) of the particle (pore), i.e., $E = E_0(1 - qP)$. Thus by calculating the stress concentration [$q = (5a/4c) + \frac{3}{4}$, where a is the spheroidal radius and c is the major axis of the elliptical cross section] of various spheroidal shapes and orientations and numerically substituting these for the stress concentrations in Hashin's equations, he extrapolated Young's modulus behavior to nonspherical pores. His graphical results for spheroidal cavities having complete alignment and random orientation (Fig. 1) are most accurate

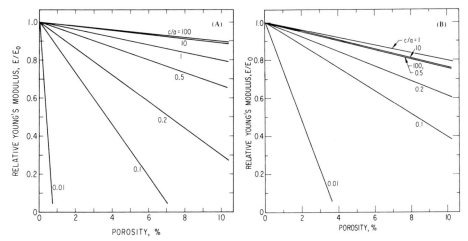

Fig. 1. Effect of pore shape and orientation on relative Young's modulus, E/E_0 (after Rossi, 1968). (A) E/E_0 for stress parallel to aligned c axis of spherical voids and (B) for randomly oriented spheroidal voids. (Original figures courtesy of Dr. Rossi and published with permission of the American Ceramic Society.)

for $v = 0.2$, since q really depends on v except at $v = 0.2$; but this assumption of q being independent of v should not normally be serious. Rossi also noted that his treatment gives good agreement for two pore extremes—spherical ($a = c$, $q = 2$) and cylindrical pores stressed parallel with their axis ($a/c = 0$, $q = 1$).

Hashin and Rosen (1964, see also Hashin, 1968) subsequently derived a theory for composites consisting of hollow cylindrical fibers in various aligned packing arrangements. These equations, with property values of the fibers set to zero for treatment of them as pores, are given in Table II. Note that Hashin and Rosen defined two-dimensional bulk and shear moduli for analysis of elastic behavior for stressing normal to the fiber axis. These moduli are for zero strain parallel with the fibers. Calculations show that this shear modulus is the same as the usual three-dimensional shear modulus G, but the two-dimensional bulk modulus equals $K + G/3$. Equations in their paper and Hashin's (1968) review are given in terms of these two-dimensional parameters, a point not particularly clear in these papers, especially the original development. Equations in Table II are given in terms of the usual, or three-dimensional, properties K and G, which give $E = E_0$ at $P = 0$, but $K > K_0$ as noted earlier and later. For stresses parallel with the cylindrical pores, these equations give a dependence on porosity identical to that obtained simply by taking E/E_0 equal to the ratio of the solid to the total cross-sectional areas. For stresses normal to the pore axes, greater porosity dependence than for spherical pores is observed as is expected and is consistent with Rossi's (1968) oriented porosity analyses discussed above.

Agarwal *et al.* (1971) have applied a finite element analysis to a body containing uniformly spaced spherical pores of equal size. Their resultant Young's modulus predictions lie slightly higher than, but close to, the BeO data of Fryzell and Chandler (1964) (Table IV). They also predict Poisson's ratio initially starting at ~ 0.23 at $P = 0$ and dropping to ~ 0.15 at $P = 0.5$. Similarly, they get good agreement between their predicted Young's modulus and those measured for glass–alumina and glass–tungsten composites by Hasselman and Fulrath (1965, 1966).

Turning next to the semianalytical relations, the exponential relationship (Eq. 8) (Table II) is one of the simplest and most widely used relations and hence most important. Duckworth (1953) apparently was the first to introduce this relation for porosity effects based on the empirical observation that compressive strength–porosity data of Ryskewitch (1953) gave a straight line on a semilog plot. Knudsen (1959) put this empirical strength relation on a semianalytical basis by considering the changing contact area between sintering grains idealized as uniform spheres coalescing together with different packing arrangements (Fig. 2). Then, assuming that strength was directly proportional to the contact (i.e., load-bearing) area, he then showed graphically that strength could be expressed by the exponential relation, with b varying from 6 to 9 for the most porous cubic through the least porous rhombohedral arrangement. Spriggs and colleagues, to whom this expression is often attributed, subsequently extensively applied this relation to not only strength, but also elastic property evaluations, as Knudsen subsequently did also. The latter apparently predicated his evaluations upon the reasonable assumption that elastic properties should be equally, if not more closely, related to the load-bearing area than is strength. Rice (1976a) has applied Knudsen's approach to bodies with cylindrical porosity and stresses normal to cylinder axes. This is fitted by the same exponential relation, but with higher values of b as expected [$b \sim 7$ for cubic packing of cylinders in contrast to $b \sim 6$ for the corresponding packing of spheres (Fig. 2)]. This dual usage of the same expression for strength and elastic porosity relations also suggests similar dual usage of other strength elastic-porosity relations where justified and useful, e.g., for special analytical purposes.

Besides the less rigorous semianalytical, load-bearing nature of the Duckworth–Knudsen exponential model, it has been criticized for its failure to approach zero as $P \rightarrow 1$. This criticism is not justified, since Knudsen's analysis clearly shows absolute upper limits of $P \sim 0.45$, 0.35, and 0.25 for idealized cubic, hexagonal, and rhombohedral packing, respectively (Fig. 2). The limits of cylinder packing are less; $P \sim 0.18$ for cubic packing. Also, it is fairly obvious that the role of pores and matrix can be reversed, leading

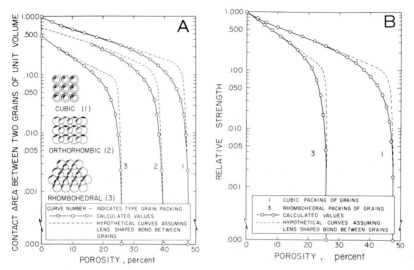

Fig. 2. Knudsen's semianalytical derivation of the exponential porosity relationships. (A) The contact area between two spherical grains that are sintering together in the various geometrical packings as sketched. Note that the top layer of spherical particles are drawn smaller simply for clarity, all are of the same size. (B) The relative strengths; while this model was derived originally for compressive strength purposes, it has been widely used for both elastic properties and tensile strengths. (Original figures courtesy of Mr. Knudsen and published with the permission of the American Ceramic Society.)

to Eq. (9), (Table II) with $b' \sim 0.5$ derived by Rice (1976) for large values of porosity. This expression is really a continuation of Eq. (28) of Table II, with both expressions being closely approximated by e^{-3P} to $P \geq 0.7$. This expression (Eq. 9, Table 2) goes to zero as $P \to 1$. It clearly does not give $E = E_0$ at $P = 0$, but it is not valid for low values of porosity (the lower bounds are $P > 0.55$ and $P > 0.75$ for cubic and rhombohedral packing, respectively). Combinations of the two models can be developed; i.e., hollow spheres representing both inter- and intragranular porosity. Such combinations are reasonably approximated by e^{-bP} to higher (e.g., $P \geq 0.7$) porosity levels than for intergranular pores alone, though with some reduction of b value.

Gatto's (1950) theory is based on considering the effects of single or uniformly spaced spherical pores on the resonance of vibrating bars, apparently then averaging various effects, but lack of clarity in the translation and probably the original article limits evaluation. Birch (1961) had empirically determined a linear additive velocity–density relation for oxides and silicates, e.g., those of geological interest. Chung (1972) subsequently showed that Birch's equation was a linear approximation to the power law equation he derived from the nondispersive wave equation. Though the expression is analytically derived, the coefficients are not, and while G can be readily derived (Table II), other elastic properties cannot be unless a second analytical expression is available. Thus, the utility of either the power or linear form of the relation for determining elastic properties is restricted. Patel and Finnie (1970) gave the equation $E = E'(1 - P)^m$ for plastic foams based in part upon empirical tests and in part on a semianalytical model. E' and m are parameters dependent on foam structure; m is typically in the range of 1 to 2 instead of a value of $\frac{2}{3}$, which one would obtain neglecting foam structure and resultant stress interactions.

Martin and Haynes (1971a) derived a semianalytical, load-bearing theory based on sterological averaging procedures for concrete and subsequently fitted data for several ceramics to the resulting equation (1971b). They considered pores of arbitrary size and shape, assuming that (1) sterological averaging procedures are accurate and that resultant equations are of the same form as those for spherical voids and (2) that $E = E_0(1 - P_A)$, where P_A is the average cross-sectional area of pores in a random plane normal to the stress axis, i.e., a load bearing nodel. From this they derive an extremely simple expression: $E = E_0(1 - kP^{2/3})$, i.e., that $P_A = kP^{2/3}$. According to their analysis, this parameter k equals AN_a/VN_v where A and V are, respectively, the average void cross sectional area and average void volume, and N_a and N_v are, respectively, the number of voids per cross-sectional area and per volume. Calculations for various simple pore geometries, e.g., uniform spheres or cylinders, show k proportional to $P^{1/3}$ and hence not a constant, so E always varies as the first power of P, contrary to their equation. In subsequent private communication, Martin concurs that the above modulus expression reduces to $E = E_0(1 - k'P)$, where $k' = k/P^{1/3}$ should truly be a constant independent of porosity. Martin (1973) has subsequently performed an analysis which he feels shows $k' = kP^{1/3}$ for any porosity.

The problem of Martin and Haynes' equations illustrates a common fallacy, the application of common sterological relations to mechanical properties. Such relations calculate averages, while mechanical properties commonly depend on extremes, e.g., being commonly substantially influenced or controlled by the minimum solid cross sectional (i.e., load bearing) area. Elastic properties, such as E, are no exception; i.e., a bar with a band of porosity across its width and thickness will effect E parallel with the bar axis far out of proportion to its contribution to the average porosity. Martin and Haynes' equation fits a fair amount of data since it is the same or similar to equations for the minimal solid area of bodies having different simple geometrical pores [Table II, Eqs. (9), (20), (28), and (29)]. Their equation is also almost identical to Ishai's (see Cohen and Ishai, 1967, and Lange, 1974a) equation for composites applied to pores, which gives

$E = E_0(1 - P^{2/3})$, derived in a similar fashion to Paul's model (discussed below and in Section II,C).

One should note, as pointed out by Hashin (1964), that the theoretical problems and inaccuracy of applying equations for elastic properties of composites increases as the difference in elastic moduli between the phases increases. Thus, Budiansky's (1970) equations will be more accurate for composites in which the dispersed phase is solid with moduli closer to those of the matrix than when the dispersed phase is void. Hashin (1964) similarly points out that, as expected, the closeness of bounds also increases as the ratio of the moduli of the matrix and dispersed phase approaches one. This is illustrated by the equations developed by Paul (1960) for metallic matrix dispersions and cermets. These equations give reasonable bounds for WC–Co bodies. However, when applied to pores his lower bound reduces to zero for all values of porosity, while the upper bound reduces to $E = E_0(1 - P)$. Clearly, the former is excessively low, and as seen by comparison to the equations of Table II, (see also data of Table IV and Section II,B), the upper bound becomes quite high as P increases. Paul's intermediate approximation for composites (see Table VI) gives $E/E_0 = (1 - P^{2/3})/(1 - P^{2/3} + P)$, which is more reasonable though still not in good agreement with data.

All of the above theories treat a single type of porosity, either a simple idealized porosity, e.g., uniformly sized and/or spaced pores, or a single, general, unspecified porosity. However, for any given porosity level the modulus will depend on the spatial shape and/or size distribution of porosity. The shape effect is readily seen from Rossi's (1968) work (Fig. 1) as well as by simply comparing expressions for spherical and cylindrical pores (Table II). The spatial distribution effect, which can also depend on size, e.g., with both large and small pores, is readily seen by recalling that most theories and data show a nonlinear dependence on P. Therefore, the elastic properties calculated from the average porosity $(\bar{P})$ will be different from those calculated by averaging the property of interest over the different regions of nearly constant porosity; i.e.

$$f(\bar{P}) \neq \sum_{}^{n} f(P_i)\bigg/n$$

where $f(P)$ is a nonlinear porosity function and P_i is the porosity in the ith section of the body having n regions of nearly homogeneous porosity. Spriggs (1962) apparently indirectly recognized parts of this in suggesting separation of open and closed porosity, as did Piatasik and Hasselman (1964), who also suggested calculating Young's modulus from its dependence on each type of intergranular and intragranular porosity. However, spatial, size, and shape distribution, which can but need not depend on the type of porosity, are the controlling variables, as pointed out by Rice (1975a,b), who notes that shape and spatial distribution effects can have either opposite or reinforcing effects.

B. Elastic Property—Porosity Data and Comparisons to Theories

1. Elastic Property–Porosity Data

An extensive but not totally exhaustive compilation of elastic property–porosity data at or near room temperature is given in Table IV. While a few studies or compilations cover only a limited range of porosity, e.g., 5–15%, most extend as high as 30–50%, and a few (e.g., glassy carbon and gypsum) to 70% porosity. As discussed in Section I, all data were fitted to the exponential relation, and some values in Table III differ from those reported by the original authors. Consideration of the relations in Table III, shows that only two of the elastic properties can exactly fit the exponential relation, but the error in fitting all four to this relation does not appear to be significant in comparison to other factors. Least squares fitting of data in the present work is apparently the major reason why the present fitting of the graphite data of Cost *et al.* (1968) and Wagner *et al.* (1972) often gives higher intercepts than the authors own evaluations. The resultant higher graphite values are closer to those calculated from single-crystal data, which would reduce the population of cracks proposed by Hasselman (1970b) to explain the single-crystal extrapolation difference in values of elastic properties. Comparative *b* values for the exponential relation and two of the next most common relations, Hasselman–Hashin's and the linear relationship, are also given in Table V showing similar general trends among them. In a number of cases in Table IV, fairly extensive compilations have been made by this author or previous authors generally showing substantial agreement between several investigators.

Overall, most data fall in a range, e.g., ± ~50% of the mean, as discussed later. However, one must first consider the scatter and variation of the data. There are, of course, some cases in which one or a few isolated data points may disagree substantially with the bulk of the data for a given material study or compilation. However, there are also some studies which disagree substantially with other data. The following variations of Young's modulus, the most widely studies property, are equal to or greater than those of shear and bulk moduli. The BeO data of Fryxell and Chandler (1964), and Udy and Bolger (1949) (see Long and Shofield, 1955) extrapolate to somewhat different zero porosity values and diverge as porosity increases (Fig. 3). Ryshkewitch's (1960) BeO data fall below all the other data at

TABLE IV

Summary of Room Temperature–Porosity–Elastic Property Data

Materials	Zero porosity moduli from single crystal values (10^6 psi)			Zero porosity values from porosity data (10^6 psi)				Porosity correction factor (b)				Source
	E	G_0	K_0	E_0	G_0	K_0	v_0	E	G	K	v	
Borides												
HfB$_2$								4.3 ± 0.3				Limited data of Clougherty et al. (1968)
TaB$_2$				69.4				3.6 ± 0.1				Claussen (1969)
TiB$_2$								3.74				Mandorf et al. (1963)
ZrB$_2$				76.0				3.1 ± 0.1				Limited date of Clougherty, et al. (1968)
ZrB$_2$				75.0				1.1 ± 0.2				Limited compilation of Jun and Shaffer et al. (1971)
Carbides												
B$_4$C				68.2	30.0			5.7 ± 0.4	4.9 ± 0.6			Liebling (1967)
NbC				75.4				3.4 ± 0.3				Speck and Miccioli (1968, see Toth 1971, p. 145)
SiC	58.2	24.6		64.6	28			4.8 ± 0.2	4.3 ± 0.2			Compiled mainly from: Carnahan (1968), Gulden (1969), and Coppola and Bradt (1972)
SiC				69.6				3.1				Prochazka et al. (1974)
ZrC				65.0				3.2 ± 0.2				Jun and Shaffer (1971)
Nitrides												
Si$_3$N$_4$				40.0	17.6		0.22	2.4 ± 0.4	3.1 ± 0.2		1.1 ± 0.1	Compiled from Coppola et al. (1972), McLean et al. (1974), and Fate (1975), limited data of Monch and Claussen (1968)
				34				4				

TABLE IV (*Continued*)

Materials	Zero porosity moduli from single crystal values (10^6 psi)			Zero porosity values from porosity data (10^6 psi)				Porosity correction factor (b)				Source
	E	G_0	K_0	E_0	G_0	K_0	v_0	E	G	K	v	
Single oxides												
Al$_2$O$_3$ (fibers)				54.2				3.9				Blakelock et al. (1970)
Al$_2$O$_3$ (fibers)				45.8				2.3 ± 0.1				Bailey and Barker (1971)
Al$_2$O$_3$	58.4	23.7	36.4	61.2	23.9			3.5 ± 0.1	3.3 ± 0.2			Compiled from Knudsen (1961), Spriggs and Brissette (1962), Binns and Popper (1966), and Neuber and Wimmer (1971)
B$_6$O + <1% B$_2$O$_3$				66.6	28.7	32.1		4.6 ± 0.3	4.5 ± 0.3	4.6 ± 0.3	0.0	Petrak et al. (1974)
B$_6$O + <1% boron				68.1	29.5	32.3		4.0 ± 0.3	4.0 ± 0.3	4.0 ± 0.2	0.0	
BeO	57.2	23.1	38.4	55.8	23.3		0.17	3.4 ± 0.5	3.0 ± 0.6		−1 ± 1	Compiled from Udy and Bolger (1949, see Long and Schofield (1955), Bentle (1962), O'Neil et al. (1966), and O'Neil (1970)
CoO				33.2				4.5 ± 0.6				Petrak et al. (1975)
Dy$_2$O$_3$				24.3	9.43		0.33	2.81	2.63		1.3 ± 0.4	Manning et al. (1969)
Er$_2$O$_3$				26.4	10.2		0.30	2.90	2.86		2.8 ± 0.2	Manning et al. (1969)
H$_2$O (ice)	1.4	0.5	1.2	2.0				4.7 ± 0.5				Tabata (1967)
Ho$_2$O$_3$				25.0	9.70		0.29	2.62	2.55		3.6 ± 0.2	Manning et al. (1969)
MgO	44.8	19.0		46.0	19.1		0.26	4.5 ± 0.2	4.9 ± 0.2		1.5 ± 0.8	Compiled from Spriggs et al. (1962) and Chung (1963)
Nb$_2$O$_5$				24.8				5.0 ± 0.2				Manning (1971)
SiO$_2$				9.8				6.6 ± 0.3				Gannon et al. (1965)
Sm$_2$O$_3$				23.6	5.63		0.34	4.4 ± 0.3	3.1 ± 0.2		0.90 ± 0.06	Hunter et al. (1974)

Material												Reference
ThO_2	36.2	14.1	30.0	34.7	16.0		0.29	3.4 ± 0.4	4.0 ± 0.2		0.9 ± 0.1	Compiled mainly from Petersen and Curtis (1970) all but low points of Spinner et al. (1963)
TiO_2	41.1	16.2		40.9	12.3			2.7 ± 0.1	2.6 ± 0.1	4.3 ± 0.2	1.0 ± 0.1	Chung and Bussem (1968)
UO_2	33.4	12.7		32.6		30.9	0.33	3.3 ± 0.2	2.7 ± 0.1			Compilation of Boocock et al. (1972)
Y_2O_3				26.9	10.1		0.28	3.4 ± 0.4				Manning and Hunter (1968)
Y_2O_3				26.1	10.1	2.81			2.70		0.0 ± 0.2	Manning et al. (1969)
ZnO	17.9	6.6	20.8	18.4				2.5 ± 0.1				Chung and Bussem (1968b)
ZrO_2				36.1				4.2 ± 0.2	4.7 ± 0.1			Smith and Crandall (1964)
Mixed oxides												
$CoO-MgO$				37.5				4.36 ± 0.30				Petrak et al. (1975)
$CoAl_2O_4$				41.3				5.39 ± 0.15				Petrak et al. (1975)
Ferrite				13.0				4.1 ± 1.6				Chen and Weisz (1972)
Mullite				32.3				3.0 ± 0.2				Penty et al. (1972)
$MgAl_2O_4$	42.6	17.0		40.6	16.3			4.2 ± 0.2	4.1 ± 0.2			Compiled mainly from Schreiber (1968)
Other materials												
Gypsum				0.64				11.5 ± 0.5				Soroka and Sereda (1968)
				0.60				6.1 ± 0.2				Soroka and Sereda (1968)
				0.56				3.2 ± 0.4				Soroka and Sereda (1968)
Sagger clay				1.0				4.2 ± 0.5				Heindl and Pendergast (1927)
Sodium borosilicate glass				11.6	4.8			1.93 ± 0.1	2.07 ± 0.1			Hasselman and Fulrath (1964)
Carbon (fibers)				1.71				6.6 ± 0.5				Ezekiel (1970)
Carbon (fibers)				1.72				9.0 ± 2.0				Butler (1973)
Graphite (POCO)				5.3	2.0		0.32	3.5 ± 0.2	3.3 ± 0.2	0.9 ± 0.4		Cost, et al. (1968)
Graphite (POCO)				4.7	2.2		0.34	5.1 ± 0.7	4.4 ± 0.3	6.2 ± 0.8		Wagner et al. (1972)
Graphite				1.9		3.6		1.7 ± 0.2			1.6 ± 0.5	Compiled from Carborundum data for Graph-i-tite G
Glassy carbon				7.8				2.4 ± 0.1				Compiled from Yamada (1968) and Hucke (1972)

TABLE V

COMPARISON OF YOUNG'S MODULUS-POROSITY FACTORS

Material	Factors			Investigator
	$E/E_0 = 1 - bP$ (b)	$E/E_0 = e^{-bP}$ (b)	$E/E_0 = 1 - \dfrac{AP}{1+(A-1)P}$ (A)	
B_6O	—	4.3	5.1	Petrak et al. (1974)
BeO	1.72 ± 0.04	2.06 ± 0.05	1.99 ± 0.04	Fryxell and Chandler, see Weil (1964)
	1.82 ± 0.02	2.21 ± 0.03	2.24 ± 0.02	
	1.80 ± 0.02	2.14 ± 0.03	2.16 ± 0.02	
	2.25 ± 0.04	2.71 ± 0.05	2.81 ± 0.04	
	1.86 ± 0.03	2.20 ± 0.04	2.19 ± 0.03	
CoO	2.8 ± 0.8	4.5 ± 0.6	6.5 ± 0.6	Petrak et al. (1975)
CoO–MgO	3.0 ± 0.4	4.4 ± 0.3	5.8 ± 0.3	Petrak et al. (1975)
Dy_2O_3	2.13	2.81	3.04	Manning et al. (1969)
Er_2O_3	2.11	2.90	3.23	Manning et al. (1969)
Ho_2O_2	2.09	2.62	2.78	Manning et al. (1969)
Sm_2O_3	2.0	4.4 ± 0.3	—	Hunter et al. (1974)
UO_2	3.1	4.4	—	Mönch and Clausen (1968)
UO_2	2.3	3.3 ± 0.2	—	Boocock et al. (1972)
Y_2O_3	2.0	3.4 ± 0.4	—	Manning and Hunter (1968)
Y_2O_3	2.02	2.81	3.1	Manning et al. (1969)
$CoAl_2O_4$	3.6 ± 0.1	5.4 ± 0.2	6.7 ± 0.2	Petrak et al. (1975)
$MgAl_2O_4$	2.5 ± 0.3	3.1 ± 0.3	3.5 ± 0.4	Petrak et al. (1975)
Mullite	—	3.0 ± 0.4	3.2 ± 0.4	Penty et al. (1972)
Sodium borosilicate glass	2.1 ± 0.1	1.9 ± 0.1	—	Hasselman and Fulrath (1964)
TiB_2	2.5	3.6	—	Clausen (1969)
B_4C	4.6	5.6 ± 0.4	—	Liebling (1967)
UN	2.2	4.0	—	Mönch and Clausen (1968)

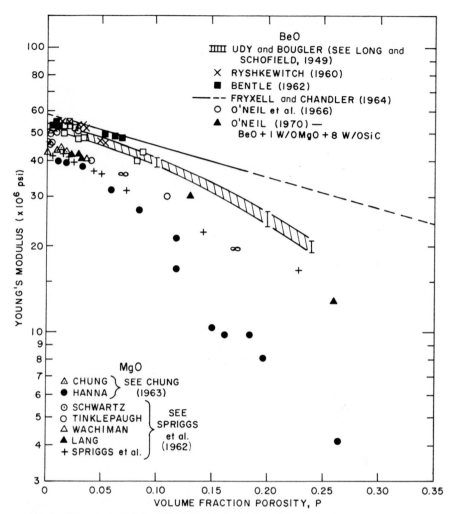

Fig. 3. Young's modulus versus porosity for BeO and MgO. Each of these sets of data illustrates some of the greatest disagreement between elastic property measurements of the same materials by different investigators.

all porosities suggesting measurement errors and hence are not included in the compilation shown in Figs. 5–8 or Table IV. Similarly, Thompson and Pratt's (1966) Si_3N_4 data are neglected, since it all falls below other data, possibly due to errors in their use of compressive testing. Also Hannah and Crandall's MgO data (see Chung et al., 1963), which fall substantially below all MgO data of other investigators at porosities beyond about 10% (Fig. 3) were neglected in the overall compilation of MgO data.

Most of the limited studies of Poisson's ratio in Table IV show it decreasing with increasing porosity at a lower rate than the other elastic properties as expected. Also, most of Bentle's (1962) BeO data gives $v = 0.2e^{-(4\pm2)P}$, i.e., a higher v_0 and b value by neglecting his three more porous sets of specimens. Petrak et al. (1974) observed essentially no porosity dependence in their B_6O samples. On the other hand, Soga and Schreiber (1968) show a slight increase in Poisson's ratio, with increasing porosity for MgO, similarly to earlier data of Soga. They also show about the same trend for earlier data of Spriggs et al., but about 30% lower than the other data. This is in contrast to the marked drop of Poisson's ratio with increasing porosity that these authors show for the data of Chung et al. Much of this variation is due to the fact that Poisson's ratio depends on measuring differences and hence is more prone to error.

Rice (1975a,b) has pointed out that the major factors in variations of elastic properties as a function of porosity are the inhomogeneous size and spatial distribution of porosity and especially variations in pore shape, as discussed in the preceeding section. That significant variations in elastic properties occurs due to inhomogeneity in the size and spatial distribution of porosity is shown by the substantial difference observed in samples made by different processing methods in the same studies, e.g., ThO_2 bodies of Spinner et al. (1963), and various gypsum bodies studied by Soroka and Sereda (1968). Similarly, Fryxell and Chandler (1964) noted that lower E_0 and G_0 values in their fine (2–5 μm) grain BeO specimens were probably due to more intergranular porosity different in size, shape, and distribution than in their larger grain bodies. As also suggested by Rice (1975a), higher b values in the Petrak et al. (1975) study of $CoAl_2O_4$ than in their study of CoO correlates with more inhomogeneous porosity in the former (Fig. 4).

Rice (1975a) has suggested that a good first-order approximation to correct for most of the variations would be to characterize the porosity sufficiently so that one could divide a body into a mosaic or regions having nearly constant porosity, calculate the elastic modulus for each region, and then average these moduli to obtain an overall body modulus. In practice, consideration of a two-dimensional mosaic based on cross-sectional examinations may be adequate. Also, tests of samples of different sizes and different orientation relative to pressing directions as well as different techniques may indicate differences in porosity that could only be discerned by fairly extensive statistical sterological studies. Clearly, further work must progress beyond the simple use of single specimens, tests, and especially only average porosity of unspecified character.

The other major and often most important factor in variable elastic properties as a function of porosity and variation in pore shape, can readily be seen from Table II. Thus, in terms of the exponential relation, the b values

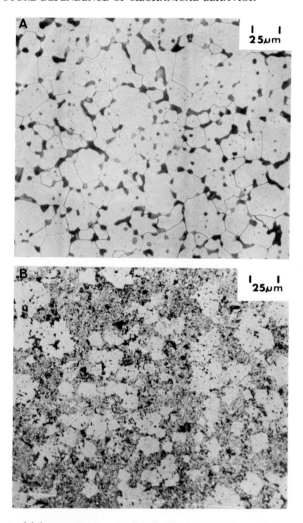

Fig. 4. Effect of inhomogeneous porosity distributions on the porosity dependence of Young's modulus. Representative photograph from two of the bodies Petrak *et al.* (1974) used for studying the porosity dependence of Young's modulus show substantial differences in the size and spatial distribution of pores (black regions): (A) CoO, $b = 4.5$; (B) $CoAl_2O_4$, $b = 5.4$. Note that the CoO, having more homogeneous porosity, has the lower b value while the $CoAl_2O_4$, having substantially more inhomogeneous porosity, has the higher b value, consistent with Rice's (1975a) discussion of inhomogeneous porosity leading to variable porosity dependence. (Photos courtesy of Drs. Ruh and Petrak, published with permission of the American Ceramic Society.)

for pores of simple geometric shapes, e.g., aligned cylindrical pores (an extreme shape), stressed parallel with their axes and cubic arrays of spherical pores give $b \sim 1.4$ and $b \sim 3$, respectively. Cubic stacking of aligned cubical pores stressed parallel with an edge, face, or body diagonal are closely fit by b values of 2, 3, and 5 over typical porosity ranges, e.g., to $p = 0.7$, 0.6, and 0.35, respectively. On the other hand, for pores between spherical grains (i.e., similar to much intergranular porosity) b is approximately 6–9, depending on sphere stacking; and for cubic stacking of aligned cylindrical pores stressed perpendicular to the cylinder axis $b \sim 7$. Note that the overall average b value is ~ 4.

Such variation of elastic properties with porosity for different processing is not unique to ceramics. Thus, for example, Pohl's (1969) data show that the b values of the exponential porosity dependence of E vary from ~ 1.6 for nodular cast iron to ~ 3.1 for sintered iron to ~ 7.3 for gray cast iron. Similarly, Artusio et al. (1966) have investigated the effect of porosity on Young's and the shear modulus of iron and copper–tin sintered materials. These authors indicate that as the size of the starting powder particles, and hence the pores, diminished there was closer agreement between experimental results and MacKenzie's theory. This is probably a result of improved homogeneity rather than an intrinsic pore size effect. Their data incidentally yield b values of about 4 and consistently fell below MacKenzie's theory.

While there are a number of variations in elastic property–porosity data, most data for each property fall in a relatively limited range. This general agreement is indicated by the summary averages and standard deviations in Table VI. An added demonstration of this basic trend is the similarity between the porosity dependence of most fibers with that of the corresponding bulk materials. The fiber data in Table VI are summarized from Table IV, plus the data of Fetterolf (1970) for Al_2O_3–mullite fibers having about 9%–25% porosity fitted through $E_0 = 40 \times 10^6$ psi. Note that E, K, and G all have a b average of ~ 4 as indicated above.

There have only been a limited number of studies on the effects of porosity on elastic properties, primarily Young's modulus, with increasing temperature. However, those that have been conducted have generally been fairly thorough. Similar observations of an overall general consistency of most data with a number of variations are indicated for the porosity dependence of the shear and bulk modulus which have been progressively less studied at high temperatures. All studies show little or no change in the porosity dependence of elastic properties in relatively pure bodies with increasing temperature, e.g., to the limits of their test temperatures (1000°C), Manning and Hunter (1969) observed essentially the same rate of decrease of Young's,

TABLE VI

SUMMARY OF

ELASTIC PROPERTY–POROSITY DEPENDENCE

	Property			
Parameter	E	K	G	ν
b^a	4.1 ± 1.8	4.0 ± 1.9	3.5 ± 0.9	1.2 ± 1.2
Number of values	46	5	21	13

Note similar behavior for fibers and bulk material, i.e., for Young's Modulus:

Material	b fibers	b bulk
Al_2O_3	2–4	3.5^b
Mullite–Al_2O_3	5	3
Carbon, graphite	6^b–9	2–5

[a] b of $E/E_0 = e^{-bP}$, $K/K_0 = e^{-bP}$, $G/G_0 = e^{-bP}$, $\nu/\nu_0 = e^{-bP}$.
[b] Extensive compilations.

bulk, and shear moduli with increasing temperatures for bodies of Er_2O_3, Ho_2O_3, and Y_2O_3 of different porosity. Similarly, Hunter *et al.* (1974) observed no differences in the rate of decrease of Young's and shear moduli with increasing porosity in their Sm_2O_3 bodies of varying porosity to their upper test temperature of about 1400°C and McLean *et al.* (1974) saw no significant differences in the rate of decrease of these moduli of hot pressed or reaction sintered Si_3N_4 bodies of different density to their upper test temperatures of 1000–1200°C. Fryxell and Chandler (1964) also apparently saw no change in the effect of porosity on Young's modulus of BeO to the limit of their tests (1400°C). Finally, Neuber and Wimmers' (1968) data on 99.5% Al_2O_3 ($0.02 \leq P \geq 0.20$) for E from $-200°$ to 1000°C in 200° intervals all show $b \sim 2.6 \pm 0.2$. This lack of temperature effects on the porosity dependence is to be generally expected since none of the bases of any of the theories would indicate any substantial changes with temperature, at least not until significant plastic deformation might occur around the pores. Even with some general yielding, significant effects would not necessarily be expected (as indicated by studies of metals with some ductility). For less pure materials, greater changes may occur, e.g., this may be the cause of the more rapid drop of Young's modulus of $\sim 40\%$ porous compared to $\sim 12\%$ porous ThO_2 bodies of Spinner *et al.* (1963) above about 1250°C. Effects of impurities will be discussed in the next section.

2. COMPARISON OF DATA AND THEORIES

Several investigators have compared their results to some of these theories; most commonly MacKenzie's, because it is one of the oldest and most well known, and recently, Fate (1975) has shown good agreement of the shear modulus of Si_3N_4 with Budiansky's theory. However, in order to make a more comprehensive evaluation, approximate standard deviations about the average values of Table IV are shown for relative elastic properties, i.e., the property at any given porosity divided by its value at zero porosity in Figs. 5–8. This provides an opportunity to test, on a comprehensive basis, the agreement of the analytical theories presented in the previous section and the great bulk of the existing data rather than tedious calculations on a case by case basis.[†]

Most theories agree reasonably well with the upper part of the range of data for Young's, bulk, and shear moduli i.e., at low P (Figs. 5–7). This is to be expected, since the inhomogeneous size, shape, and distribution of pores in actual bodies become more important as P increases, and is at best only partially accounted for in any of the theories. As noted earlier, such inhomogeneities will usually increase the porosity dependence, hence commonly giving lower values, as shown in Figs. 5–7. Budiansky's equations have been plotted assuming that $v = v_0(1 - P)$, which as seen from Fig. 8, is somewhat of an over estimate, but it generally makes a limited difference. Overall, the combination of Hashin's low and high concentration theories gives the best agreement, with Dewey's, Budiansky's, and MacKenzie's also being fairly good, but all are generally higher than the average. Note also that except for theories for special, e.g., cylindrically, shaped pores, all give the same general result over much of the porosity range (i.e., $b \sim 3$) as for the simple spherical load-bearing models (Eqs. 9 and 28, Table II).

Of the various theories for Poisson's ratio (Fig. 8), only Budiansky's theory (which gives a cubic equation having one real root) falls within the range of data (it also goes to zero at $P = 1$). Hashin's equations for high concentrations of spherical pores gives v independent of P, as does Hashin and Rosen's theory for stress parallel to cylindrical pores. Similarly, MacKenzie's equations can be solved for v if only first-order P terms are used, giving v approximately independent of P for $P < 0.4$, where it is valid. Beyond this, it first increases to $+\infty$ before approaching $-\infty$ at $P = 1$. Hashin and Rosen's theory for stress perpendicular to the axis of cylindrical

---

[†] Note that several of the theories considered cannot be analytically solved for each of the four elastic properties, therefore, where necessary, $v_0 = 0.25$, $G_0 = 24 \times 10^6$ psi or $K_0 = 36 \times 10^6$ psi have been assumed, as generally found from alumina. Assumption of these specific numerical values should not introduce significant errors, since only the ratio of the modulus at a given porosity to the modulus at zero porosity is considered.

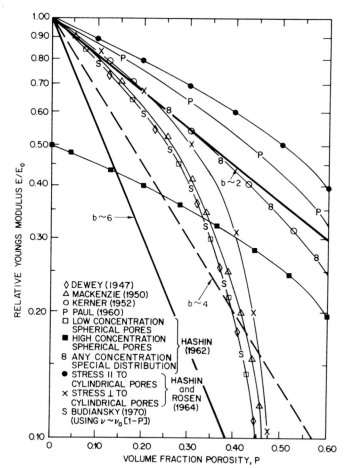

Fig. 5. Relative Young's modulus E/E_0 as a function of volume fraction porosity. The heavy dashed and solid straight lines are the approximate average and $\pm \sim 1$ standard deviation of data of Table IV. Ishai's two-phase theory (Table VII) used for porosity is identical with the line $b = 2$. Note that over the range that is usually a primary concern, i.e., to $P \sim 0.4$, all theories applicable to this range fall above the typical average porosity dependence of Young's modulus. This is attributed to inhomogeneous size, shape, and spatial distribution of real porosity leading to a greater porosity dependence than predicted by the models typically based on simpler more idealized porosity. Note that the combination of Hashin's low and high pore concentration models give fairly good agreement over all porosity.

pores gives v increasing substantially with P, e.g., about double v_0 at $P = 0.5$, then increasing faster. Such poor predictions of v are not surprising, since it is a small quantity dependent on differences of the other elastic properties and hence quite sensitive to errors in them. Based on the agreement with v,

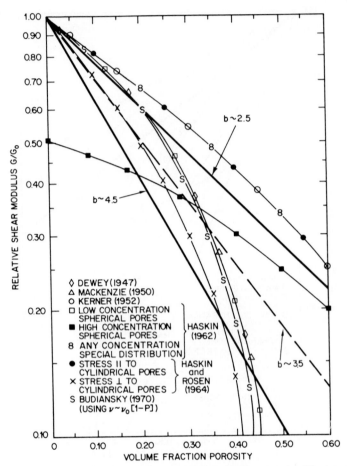

Fig. 6. Relative shear modulus G/G_0 versus volume fraction porosity. The heavy dashed and solid straight lines are the approximate average $\pm \sim 1$ standard deviation for all data of Table IV. Note that the theories fall in approximately the same order of agreement as for Young's modulus. Again, combination of Hashin's low and high pore concentration equations agrees fairly well with data over the full range of porosity. Note also that overall, the theories agree more closely with experimental data than for other elastic properties.

Budiansky's theory is overall the most self-consistent, but Dewey's, Hashin's low concentration, and MacKenzie's theories are still useful at low porosity.

In summary, several analytical theories, e.g., Hashin's for low and high concentrations of pores, Budiansky's, Dewey's, and MacKenzie's give fairly good but not excellent agreement with experimental data for Young's shear and bulk moduli. The situation is clearly much poorer for Poisson's ratio. Thus, further improvement in theories is needed, especially for Poisson's

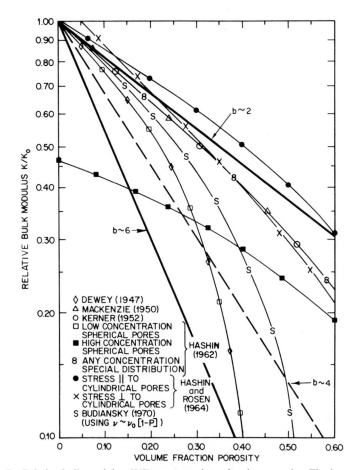

Fig. 7. Relative bulk modulus K/K_0 versus volume fraction porosity. The heavy dashed and solid straight lines are the approximate average and ± 1 standard deviation for all data of Table IV. While the combination of Hashin's low and high concentration of pore models gives fair overall agreement, the agreement of theories with experimental data is generally poorer with Hashin and Rosens' theory for stress perpendicular to aligned cylindrical pores—higher, since it gives $K = K_0 + G_0/3$ at $P = 0$.

ratio, with the basis and results of the past theories being a good guide to further development.

The major theoretical need is to account for variable pore shape, size, and spatial distribution. Much of this can be done based on existing theories. Note that the exponential function is analytical founded for some important pore shapes and that this and other simple load-bearing theories for particular pore shapes agree as well or better with data for those approximate shaped pores than do the more complex stress based theories.

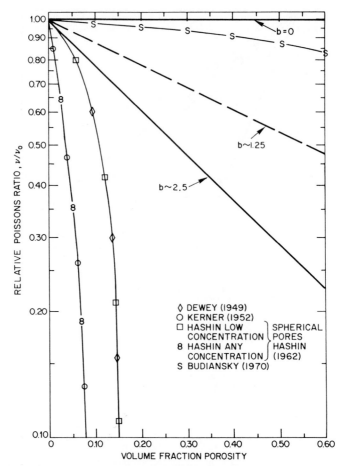

Fig. 8. Relative Poisson's ratio v/v_0 versus volume fraction porosity. The heavy dashed and solid straight lines represent, respectively, the approximate average and ± 1 standard deviation of the data from Table IV. Only Budiansky's theory falls within the general data trend, while others give v dropping more rapidly, independent of porosity, more complicated behavior, or disagreement with the data, as discussed in the text.

Note that there is still an important role for simple empirical or especially semianalytical relationships, particularly the exponential function e^{-bP} generalized to approximate the behavior of any porosity. They can generally fit experimental data better for extrapolation purposes, and simple expressions such as e^{-bP} are extremely useful in estimating the porosity dependence of behavior related to several parameters, e.g., thermal shock criteria such as $S/\alpha E$ as discussed later. Also, the mean trend of E with P is the same as for both fracture energy and tensile strength, as discussed in later sections, so many of theoretical and especially experimental observations there can be a useful guide. In view of similar and imperfect agreement of theories with data, along with the typical variable porosity of actual bodies in contrast to the idealized or unspecified pore character of the various theories, there is little or no significance to one theory fitting a particular set of data somewhat better than another theory.

The major overall experimental need is for much better characterization of porosity, i.e., shape, size, and spatial distribution. Rossi's (1968) numerical analysis of the effects of variable pore shape and orientation are an excellent guide to shape effects. Indirect characterization, i.e., by comparing moduli results from different types of tests, and different orientations and sizes of samples, is very useful along with extensive direct quantitative (i.e., sterological) characterization. In the interim, while theories are improved or developed to handle variable shape and spatial distribution of porosity, Rice's (1975a) suggestion of determining the statistical distribution of regions having nearly homogeneous porosity based on ceramographic examination and calculating an average body modulus from the calculated moduli of each section is recommended. The real need is to build on the simpler average porosity experiments of the past, not to continually simply duplicate them on another material or body.

C. Effect of Impurities or Additives on Elastic Properties

The effect of impurities or additives on elastic properties generally depends primarily on whether these constituents are in solution or form a second phase. For example, species in solid solution probably have the least effect, since they can then be homogeneously distributed, and the fact that they go into solid solution will often mean that their elastic properties differ less from the matrix than for second phases. With the exception of Hashin and Rosen's (1964) complex equations for a matrix reinforced with aligned fibers, the primary equations for calculating elastic properties of two-phased ceramics are listed in Table VII. Except for Paul's (1960) and Ishai's equations, the type of equation and their assumptions, including those concerning the distribution of the second phase, are identical to those of Table II, where the equations had been reduced to treat porosity by allowing the properties of the second phase to go to zero. It should also be noted that Rossi (1968) used his perturbation technique to numerically predict effects of particle shape on Young's modulus as a function of particle and matrix elastic properties. It is important to note that the equations for elastic properties of two-phased bodies should typically fit data substantially better than the counterparts of these equations used for porosity, since the difference in matrix and second-phase properties, which in the case of porosity is at a

TABLE VII

ELASTIC PROPERTIES OF TWO-PHASE CERAMICS

Investigator	Equation[a,b]
Kerner (1952)	$K = \dfrac{(3K_1 + 4G_0)K_0(1 - \phi) + K_1\phi(3K_0 + 4G_0)}{(1 - \phi)(3K_1 + 4G_0) + \phi(3K_0 + 4G_0)}$
	$G = G_0\dfrac{15G_1\phi(1 - v_0) + (1 - \phi)[(7 - 5v_0)G_0 + (8 - 10v_0)G_1]}{15G_0\phi(1 - v_0) + (1 - \phi)[(7 - 5v_0)G_0 + (8 - 10v_0)G_1]}$
Paul (1960)	$E \sim E\left[\dfrac{E_0 + (E_1 - E_0)\phi^{2/3}}{E_0 + (E_1 - E_0)\phi^{2/3}(1 - \phi^{1/3})}\right]$
	$\left(\dfrac{1 - \phi}{E_0} + \dfrac{\phi}{E_1}\right)^{-1} \leq E \leq [E_0(1 - \phi) + E\phi]^c$
Hashin (1962, 1968) (spherical inclusions)	(A) Small ϕ
	$K = K_0 + (K_1 - K_0)\dfrac{(3K_0 + 4G_0)\phi}{3K_1 + 4G_0}$
	$G = G_0 + (G_1 - G_0)\dfrac{5(3K_0 + 4G_0)\phi}{9K_0 + 8G_0 + 6(K_0 + 2G_0)(G_1/G_0)}$
	(B) Large ϕ
	$K = K_1 - \dfrac{(K_1 - K_0)[2(1 - v_0) + (1 + v_0)(K_1/K_0)](1 - \phi)}{3(1 - v_0)}$
	$G = G_1 - \dfrac{G_1 - G_0[7 - 5v_0 + 2(4 - 5v_0)(G_1/G_0)](1 - \phi)}{15(1 - v_0)}$
Hashin and Rosen (1964) (aligned fiber inclusions)	Complex equation for bounds on axial and transverse properties (see Hashin, 1968)
Ishai (see Lange, 1974a)	$E = E_0\left[1 + \dfrac{\phi}{E_1/(E_1 - E_0) - \phi^{1/3}}\right]$
Budiansky (1970)	$\dfrac{1 - \phi}{1 - a + a(K_0/K)} + \dfrac{\phi}{1 - a + a(K_1/K)} = 1 \qquad a = \dfrac{1}{3}\left(\dfrac{1 + v}{1 - v}\right)$
	$\dfrac{1 - \phi}{1 - b + b(G_0/G)} + \dfrac{\phi}{1 - b + b(G_1/G)} = 1 \qquad b = \dfrac{2(4 - 5v)}{15(1 - v)}$

[a] Subscripts 0 and 1 refer, respectively, to matrix and second phase, ϕ = volume fraction second phase.

[b] The basis and assumptions for these equations are the same as for their special case when the second phase consists of pores (see Table II).

[c] Upper bound in exactly this form only when $v_1 = v_0$.

maximum, increases any discrepancies between theoretical assumptions and the real nature of the body.

Because Paul's equations are simple yet fairly useful and have not been discussed elsewhere in this chapter, they are covered somewhat more extensively here. First, consider Paul's bounds, which are based on simple energy considerations. His upper bound actually depends on the ratio of Poisson's ratio of the matrix to that of the dispersed phase. However, this very cumbersome expression greatly simplifies to the upper bound shown in Table VII when the two Poisson's ratios are equal, a situation that is not too uncommon, so that errors using the simplified expression of Table VII should not be too great. Note that upper and lower bounds for bulk modulus and shear modulus can be obtained for Paul's approach by simply replacing E in his bounds by K or G. Paul derives an approximate expression that lies about half way between his upper and lower bounds for the case when the body can be divided up into uniform cubes containing a symmetrically located section of the second phase. In the special case when the inclusions are aligned prismatic fibers, whether solid or hollow, his approximate equation gives the identical result as his upper bound. On the other hand, if the inclusions are in the form of aligned slabs perpendicular to the direction of applied stress, then this approximate equation is exactly the same result as his lower bound. Ishai's derivation is similar to Paul's.

Comparison of the various equations listed in Table VI with data for two-phased ceramic systems is far more limited than the comparison of data for porous materials with equations relating elastic properties and porosity. Hashin (1968) has shown that his calculated bounds for Young's modulus of WC–Co bodies give a much narrower range between them than do Paul's; e.g., Paul's high and low bounds are about $\pm 15\%$, respectively, above and below the true modulus at 50–50 concentration, while Hashin's bounds are about $\pm 5\%$. However, Paul's approximate solution, being very close to intermediate between his upper and lower bounds, is close to the actual experimental data. Ishai's model falls somewhat higher than Paul's but also agrees well with composite data (e.g., see Lange, 1974a; Cohen and Ishai, 1967). Rossi et al. (1972) have shown that Rossi's (1969a) perturbation technique agreed with experimental results of MgO with $\sim 1\%$ graphite flake, giving Young's modulus parallel with the partially aligned flakes about 20% lower than normal to the flakes.

Knudsen's (1959) exponential equation relating elastic properties or strength with porosity can also be adapted for empirical correlation of volume fraction of second-phase inclusions and Young's modulus or other elastic properties. Thus, b values of 5.7 ± 0.9 and 3.7 ± 0.9 are respectively obtained using the apparent and estimate true porosities for Al_2O_3 data of Binns and Popper (1966) containing decreasing amounts of silica as density increases if all data is fitted to $E_0 = 61.2 \times 10^6$ psi. From these b values, one can estimate the exponential dependence of Young's modulus on the silica content.

As temperatures increase, properties of the second phase (e.g., silicate phases) may decrease much more rapidly than the matrix phase, leading to greater effects of the second phase, e.g., as previously suggested for the 40% porous ThO_2 bodies of Spinner et al. (1963). Similarly, as second phases become very soft (e.g., silicate phases), measurement of elastic properties, especially by applying a significant stress as in normal strength measurements, can lead to significant errors.

Cohen and Ishai (1967) used Ishai's simple composite formula to express the porosity dependence of a matrix material; then they substituted this expression for the matrix modulus in their composite equation. This gives them:

$$E = E_0(1 - P^{2/3})\left[1 + \phi\left(\frac{E_1}{E_1 - E_0(1 - P^{2/3})} - \phi^{1/3}\right)^{-1}\right]$$

They report good agreement with experimental data for epoxies containing voids and sand.

III. Microstructural Dependence of Fracture Energy

A. Background

The Griffith (1921) theory of brittle fracture is predicated on the concept that a crack will propagate in a stressed body when the strain energy in the body can provide the energy necessary to form the new surface area formed by the advancing crack. This fracture "surface energy" is commonly denoted by γ in the familiar form of the Griffith equation:

$$S = Y(E\gamma/a)^{1/2} \tag{2}$$

where S = fracture stress; Y = a geometrical factor determined by flaw and specimen geometry ($Y \sim 1.12$ for a surface half-penny crack and ~ 1.25 for an internal penny crack); E = Young's modulus; a = flaw size, e.g., radius of surface half-penny crack or internal penny crack. Many have assumed that γ was the surface energy, i.e., the energy needed to form a unit area of new surface by such processes as sublimation, chemical dissolution, etc. While there are a few earlier studies of fracture energy and associated measuring techniques (e.g. including Griffith, 1921; Obreimow, 1930; and Orowan, 1933), it was not until nearly 40 years after Griffith that practical and widely used fracture energy measuring techniques came into use. Gilman's (1960) development of the double cantilever beam technique and its rapid adoption were probably a major turning point. Subsequently in the 1960s a significant and increasing amount of work on measuring fracture energies was undertaken, spurred in part by the growing development of fracture mechanics (e.g., Irwin, 1957). During this time, other useful fracture energy measuring techniques, such as the notched beam (Davidge and Tappin, 1968) and work of fracture (Nakayama, 1965; Tattersall and Tappin, 1966) techniques, were developed, e.g., see Evans (1974).

Fracture energy studies, which continue at a high if not still increasing level of effort, have provided a great deal of useful information. Such work is the underpinning for the fracture mechanics approach to design of ceramic components for structural loads (e.g., see Wiederhorn, 1974, and Davidge *et al.*, 1973). It is now well established that only under special conditions in some single crystals that fracture energy and surface energy are equal (e.g., Wiederhorn, 1970). Fracture energies are commonly several times thermodynamic surface energies ($\sim$ threefold in glasses, see Wiederhorn, 1966). Further polycrystalline fracture energies are typically three- to tenfold times single crystal fracture energies. Because of such differences between fracture and thermodynamic surface energies and evidence of sub-surface processes in some fracture, it is preferable to delete the term surface, referring only to fracture energy as many investigators do.

Despite the above advances, much is unknown about fracture energy. The reasons for the differences between fracture and surface energies, e.g., in glasses, remain unknown. Similarly, a complete understanding of the reasons for the difference between single and polycrystalline fractures is not at hand. Roughness of polycrystalline fracture surface is estimated to account for only about two- to at most a fourfold increase over that of single crystals (e.g., Wachtman, 1972). Roughness is a factor, since fracture energy is computed on the basis of projected fracture areas, and a rougher surface has more actual surface area. In some cases, especially in softer materials such as MgO, plastic deformation associated with a more tortuous crack path in a polycrystal may account for much of the polycrystalline fracture energy [e.g., Evans (1970a) estimates $\frac{2}{3}$ in MgO]. However, it is quite doubtful that such processes are significant in much harder materials, such as Al_2O_3, SiC, and B_4C, at least until high temperatures are reached.

Another major deficiency in our understanding of fracture energy has been the general lack of any theoretical basis for understanding the effect of microstructure on fracture energy. There is also some uncertainty in the experimental understanding of microstructure–fracture energy relations. Because of the previous confusion, the recentness of some understanding (e.g., Rice and Freiman, to be published) and the lack of any previous review of the microstructural dependence of the fracture energy, this will be a more comprehensive review.

B. Porosity Effects on Room Temperature Fracture Energy

Increasing porosity (P) reduces the actual cross-sectional area of the material to be fractured and hence should generally reduce fracture energy. Few investigators have considered this, and those that have have commonly assumed that $\gamma \propto (1 - P)$ based on typical sterology. For example, Simpson (1973) used this linear porosity correction for stress intensity and $E = E_0 e^{-bP}$ in calculating fracture energy from work of fracture measurements. However, Rice and Freiman (to be published) have pointed out that sterology gives the average area, while the crack will seek the path having the minimum solid area, i.e., the maximum, not the average cross-sectional area of porosity. They point out that the idealized model of sintering spherical grains that Knudsen developed for strength and used for elastic moduli (Section II,A) closely approximated by e^{-bP} gives the minimum solid area. They further showed that this model should still be a good approximation even with substantial variation in grain and pore size, so fracture energy data should commonly follow an e^{-bP} dependence similar to Young's modulus and strength. Similarly, load-bearing theories for elastic properties (Table VII) or strength (Table X) are generally applicable to fracture energy.

Compilation and analysis of the limited $\gamma-P$ studies, though complicated by variability clearly show γ commonly fitting the exponential relation similar to Young's modulus (e.g., compare Tables IV, VI, and VIII). As

TABLE VIII

POROSITY DEPENDENCE OF FRACTURE ENERGY AT ROOM TEMPERATURE

Material	Approximate grain size (μm)	Porosity range (%)	γ_0 (J/m^2)	b^a	Reference
Al$_2$O$_3$	1	3–10	75	18.9 $\pm$ 0.9	Simpson (1973)
	1	1–9	12	0.4 $\pm$ 0.9	Coppola and Bradt (1973a)
	3	5–50	22	2.2 $\pm$ 0.1	Evans and Tappin (1973)
	10–50	1–43	30	0.3 $\pm$ 1	Pabst (1974)
HfTiO$_3$	7	12–27	35	2.1 $\pm$ 0.2	Hoagland et al. (1974)
SiC	20	0–15	15	2.3 $\pm$ 0.3	Matthews et al. (1973)
	100	0–16	22	2.6 $\pm$ 0.9	Coppola and Bradt (1972); Matthes et al. (1973)
Si$_3$N$_4$	1.5	0–33	50	7.1 $\pm$ 0.5	Evans and Davidge (1970); Coppola et al. (1972); Lange (1973a)
Graphite	—	16–32	101	3 $\pm$ 1	Andersson and Salkovitz (1974)

^{a} b of $\gamma = \gamma_0 e^{-bP}$.

examples of different degrees of variability consider Al_2O_3, Si_3N_4 and SiC. Most data for Al_2O_3, the most extensively studied material (Fig. 9) shows γ decreasing with increasing P. Binns and Poppers' (1966) data for various commercial Al_2O_3 bodies is reasonably consistent with other data, but covers too narrow a range of porosity combined with too much variation of grain size and purity to define any clear trend. Coppola and Bradt's (1973) data for hot-pressed Al_2O_3, also over a limited range, is essentially independent of the mostly intergranular porosity. The fracture mode was about 20% transgranular, in contrast to Simpson's (1973) nearly total intergranular fracture (also hot-pressed materials). This may account for the difference between these two investigations, though, the initial low Young's modulus and strengths of Coppolas and Bradt's Al_2O_3 prior to annealing leaves some question of gaseous impurities or other weakening effects in their material. Note that the low b value for the Al_2O_3 data of Evans and Tappin (1973) is consistent with the purposely introduced approximately spherical porosity. The low b value for Pabsts' (1974) data also suggests approximately spherical pores, which is consistent with the high porosities achieved. Analysis of the data of Claussen *et al.* (1975) for a variety of commercial aluminas shows his data fit the exponential relation well, giving $b = 5.5 \pm 0.8$. In view of the similar dependences of γ and E on P and that $K_{IC} = (2E\gamma)^{1/2}$, his data shows γ has a similar porosity dependence as other data. Thus, his b value suggests substantial intergranular porosity. This

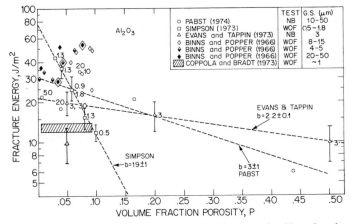

Fig. 9. Fracture energy of Al_2O_3 versus volume fraction porosity. Note that the numbers near some of the data points indicate the approximate grain sizes of those bodies. While there is substantial scatter, most data is reasonably consistent except for Cappola and Bradt's and part of Simpson's data. For this reason, as well as their extreme slopes and limited range of data, their b values have not been included in the general evaluation of porosity dependence of fracture energy.

clearly is the case for Si_3N_4 (Fig. 10A) consistent with the high b values. In contrast, to Si_3N_4 (Fig. 10A), SiC (Fig. 10B) shows no clear trend, with porosity due in part, if not completely, to greater variation in phase composition, grain size, and pore location. However, comparison of bodies with the same grain size (e.g., 20 or 100 μm) that cover a span of porosity do indicate a limited decrease of fracture energies with increasing porosity, again consistent with the more spherical like porosity expected in larger grain bodies, e.g., due to intergranular porosity.

Two porosity effects can alter expected $\gamma-P$ trends. At low to moderate porosity, pore location can be important. Thus, as an uncommon example, most Lucalox has some trace porosity clustered within grains, but fracture is typically intergranular in this material, so no direct crack–pore interactions take place. More generally, other factors being equal, cracks preferentially follow the most porous path; e.g., intergranular failure occurs with predominately intergranular pores, transgranular failure with intragranular pores. Changes in pore location are often associated with other changes; e.g., pores are often trapped within grains by the onset of exaggerated grain growth, which increases grain size, and often results in inhomogeneous grain sizes and tabular-shaped grains, e.g., in Al_2O_3 and SiC. Further, in the case of SiC, exaggerated grain growth is associated with the β to α phase transformation. Change in pore location determines whether fracture "tails," the result of a crack failing to rejoin on the same plane as it completes its passage around a pore, will form. As reviewed by Rice (1974a), such "tails" generally form when a crack passes around pores larger than the grains for any fracture mode, pores within grains (i.e., for transgranular failure), and pores on grain boundaries where the fracture changes from intergranular to transgranular at the pore. Such tails do not form at grain boundary pores with intergranular fracture. Fracture tails make small increases in fracture energy through the increased surface area they represent and possibly substantially larger increases through elastic and plastic processes. Elastic energy can be lost through vibration and friction processes when the overlapping web forming most tails (e.g., with transgranular failure) fractures. Clark and Wilks (1966) have suggested that the energy in forming such tails or fracture steps can be quite high (e.g., 10 J/m^2) due to shearing processes that may be involved. However, whether fracture tails or steps significantly increase fracture energy due to such effects is uncertain.

Plastic processes may increase fracture energy effects of fracture tails in softer materials. Forwood (1968) estimated from changes in crack velocity that the fracture energy of NaCl crystals containing voids about 30 μm in size about 30 μm apart was increased about a hundredfold. While Forwood has apparently since concluded that plastic deformation was not the cause of the fracture steps (see Groves, 1971) and a hundredfold increase seems

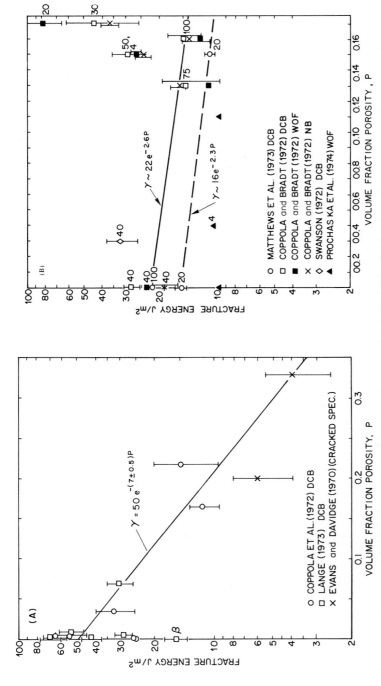

Fig. 10. Effect of porosity on fracture energy of Si_3N_4 (grain size 0.1–1.5 μm) and SiC. (A) Si_3N_4, showing a clear trend of γ to decrease with increasing P. (Note Lange's weaker material made from β powder neglected in computing the porosity dependence). (B) SiC, note numbers near symbols are approximate grain sizes. Porosity dependence for 20 μm and 100 μm grain sizes from DCB tests is indicated by lines and equations.

unusually high, it does indicate a real increase in fracture energy due to the voids, with some of the increase probably being due to plastic flow in the tails preceeding their fracture. Thus, Hing and Groves (1972a) introduced 2% Fe or Ni particles into MgO crystals, which significantly increased the density of fracture tails and did increase fracture energies, though only about twofold or so. Rice (1972a) has reported significant reductions (e.g., tenfold) in crack branching in MgO with about 1% porosity, correlating with pores within grains and the generation of fracture tails associated with the significant portion of intragranular porosity. More generally, the occurrence of plastic deformation from such tail formation or related crack–pore interactions in harder ceramics is quite uncertain, since little study of this has been made. Bursill and McLaren (1965) reported slip along fracture of cleavage steps (which are similar to fracture tails) in zircon but not quartz.

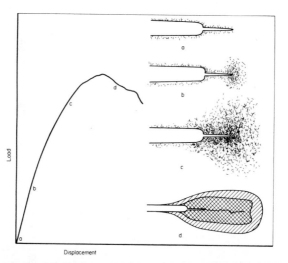

Fig. 11. Parasitic cracking in porous rocks (after Hoagland *et al.*, 1973). Sketches of microcrack development around a machined slot in fracture energy test sample of porous rock, and a photo of such a slot (left) and crack (darker zones near end of slot are due to more potting resin absorbed in microcracks) in sandstone, $\gamma \sim 300–1600$ J/m^2. Sketch and photo courtesy of Dr. Hoagland, published with permission of *Rock Mechanics*.

When there is substantial porosity, pores can cause branching or parasitic cracking (the opening up of extra cracks along or around the main crack), e.g., as could occur due to a crack running into a pore on a triple line. Hoagland *et al.* (1973) have shown by both acoustic emission and direct optical examination (e.g., Fig. 11) that such parasitic cracking is the cause of very large fracture energies in porous limestones and sandstones (e.g., fracture energies of $50-230$ J/m^2 and $465-1580$ J/m^2, respectively). Such effects are presumably a major factor in the high fracture energy or toughness of porous refractory brick (e.g., Larson *et al.*, 1974; Kuszyk and Bradt, 1975), giving DCB or WOF fracture energies (but not necessarily NB fracture energies) two to five times higher than those of dense bodies. How important the relative roles of pores and cracks are in generating parasitic cracking has yet to be determined.

C. Grain Size and Shape Effects on Room Temperature Fracture Energy

Until very recently, there was much confusion about fracture energy–grain size data (i.e., whether fracture energy increased, decreased, or remained constant as grain size increased) and there was no theory for guidance. However, a recent survey and experimental tests by Rice and Freiman (to be published) appears to have removed much of the confusion. They report that the primary factors in the grain size dependence of fracture energy is whether the material is cubic or noncubic. There is also some reinforcement and extension of Simpson's (1974) earlier observation that the notched beam (NB) technique generally gives γ independent of, or slightly decreasing with increasing grain size for any crystal structure. Limited data using fracture mirror and crack branching data from normal strength tests to measure fracture energies, also indicates the same trend. This may result, since both the NB and normal strength tests lack grooves to guide the crack, and failure may initiate from flaws small in comparison to specimen dimensions. Thus, failure usually occurs from a single more serious flaw, which as grain size increases to become comparable to or larger than the flaw size, fracture energies controlling failure decrease to or near single crystal values Section IV,C). Other fracture energy techniques—e.g., double cantilever beam (DCB), double torsion (DT), and work of fracture (WOF)—all use machined slots, which could enhance microcracking effects in materials prone to such cracking (discussed below). While more work is needed to confirm the above or other possible mechanisms (e.g., doing tests on grooved samples with possible machining damage removed by annealing, chemical polishing, or sputtering), there often appear to be very real differences between these two sets of techniques. This will be seen from the data discussed below and is consistent with the NB tests giving fracture energies agreeing much more closely with strengths of porous refractories as discussed in the previous section.

Turning next to the effect of crystal structure, Rice and Freiman report that DCB, DT, and WOF techniques show no significant grain size dependence of fracture energies in cubic materials similar to the NB and fracture mirror or crack branching measurements. However, in noncubic materials, they report that the DCB, DT, and WOF techniques give fracture energy first increasing with increasing grain size, then reaching a broad maximum, and finally decreasing. The relative amount of fracture energy increase with increasing grain size is indicated to increase with the degree of anisotropy, primarily thermal expansion anisotropy (TEA), but elastic anisotropy (EA) may also contribute. The grain size of the fracture energy maximum decreases as the anisotropy increases. These effects are attributed to TEA and possibly EA stresses aiding microcracking, but twinning, which is often favored in noncubic materials may also contribute in some cases. The maximum of fracture energy is attributed to the opposite effects of the increasing ease of microcracking and its increasing volume around the crack over which it occurs as grain size increases. Thus, as the grain size becomes large enough for spontaneous cracking (see Section IV,G), fracture energy should essentially go to zero. Clearly, porosity can decrease such anisotropy effects, and impurities can increase or decrease them.

A particularly clear demonstration of the effect of crystal structure on fracture energy is shown by Pohanka et al. (submitted for publication) DCB measurements of $BaTiO_3$. Above the Curie temperature $(T_c \sim 120°C)$ where the material is cubic, fracture energy was independent of grain size, while below T_c where it is tetragonal, fracture energy was the same at fine $(\sim 1 \mu m)$ grain size, but increased nearly threefold as grain size increased to $\sim 100 \mu m$. Recent work shows that twinning, which is in part due to internal stresses is a factor in this increase.

Both the above test and crystal structure effects are seen in the following literature summary; materials are considered approximately in order of decreasing anisotropy. The WOF data of Kuszyk and Bradt (1973) for nearly dense $(\sim 1.6\%$ porous) highly anisotropic $MaTi_2O_5$ clearly shows the above described trend (Fig. 12). The DCB tests of Freiman et al. (1974a) for a crystallized barium silicate glasses show fracture energy increasing from ~ 3 to over 12 J/m^2 as grain size goes from submicron level to a few microns; tests to larger grain sizes were apparently not done because of specimen cracking. Similarly, Barry et al. (1972) note that fracture energies of 10–30 J/m^2 for crystallized glasses (which typically have anisotropic crystalline phases) they studied generally increased with grain size, but differentiation between NB and DCB techniques used was not made. Lynch and Bradt's (1973) BeO WOF data show fracture energy increasing

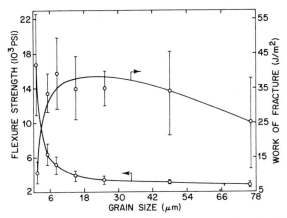

Fig. 12. Fracture energy and flexure strength of $MgTi_2O_5$ as a function of grain size ($P \sim$ 0.016) (after Kuszyk and Bradt, 1975). Note that fracture energy clearly rises to a maximum and then decreases with increasing grain size in this material having high thermal expansion anisotropy, while strength continuously drops with increasing grain size. These effects are attributed to the internal stresses and expected related microcracking from the high thermal expansion anisotropy of this material. Published with permission of the American Ceramic Society.

from about 30 J/m² at $G = 12$–$22 \ \mu$m to over 35 J/m² at $G \sim 30 \ \mu$m (which is well below the expected grain size for the maximum fracture energy) is in good agreement with Swanson's (1972) DCB data of $\sim 32 \pm 2$ J/m² at $G = 3 \ \mu$m (Fig. 13).

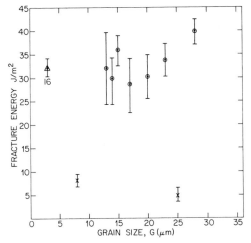

Fig. 13. Fracture energy versus grain size of BeO and UO_2. Note that the fracture energy BeO [○, Lynch and Bradt (1973) (WOF); △, Swanson (1972) (DCB)] increases with increasing grain size, while there is no statistically significant trend for the two data points for UO_2 [×, Evans and Davidge (1969) (NB).

We next consider Al_2O_3, the most studied materiel. As seen in Fig. 14, though scattered, all DCB data (no WOF data was found) generally agree and show fracture energy increasing with increasing grain size. On the other hand, Evans and Tappin's (1972) and Simpson's (1973) NB tests show fracture energy remaining nearly constant or decreasing some with increasing grain size. Subsequently, Simpson (1974) used both NB and DCB tests on the same Al_2O_3 bodies, with the former test giving no grain size dependence, while the latter gave a definite increase with increasing grain size. Similarly, NB tests by Claussen et al. (1975) of commercial and hot-pressed Al_2O_3 bodies show K_{IC} and, hence, γ, independent of grain size or slightly decreasing as grain size ($1-20$ μm) increases. Evans reports (private communication) that DT tests of Lucalox shows fracture energy increasing with grain size. Tests by Rice and Freiman indicate the fracture energy of Al_2O_3 does decrease at larger grain size and that the fracture energy of TiO_2, and Nb_2O_5 pass through a maximum as grain size increases.

Limited DCB measurements by Rice and Freiman indicate that the fracture energy of hot-pressed B_4C (low anisotropy) is essentially constant

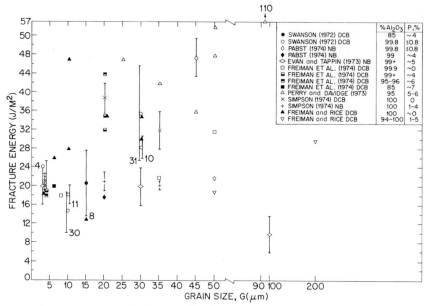

Fig. 14. Fracture energy versus grain size data for Al_2O_3. Sub- or superscripts on Swanson's data refer to the number of tests averaged. Note that while there is substantial scatter in part due to the variable porosity, all measurements by the double cantilever beam technique show fracture energy increasing with increasing grain size. However, tests using the notch beam technique generally indicate fracture energy independent of grain size or decreasing somewhat with increasing grain size.

at ~ 20 J/m^2 over the grain size studied (~ 5–150 μm). Similarly, they found no clear grain size dependence to the fracture energy of hot-pressed or CVD SiC (mostly cubic over $G \sim 10$–100 μm). Their fracture energy of SiC of about 25 J/m^2 is in reasonable agreement with Simpson's (1974) limited NB tests of SiC showing a possible decrease in γ and limited DCB tests by Matthews *et al.* (1973) showing a possible limited increase in γ with grain size over the same grain size range.

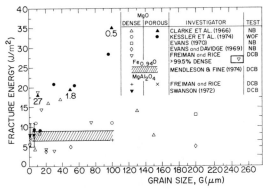

Fig. 15. Fracture energy as a function of grain size for MgO, FeO, and MgAl$_2$O$_4$. Note that in all of these cubic materials that the fracture energy, while scattered, is essentially independent of grain size unless some limited porosity is present, in which case fracture energy can increase with increasing grain size. Note that the higher value for the largest (250 μm) grain size sample of MgO (sample of Freiman and Rice) is also probably due to some limited porosity included within the grains.

Data for UO$_2$, MgAl$_2$O$_4$, FeO, and MgO, all cubic materials, generally show no significant dependence of fracture energy on grain size. Evans and Davidge's (169) NB tests of UO$_2$ at $G \sim 8$ μm and $G \sim 25$ μm indicate a possible limited decrease in γ from ~ 8 J/m^2 at 8 μm to ~ 5 J/m^2 at 25 μm (Fig. 13). Data for the other materials show fracture energy generally independent of grain size, except for MgO having some porosity (Fig. 15). Kessler *et al.* (1974) found that while previous studies had indicated different possible trends of fracture energy of MgO with grain size, fracture energy of dense MgO was essentially independent of grain size. Increased fracture energy with grain size in somewhat porous bodies correlated with porosity changing from predominately intergranular to substantially intragranular as grain size increased. This is consistent with porosity effects discussed in the previous section. Such pore location effects may also be a factor in the general increase of fracture energy of magnesite–chrome refractories with grain size (e.g., Kuszyk and Bradt, 1975). Recent DCB tests by Rice and Freiman (to be published) corroborate the results of Kessler *et al.*

Lange (1973a), in one of the few observations on grain shape effects, reports that the fracture energy of hot-pressed Si_3N_4 is lower for bodies made from β phase than from α phase powder. He attributed this difference to the tabular grain shape and the attendant increase of fracture roughness, area, and energy resulting from use of α powder as opposed to equiaxed grains and flatter fracture using β powder. However, Rice and Freiman (to be published) have shown that CVD SiC with a much greater grain aspect ratio has a fracture energy similar to that of hot-pressed SiC with equiaxed grains. Also, the CVD material had a definite texture, but no clear anisotropy of fracture energy. This difference between SiC and Si_3N_4 may be due to transgranular fracture in the former and mostly intergranular fracture in the latter. These differences may in turn be due to internal stresses, since Si_3N_4 is noncubic and generally has a greater amount of grain boundary phase than SiC.

While clear demonstrations of parasitic cracking is often difficult to obtain, there is some evidence of this. Noone and Mehan (1974) have shown evidence of such cracking in Lucalox, and Green et al. (1973) have shown that high (e.g., 150 J/m^2) fracture energies in partially stabilized ZrO_2 are due to a network of cracks in the material. Similarly, Freiman et al. (1974) showed that large (typically tubular) grains in their barium silicate glass ceramics had microcracks, which they attributed to stresses from the anisotropic expansion. Further, the ratio of fracture energy and Young's modulus for cubic materials (and for fine-grain materials that are not too anisotropic) all fall in a limited range suggesting a relationship between γ and E, as porosity does. However, this ratio increases with the degree of anisotropy and grain size in noncubic materials, with graphite, the most anisotropic material studied and one known to contain microcracks, showing the greatest deviation (e.g., see Freiman and Rice, to be published; Mecholsky et al., 1976). Finally, the concept of microcracking from TEA stresses is consistent with observed and predicted spontaneous cracking from such stresses (Section IV,G).

The mode of fracture, which can depend on intrinsic or extrinsic effects, may affect fracture energy. This may in fact be a factor in impurity and porosity effects, since pores and impurities are often at grain boundaries and encourage intergranular fracture. However, possible effects of the mode of fracture do not appear explainable on the basis of basic fracture energies or paths. For example, if the fracture energy of grain boundaries is assumed to be half that of single crystal fracture (an often quoted, but poorly substantiated result), a change from inter- to transgranular failure clearly cannot explain all of the observed grain size changes. Further, simple geometrical models, e.g., a one-dimensional model considering grains as hexagons, indicate a greater area generated by intergranular failure, thus, in part compensating for possible inter- versus transgranular fracture energy differences. Also, such models that show the area differences between the two modes of fracture are independent of grain size. Thus, possible inter- versus transgranular effects may depend more on cleavage or fracture step effects similar to those suggested for fracture tails around pores. Further, while many have assumed that a change from inter- to transgranular failure was intrinsic, Rice (1974a) has recently questioned this, at least in materials without crystalline anisotropy and in the presently attainable grain size range. Thus, further study is needed on the effect of grain boundaries and hence the mode of fracture on fracture energy.

D. Effect of Impurities and Additives on Room Temperature Fracture Energy

The effect of additives and impurities on fracture energy depends on whether they are in solution or occur as second phases. Unless the amount or type of additives or impurities in solid solution significantly change Young's modulus, which is uncommon, they should generally have no significant effect on fracture energy unless the matrix is soft enough for plastic deformation to be a significant factor. Even then, strain rate and dislocation density may play a large role. Thus, for example, Freiman et al. (1975) showed varying trends of fracture energy in alloyed KCl crystals with increasing alloy content.

They attribute these variations to competition between dislocation crack blunting and dislocation stress focusing or feeding at crack tips to aid crack growth. Thus, at too high an alloy content or too high a strain rate, dislocation processes are not operative on a significant scale, so measured fracture energies are quite close to the surface energy (~ 0.2 J/m^2). However, the fracture energy of alloyed or unalloyed polycrystals was consistently high (2–4 J/m^2) due to crack–grain boundary dislocation effects.

The effect of second phases on fracture energy depends on their chemical nature, shape, size, and location. Lange (1970) has developed the following equation for the fracture energy of a composite containing uniformly spaced particles or other objects (e.g., fibers and possibly pores) that can pin cracks temporarily:

$$\gamma = \gamma_0 + \Gamma/2D \tag{3}$$

where γ_0 = matrix fracture energy, D — particle spacing, and Γ = line tension of the crack front ($= \frac{2}{3} r \gamma_0$ for a circular, i.e., internal, penny crack of radius r). Two reservations about this model should be noted. First, it is based on line tension concepts that are not necessarily valid for cracks. Second, it does not account for the choice of inter- versus transgranular fracture paths in a polycrystalline body containing intragranular particles. Nevertheless, as will be shown later, this equation provides a reasonable basis for interpreting some ceramic composite data, e.g., for glass matrices or polycrystalline bodies with particles larger than the matrix grains.

However, first consider some location effects. Grain boundary phases are normally assumed to result in a reduction of fracture energy. This is apparently based on some observed effects of such phases on strength and fracture of some ceramics and metals, since almost no direct fracture energy studies have been made. Kessler et al. (1973) did show that MgO, as hot pressed with LiF, has a fracture energy of ~ 5 J/m^2, while specimens annealed (e.g., at 1100°C) to remove essentially all remaining LiF gave normal MgO fracture energies of ~ 10 J/m^2 with a change from total to partial inter-granular failure. The demonstration by Johnson et al. (1974) that the LiF is at grain boundaries of the as-hot-pressed MgO but not the annealed MgO clearly indicates that the reduction of fracture energy is due to a grain boundary phase. Simpson and Merrett (1974) have provided indirect evidence for weakening and reducing fracture energy of hot-pressed Al$_2$O$_3$ due to anion impurities, presumably at grain boundaries as suggested by Rice (1969a). They showed that their results overlapped with those of Coppola and Bradt (1973a), so that the independence of Coppola and Bradt's fracture energy from porosity may be due to anion impurities as noted earlier.

There can be important exceptions to the expected general reduction of fracture energy by a grain boundary phase. First, intragranular pores or second phases that encourage transgranular failure may reduce or eliminate effects of grain boundary phases. On the other hand, grain boundary phases in anisotropic materials may allow some relaxation of internal stresses from expansion anisotropy, especially in materials with some microplastic phenomena such as twinning. Such relaxation could partly or more than compensate for the normally deleterious effect of grain boundary phases on fracture energy. Finally, some grain boundary phases, namely metals, can increase fracture energy. For example, Mendleson and Fine (1974a) showed that Wustite with an Fe layer along the grain boundaries fractures inter-granularly giving DCB fracture energies of 20–40 J/m^2 compared with about 8 J/m^2 without the Fe phase. Similarly, Virkar (1973), have hot-pressed dense bodies of unstabilized ZrO$_2$ with (10–40 vol % of) Zr as a grain boundary layer. Fracture energies (DCB) of bodies with 23 vol % Zr ranged from 38 J/m^2 to 65 J/m^2, depending on whether the bodies were slow cooled to reduce stress from the monoclinic–tetragonal phase trans-formation or fast cooled to minimize stress relaxation, respectively. Such composites are, of course, similar in nature to metal-bonded (e.g., Co) WC. However, they can give differing results. Mendelson and Fine showed that fracture energy varied linearly with the reciprocal of the grain size of their FeO–Fe composites; i.e., fracture energy was consistent with Lange's (1971) line tension model. Thus, this data was dependent on the spacing between metal films and not their thicknesses. In contrast to this, Lueth (1974) showed excellent correlation between fracture energy of Co-bonded WC and the Co film thickness. The data were essentially independent of film spacing, i.e., reciprocal grain size. These differences may be related to the ductility of the films or other factors, but more work is clearly needed.

Another common exception is the case of particulate composites, since the second-phase particles are generally inter- rather than intragranular. Thus, for example, Lange (1971) generally increased both the strength and fracture energy of a sodium borosilicate glass by addition of various sized Al_2O_3 particles. The fracture energies varied linearly with the inverse of the Al_2O_3 particle spacing in agreement with his line tension model (Eq. 3). Lange (1973b) also added SiC particles to Si_3N_4, which more commonly decreased strength and fracture energy, as shown in Fig. 16. He attributed the continual decrease of fracture energy with all but the volume fraction addition of the largest SiC particles and the failure to follow the line tension model to SiC particle–Si_3N_4 grain size effects. Thus, he noted that the

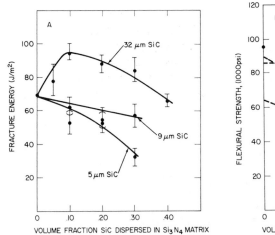
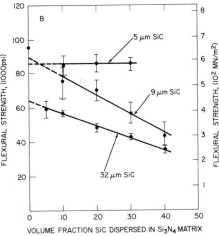

Fig. 16. Fracture energy and flexure strength as a function of size and volume fraction of SiC dispersed particles in hot-pressed Si_3N_4 (after Lange, 1973b). Note that the fracture energy and strength trends do not agree, first because of reversals between the relative effects of different sized particles on strength and fracture energy. Second, even when strength and fracture energy have the same trends, strength shows a more pronounced change with volume fraction SiC than fracture energy when the reverse should be true, because strength varies only as the square root of fracture energy. These effects are attributed to variations between the flaw size and the particle spacing (see Table XV). Published with the permission of the American Ceramic Society.

fracture energy of SiC is one-half to one-third that of Si_3N_4 and that the 5 and 9 μm SiC particles are not much larger than the Si_3N_4 grains. He therefore felt that only the larger SiC particles were large enough relative to the Si_3N_4 grains to perturb the crack enough to dissipate enough energy by generating fraction tails, parasitic cracking, etc. to overcome the lowering of fracture energy because of the replacement of Si_3N_4 by SiC along the fracture surfaces. However, his reasoning may not be correct, since the above $SiC-Si_3N_4$ fracture energy differences are based on poly-crystalline behavior, which may not reflect the fracture energy behavior for single crystals, hence, single grains; e.g., note the deviations of Si_3N_4 from the $\gamma-E$ trends discussed in the previous section. Rankin et al. (1971) report that additions of 5 vol % Mo particles to Al_2O_3 increased the fracture energy (to $\sim 55 \ J/m^2$) over that of Al_2O_3 of comparable grain sizes. However, strengths of Al_2O_3 with and without Mo were the same for comparable grain sizes.

Hing and Groves (1972b), using a notched beam technique, found that the energy for fracture initiation (γ_i) in hot-pressed MgO containing 10–40 vol % Co, Fe, or Ni particles ranged from 10 to 60 J/m^2. Generally, γ increased with the volume fraction of the metal particulates. The energy for crack propagation varied similarly up to 260–740 J/m^2, i.e., an order of magnitude larger. These effects, especially the latter, were attributed to plastic deformation of the metal particles. This is in contrast to Hing and Groves' (1972a) studies of single crystals of MgO noted earlier in which 2 vol % of Fe or Ni particles increased fracture energy only 1–2 J/m^2 (and little or no increase with up to 40 vol % magnesioferrite. Note also that the polycrystalline strength, while increasing with metal particulate content (due to decreasing grain size) was the same as the strenth of good hot-pressed MgO of comparable grain size without additives.

More recently, Bansal and Heuer (1974) reported that heat treatment of Al_2O_3-rich $MgAl_2O_4$ crystals to precipitate the Al_2O_3-rich phase increased stress intensities $\sim 70\%$, and hence fracture energies $\sim 300\%$. Strengths were similarly increased, following the same trend as hardness with heat treatment. Similarly Porter and Heuer (1975) and Porter et al. (1976) have indicated increases of fracture toughness by precipitation phenomena in ZrO_2 partially stabilized with MgO. Particularly spectacular is the report of Garvie and co-workers (1976) of developing partially stabilized ZrO_2, i.e., cubic ZrO_2 with monoclinic and tetragonal ZrO_2 precipitates having work of fracture of up to 500 J/m^2, yet having strengths of 60,000–100,000 psi. Such precipitates are, of course, typically less than a micron in diameter, and having separation of a few microns or less. Similarly Claussen (1976a) has reported significant (e.g., 100%) increase in K_{IC} (i.e., $\sim$ fourfold increases of γ) for Al_2O_3 with small (about 1 μm or less) particle of unstabilized ZrO_2 in

Al_2O_3 (stabilized ZrO_2 is not effective). Preliminary work by the author and colleagues verify that unstabilized, but not stabilized, ZrO_2 toughen Al_2O_3. As Claussen (1976a,b) and Rice (submitted for publication) point out, these effects are most attributable to microcracking; e.g., a line tension or similar crack front particle interaction cannot explain why intergranular failure does not occur to avoid precipitates in the toughened grains, while microcracking around particles in the vicinity of the crack front can.

While a complete discussion of ceramic composites is beyond the scope of this chapter, a brief outline of fiber composites is in order. Hing and Groves (1972b), showed that the fracture energy of dense MgO containing Ni fibers (89 μm diameter) increased from about 2000 J/m^2 to about 6000 J/m^2 as the volume fraction of fibers increased from 11% to 36%. (Note that although these energies are one to two orders of magnitude higher than in their MgO–metal particle bodies, the strengths with fibers were somewhat lower than with particles or no additives at all). Sambell *et al.* (1972a) added random or oriented discontinuous carbon fibers to MgO, Al_2O_3, soda lime glass, borosilicate glass, and lithia aluminosilicate glass ceramic bodies. Fracture strength decreased with discontinuous fibers that were randomly oriented, increased some with partial alignment of discontinuous fibers, and increased still more for continuous aligned fibers. Fracture energy typically increased ten to a hundredfold, e.g., at 20 vol % except for Al_2O_3 (no charge) with random discontinuous fibers (but all strengths decreased 30–90%) (Fig. 17). However, discontinuous carbon fibers that were aligned in the glass increased both fracture energy and strength ($\sim 40\%$ with 20 vol % fibers). Pursuing this, Sambell *et al.* (1972b) showed that incorporation of about 40 vol % aligned, continuous carbon fibers in the glasses and the glass ceramic increased fracture energies about a thousandfold (i.e., to 3000–4000 J/m^2) while increasing strengths four- to sixfold.

Similarly, various investigators have added refractory metal wires to ceramics. Simpson and Waslyshyn (1971), for example, note that the fracture energy of Al_2O_3 hot pressed with 80 pm diameter chopped Mo fibers increased approximately linearly with fiber content, being about 250-fold higher than pure Al_2O_3 at 12 vol % fibers. However, Simpson (private communication) notes that such systems may not be suitable for a number of applications, since cracks tend to run through the Al_2O_3, leaving only the fibers holding the sample together but exposed to the environment. Such effects would appear to preclude such toughening approaches for application in high temperature oxidizing environments. However, Brennan (1974, 1975) reports that Si_3N_4 hot pressed with Ta wires (~ 100 μm diameter), e.g., 25 vol %, results in an increase of import toughness, e.g., at 25 vol %, Charpy Impact, was up nearly sixfold over that of Si_3N_4 alone

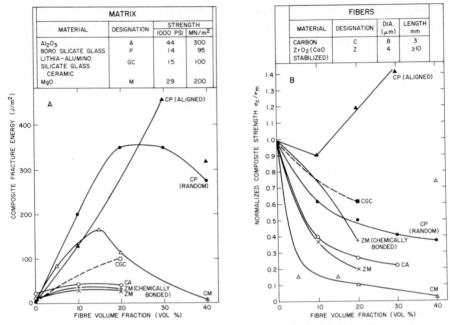

Fig. 17. Fracture energy and strength of fiber-reinforced ceramics. (A) Fracture energy generally rises with increasing fiber content, then drops due to failure to maintain adequate fiber spacing, especially with random fibers. (B) Strength drops with increasing volume fraction of fibers, except for higher contents of aligned fibers, due to maximum fiber spacing being greater than the flaw size and stresses from the fibers. Data after Sambell *et al.* (1972a), published with permission of *Journal Materials Science.*

TABLE IX

FRACTURE ENERGY OF METAL CERAMIC EUTECTICS[a]

System	Metal content (wt %)	Fabrication[b]	K_{IC}[c] (MN/m$^{3/2}$)	Work of fracture[c]	Comments
ZrO$_2$–Ta	14–17	DS	3.3–7.0	138–573	—
ZrO$_2$–W	15–17	DS	5.9	450	—
Cr$_2$O$_3$–Mo	7	DS	1.15	80	Parallel to cell direction
	7	DS	3.32	250	Across cell direction
	7	H	5.4	50–60	Grain size <71 μm
	7	H	3.7	150	Grain size 5–20 μm
Cr$_2$O$_3$	—	H	3.9	19	—

[a] Data of Claussen *et al.* (1974).
[b] DS—directionally solidified; H—hot pressed, Cr$_2$O$_3$ from powder, others from crushed, directionally solidified eutectics.
[c] Average of at least four measurements.

without any apparent damage. This indicates a substantial increase, but less than the total increase in Charpy toughness of nearly thirtyfold in the Si_3N_4–Ta body, but with attendant cracking. Note, however, that while the Ta wires had no effect on strength at 1300°C, they reduced room temperature strength by one-third, e.g., from ~120,000 psi to ~80,000 psi. How much the above differences in Al_2O_3 and Si_3N_4 represent differences in the toughnesses of these respective matrices and how much to different matrix-fiber compatibility is unknown.

Claussen et al. (1974) studied the fracture of metal–ceramic eutectics based mostly on ZrO_2 or Cr_2O_3. The in situ metal fibers were commonly about 1 mm in diameter and spaced a few microns apart. Fracture energies were increased about four- to twentyfold (Table IX), but stress intensities were about the same or less than the pure oxide, apparently due in part to lower Young's moduli of the composites (values not given).

E. Microstructural Dependence of Fracture Energy at Elevated Temperatures

Only limited experimental and theoretical work has been done on fracture energy at elevated temperatures. Stevens and Dutton (1971) have developed a model of crack growth by diffusion processes, which may be applicable before plastic deformation by slip becomes important. However, the rate of change from low-temperature to high-temperature processes and their interaction at intermediate temperatures are not at all clear. Thus, for example, the relative effects of weakening of grain boundaries to lower fracture energy on one hand and to increase blunting and possible parasitic cracking effects from grain boundary sliding on the other are not at all clear. Some answers may result from slow crack growth studies (in Section IV,C) that have been initiated.

Only SiC, Si_3N_4, and UO_2 have been studied in any detail for the temperature dependence of their fracture energies. Evans and Davidge (1969a) showed fracture energies of their 8 and 25 μm grain size UO_2 bodies to progressively increase from about 8 and 5 J/m^2 at 22°C to about 22 and 11 J/m^2, respectively, at 1600°C. The finer grain size material exhibited ~30% transgranular failure at room temperature, with this dropping to <5% by 1100°C. The 25 μm grain size UO_2 was >60% transgranular failure at all temperatures.

Coppola, and co-workers' (1972) WOF results for Si_3N_4 are from the same materials and study as their room temperature DCB data (Fig. 10A) and are in agreement with that data. Generally, the data show a slight decrease in fracture energy, reaching a minimum at temperatures of 400–800°C, then at higher temperatures, e.g., 1000–1200°C, increasing to about their room temperature level. The effects of porosity on the fracture energy of Si_3N_4 are relatively constant over this temperature range. The major variations in this trend were in hot-pressed materials, which the authors attributed to hot-pressing additive effects. The increases in fracture energy at higher temperatures were attributed to plastic flow. Hartline *et al.* (1974), have presented further evidence of such plastic flow effects, e.g., showing lower high-temperature WOF in Si_3N_4 with increasing plastic deformation which are also corroborated by results of Evans and Wiederhorn (1974).

Extensive measurements by Bradt *et al.* (1974) on SiC show trends basically similar to those of Si_3N_4; i.e., some changes, often in different directions, in different materials (Fig. 18). These changes are most likely related to the various phases (e.g., at grain boundaries) and their behavior with increased temperature. This is supported by results of McLaren *et al.* (1972), who found a marked drop in the fracture energy of Refel SiC when the included Si melts. Again, there is a general inverse trend between fracture energy and strength.

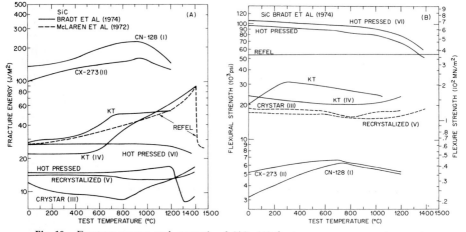

Fig. 18. Fracture energy and strength of SiC; (A) fracture energy versus temperature, (B) strength versus temperature. Note the approximate inversion of strength and fracture energy between the different sets of materials. Most data after Bradt *et al.* (1974), adapted with permission of University of South Carolina Press.

Much of the remaining elevated temperature fracture energy data is for composites—again, note that strength and fracture energy often have opposite trends (Fig. 19). Hulse and Batt (1974) attribute the large increase in fracture energy of their two all cubic composites to plastic deformation. However, Kennard's results for $MgO–MgAl_2O_3$ directionally solidified composites having the same $\langle 111 \rangle$ growth texture showed a much lower

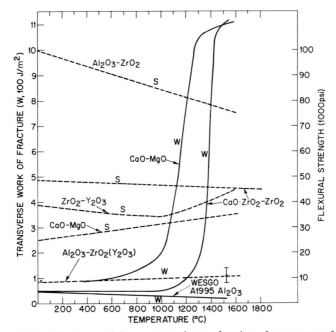

Fig. 19. Work of fracture and flexure strength as a function of temperature for various solidified ceramic eutectics (from the data of Hulse and Batt, 1974). Note the high fracture energies, particularly of these materials involving non-cubic phases. The general trends of strength and fracture energy are approximately consistent, indicating flaw sizes are greater than the size of each phase region, e.g., typically lamali spaced ~ 1 μm apart; however, at high temperatures, the quantitative changes in strength and fracture energies are not consistent.

level of fracture energy and little increase with temperature. Whether this is due to lower plasticity in the $MgAl_2O_4$ phase in comparison to those of the other composites is uncertain.

F. Discussion and Summary of Fracture Energy

The preceding survey shows several important aspects and problems of the testing and microstructural dependence of fracture energy. First and foremost is the fact that fracture energy often does not correlate with strength. This can be due to changes in flaw sizes, but it commonly results from differing effects when the crack size in the fracture energy test is large in comparison to the strength controlling microstructure, but strength-controlling flaws are not, as will be discussed further in Section IVc and d. The most common correlation of fracture energy, strength, and micro-structure is their general decrease with increasing porosity similar to that for Young's modulus. Thus, in general, fracture energy decreases nonlinearly with increasing porosity due to the nonlinear dependence of the maximum pore, i.e., minimum solid cross-sectional area for almost all pore shapes. Most data can be fitted to an exponential dependence, e^{-bP}, with different b values associated with different types of pores. But these b values can also be affected by pore location and inhomogeneous spatial distribution of pores. There are two possible departures from the above porosity trends. Bodies having limited, primarily intergranular porosity can have fracture energy increased due to pore–crack interactions, i.e., formation of fracture tails. However, this is important mainly, if not exclusively, in softer materials such as MgO where plastic processes at such tails can be important. Bodies having substantial amounts of porosity generally show reduced decreases and often substantial increases in fracture energy in contrast to the trends at lower porosity by most techniques, e.g., DCB. The NB technique does not appear to show nearly as much, if any of this high porosity effect which is probably enhanced by large grains, second phases, and crystalline anisotropy.

The WOF, DCB, and DT tests indicate that the fracture energy of dense cubic or nearly cubic materials is independent of grain size. This is partly supported by eutectic results. The combination of grain size and eutectic composite results for noncubic (i.e., anisotropic) materials indicates that fracture energy initially increases, passes through a broad maximum, then decreases with increasing grain size (or phase spacing). The extent of the increase and the size of the grains for these changes should depend on the level of internal stress and hence, respectively, directly and inversely with the degree of anisotropy (primarily TEA). On the other hand, the NB test and possibly determination from fracture topography, i.e., fracture mirrors or crack branching, apparently indicate that fracture energy is independent of, or decreases somewhat, with grain size in all materials.

The differences between the NB, and possibly fracture topography results, in contrast to those of DCB, WOF, and DT techniques, are uncertain and requires further study. They may be due to crack size effects. Normal strength tests (the basis of fracture topographic techniques) and NB specimens apparently fail from the most severe flaw; e.g., the most porous region, making them more sensitive to microstructural effects. The other tests generally: (1) use machined grooves that could cause microcracking along the groove and (2) have smaller fracture areas, making it easier to have a crack initiating fracture that crosses the complete test area and hence averages microstructural effects. Parasitic cracking is felt to be the most important effect increasing fracture energy in both quite porous and anisotropic bodies. While internal stress effects are progressively relaxed as temperatures rise, parasitic cracking may not necessarily decrease. It may in fact increase due to grain boundary sliding but be complicated by an increasing bluntness of cracks. However, much more work is needed to determine the extent and role of parasitic cracking in determining fracture energy.

Impurity and additives can have a variety of effects on fracture energy, complicating analysis of their and other microstructural effects. They, especially second phases, can lower or substantially increase fracture energy. Increased fracture energy due to crack pinning inclusions is accounted for in some cases (e.g., particles in glass matrices) by Lange's line tension model, but the size and fracture energy of the particulates in comparison to the fracture energy and matrix grain size may have to be considered. Also, in polycrystalline bodies where the pinning particles are smaller than the grains, this model gives no information about the possibility of intergranular failure to totally avoid intragranular particles. Of particular significance is the indicated toughening of ceramics (e.g., ZrO_2 and Al_2O_3) by inclusion of fine, uniformly dispersed second-phase particles having significant differences in expansion behavior from the matrix. Such particles can apparently lead to a microcrack, hence energy dissipation, zone around the tip of a crack. If the size of the particles is sufficiently small and their volume fraction high enough so that the spacing between the particles is substantially smaller than the flaw size yet greater than the particle size, strength need not be low.

IV. Microstructural Dependence of Tensile Strength and Fracture

A. Background and Strength–Porosity Theories

1. Pores as Stress Concentrators

Tensile strength is generally the single most important mechanical property, since the level and/or reliability of tensile strengths are commonly a major, if not determining factor in the use of ceramics. Despite this importance, there has been somewhat less development of strength–porosity than of elastic property–porosity relations, especially of a fully analytical nature. This is probably due in part to the more complex nature of tensile strength, i.e., its interactive dependence on porosity, impurities, grain size, surface finish and test conditions. However, recently substantial progress has been made in understanding the effect of porosity on strength. Most of the strength–porosity models that have been developed are shown in Table X along with a summary of their assumptions. These are briefly discussed below along with some similar models. Also note that many if not all of the elastic property–porosity relations could be used for strength in view of the similar strength and elastic property changes with porosity. In particular,

TABLE X

Strength–Porosity Models

Investigator	Type[a]	Assumptions	Equation
Balshin (1949)	SA (LB)	Derived for metals, apparently assuming some plastic deformation and work hardening, and probably uses fact that most data give a straight line on a log–log plot	$S = S_0(1 - P)^m$
Schiller (1958)	SA (essentially LB)	No pore interaction; sterological relations for uniformly spaced pores hold for randomly spaced pores	
		(A) Spherical pores	$S = S_0[1 - (P/P_c)^{1/3}]$
		(B) Cylindrical pores	$S = S_0[1 - (P/P_c)^{1/2}]$
Schiller (1958)	SA (essentially LB)	General porosity can be represented by generalizing the above equations; generalized equation based on first-power terms of a series expansion; $P > 0$	$S = S_0[1 - (P/P_c)^{1/n}]$ $S = \beta \log P_c/P$ $S = 0 \quad \text{at} \quad P = P_c \lesssim 1$
Knudsen (1959)	SA (LB)	Strength is directly proportional to contact area normal to the stress between uniform sintering spheres	$S = S_0 e^{-bp} \qquad P > \sim 0.5$
Rice (1976)	SA (LB)	The position of pores and solids can be exchanged in Knudsen's model	$S = S_0[1 - e^{-b(1-P)}] \qquad P > \sim 0.5$
Eudier (1962)	A (LB)	Uniform spacing of equal size spherical pores	$S = S_0(1 - 1.19P^{2/3})$
Bache (1970)	SA (LB)	Contact area between identical spherical grains is limited and stress in the body = stress in each contact area	$S = S_0(d/G)^{2/3}(\gamma E/G)^{1/2}$ d = diameter of contact area

Reference	Type	Description	Equations
Harvey (see Bailey and Hill, 1970)	SA (SC)	Uniform spherical pores	$S = S_0\left(\dfrac{(1-U)^3}{U^3 + (1-U)^3}\right)$ $U = (3P/4\pi)^{1/3} = R/\lambda$ R = pore radius λ = separation between pores
Carniglia (1972a)	A (LB)	Flaw size $\gg$ pore spacing (mainly for intragranular pores)	$S = S_0[(1-P)E/E_0]^{1/2}$
Carniglia (1972a)	SA (LB–SC)	Flaw size $\ll$ pore spacing; $E/E_0 = e^{-bP}$ ($\xi \geq 1$).	$S = \dfrac{S_0[e^{-bP}(1-P)]^{1/2}(P - \frac{1}{3} - 1)^\xi}{\left[P^{-1/3} - \dfrac{11}{16} - \dfrac{P^{2/3}}{8} - \dfrac{3P^{4/3}}{16}\right]^\xi}$
Carniglia (1973)	—	Numerical approximation for above equation with $P \leq 0.4$	$S = S_0 \exp(0.26 + 0.22b + 0.295\xi)^P$ $\xi = 0$ for intragranular porosity, otherwise $1 \leq \xi \geq 5$
Hrma and Satava (1974)	SA (LB)	Strength is directly related to the contact area between particles, and sterological relations for spherical particles can be generalized to irregular particles	$S = \xi(f - fc)$; ξ = pore shape factor $f = (1-P)^{4/3}/P^{1/3}$ $fc = (1-P_c)^{4/3}/P_c^{1/3}$ $S = 0$ at $P = P_c$
Rice (submitted for publication)	A (SC)	Pores plus part of surrounding grains are flaws	$S = \dfrac{\pi}{2}\left(\dfrac{\eta}{\tan\eta}\right)^{1/2}\left(\dfrac{E\gamma}{R + L}\right)^{1/2}$ $\eta = \dfrac{\pi R}{2R + \lambda}$ L = a fraction of grain size (G), dependent on R and G λ = pore spacing $= 4R(1 - P)/3P$ for spherical pores

[a] SA = semianalytical, A = analytical, LB = load-bearing, SC = stress concentration.

259

the exponential relation e^{-bP} is the most widely used relation for strengths as it is for elastic properties (Section II,A).

Bal'shin's (1948, 1949) semianalytical, essentially load-bearing model appears to be based primarily on empirical observation that data often give a straight line on a log–log plot, since he has used the same expression for other properties, e.g., hardness and elastic moduli. Also, m is empirically determined (typically 3 to over 5 for metals according to Bal'shin). Use of log–log plots minimizes apparent data scatter, e.g., Thompson and Pratt's (1966) plotting of Si_3N_4 data, but does not really lead to any real increase in accuracy of extrapolation or comparison of data. Since this model appears to depend upon some plastic deformation and apparently includes corrections for work hardening, Thompson and Pratt's obtaining $m = 3.4$ for the room temperature (i.e., brittle) behavior of Si_3N_4 in agreement with metal values may be fortuitous simply, due to the fact that a wide variety of data will appear as a straight line on a log–log plot. Pattel and Finnie (1970), based in part upon empirical tests and in part on a semianalytical model of plastic foams, gave an equation of the same form as Bal'shin's. They noted that the exponent m was dependent on foam structure and was typically in the range of 1 to 2 instead of $\frac{2}{3}$, which one would obtain neglecting foam structure and resultant stress interactions. Whether this equation, which is based in part on plastic buckling of parts of the foam structure, is legitimately applicable to brittle ceramics remains to be determined. Also, whether either the identical form or differing m values for this and Bal'shin's model is significant is uncertain.

Schiller (1958) assumed that the stress around a pore is a function of the stress in the absence of the pore and the ratio of the average pore radius to the average pore spacing. Constants of his generalized, approximate equation are determined from the boundary conditions, $S = S_0$ at $P = 0$, $S = 0$ at $P = P_c$. In the discussion following Schiller's paper, Evans notes a much simpler empirical derivation not dependent on any sterological considerations, while Tabor presents a simple load-bearing sterological derivation that raises questions about the value of the exponent m. Despite this uncertainty and the question of its applicability to systems with close, interacting pores, e.g., very porous bodies, it does show good correlation with data for some highly porous materials, e.g., Fig. 20, which shows $P_c < 1$, as discussed later.

Two other investigators have followed Schiller's approach, lending some further support to it. In the discussion following Schiller's paper, Millard indicates the derivation of an equation of the form $S/S_0 = [1 - (P/P_c)^{m_1}][1 + k(P/P_c)^{m_2}]$, where the first bracket represents the reduced load bearing area and the second stress concentration effects. He notes the second bracket can often be reduced to $(P/P_c)^{m_2}$, but gives no de-

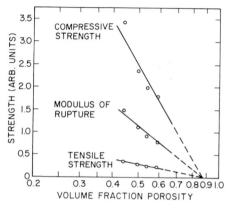

Fig. 20. Strength versus porosity for plaster (after Schiller, 1958). Note that all three strengths project to zero at $P < 1$. Modified figure published with the permission of Interscience Publishers.

tails of the derivations. He cites experimental evidence that some porous materials follow the simple form of the equation

$$S/S_0 = (1 - P/P_c)(P/P_c)^{-1} = P_c/P - 1.$$

Hrma and Satava (1974) (Table X) treated a body built up of spherical particles in a similar fashion to Schiller's treatment of spherical pores. This is another example of a model and its inverse (i.e., letting the solid and pore phase be interchanged) giving similar results, as noted in discussion of elastic property–porosity relations. A critical volume fraction porosity of less than 1 is required for the body to begin to have strength, since there must be a sufficient number of solid particles before they become connected to provide strength, similarly to percolation theory as noted in the discussion of elastic behavior. The pore shape factor ξ could be calculated if factors to relate a particle "radius" to its surface and volume and the strength of particle contacts and their orientations relative to the applied stress were known. This is generally not practical, so their equation must be basically regarded as only semianalytical.

Eudier (1962) derived his equation for sintered metals assuming that strength is determined solely by the minimal cross-sectional areas of solid material. Similarly, Brown et al. (1964) assumed that strength is directly proportional to the ratio of the projected area of the solid material between pores onto the plane of fracture divided by the total cross-sectional area of a flat plane of fracture, i.e., $S = S_0(1 - \sum\alpha_i P_i)$ where the α_i are the geometrical projection factors for each various pore shape (see Table XI) and P_i is the volume fraction of each type of pore.

TABLE XI

GEOMETRICAL PORE SHAPE FACTORS OF BROWN *et al.* (1964)

Pore shape and orientation	x_i	a_i	v_i	δ_i
Sphere F ← ◯ → F	d	$\frac{1}{4}\pi d^2$	$\frac{1}{6}\pi d^3$	1.50
Oblate spheroid	c	$\frac{1}{4}\pi a^2$	$\frac{1}{6}\pi a^2 c$	1.50
Oblate spheroid	a	$\frac{1}{4}\pi ac$	$\frac{1}{6}\pi a^2 c$	1.50
Ellipsoid Force parallel to any axis	a	$\frac{1}{4}\pi bc$	$\frac{1}{6}\pi abc$	1.50
Cube ← ▢ →	s	s^2	s^3	1.00
Cube ← ◇ →	$s\sqrt{2}$	$s^2\sqrt{2}$	s^3	2.00
Cube	$s\sqrt{3}$	$s^2\sqrt{3}$	s^3	3.00
Cylinder ← ▭ →	l	$\frac{1}{4}\pi d^2$	$\frac{1}{4}\pi d^2 l$	1.00
Cylinder	d	ld	$\frac{1}{4}\pi d^2 l$	1.27

Bache's (1970) model (Table X) for a body consisting of identical spherical particles follows Griffith's approach, calculating the changes in energy as a function of crack growth in the limited contact area of one set of particles and using an approximation based on the fact that the contact area is small in comparison with the particle diameter. [He also obtains an expression relating strength to the volume fraction of a bonding phase between spherical grain (Section IV,E)]. While Bache does not directly relate the parameters in his model to the porosity of the body, this can be done by using Knudsen's semianalytical relations, which give $(d/G)^2$ proportional to e^{-bP}. Thus, for example, using cubic stacking of particles ($b = 6$), which would be most applicable to Bache's model, one obtains $S = e^{-4.5P}(\gamma E/G)^{1/2}k$.

Harvey (Bailey and Hill, 1970; Bailey, private communication) derived his equation for spherical pores assuming that the stress in the vicinity of a pore (σ_s) is related to the applied stress (σ_a) and the distance from the pore center (l) has the form $\sigma_c = \sigma_a[N/l^3 + Q]$. Next, assuming that failure occurs due to the concentration of the stress from one pore on the surface of an adjacent pore adds a factor 2 in front of σ_a. Constants

$$N \, (= R^3/2) \quad \text{and} \quad Q \, (= \tfrac{1}{2})$$

are obtained by recognizing that at $l = \infty$, $\sigma_c = \sigma_a$, and that at $l = R$ (i.e., at the pore surface), $\sigma_c = 2\sigma_a$. Then, recognizing that the maximum stress occurs on a line between the pore centers so $l = \lambda - R$ where R is the radius of the second pore from whose surface failure occurs, and that $\sigma_a = S$ and $\sigma_c = S_0$, and defining $U = R/\lambda$, the equation shown in Table X is obtained.

Carniglia (1972a) analytically derived his equation for flaw size $\gg$ pore spacing (Table X) by assuming that stress concentrations are not significant; i.e., load bearing conditions dominate. He equated the total strain energy in porous bodies expressed in terms of the (1) applied stress and Young's modulus of the body and (2) average stress and Young's modulus of the matrix. His general expression (Table X) for flaws $\ll$ pore spacing is obtained using his analytical load-bearing expression $E/E_0 = e^{-bP}$, base 10 logarithms, and considering some analytically solvable geometrical arrays of one or more spherical pores and a planar crack of variable distance from the pore. In using the latter, he assumes (1) that the expression for the effect of one pore and nearby crack geometry can be generalized to all such geometries, since the few important geometries considered gave similar results, and (2) that nonspherical pores can be treated by introducing a single shape parameter ξ which becomes greater than 1 as pores deviate from spherical shape ($\xi = 1$).

2. PORES AS FLAWS

In the last few years, a basically different approach, that of treating pores themselves as an integral part of flaws rather than as simply a means of concentrating stress on flaws away from the pore, has been taken. While Hasselman (1969a) had considered some aspect of pores as flaws, Evans, Davidge, and colleagues (Evans and Davidge, 1969a; Evans and Davidge, 1970; McLaren et al., 1972) were major developers of this concept for large pores. They considered the pore diameter, plus one grain depth on either side of the pore, to be a flaw based on the concept that a crack would propagate along grain boundaries emanating from the pore wall into the body until the next layer of grains was encountered. Crack propagation around or through grains of this second layer was then considered to be the strength–controlling step. Evans and Tappin (1972) suggested use of Bowie's (1956) (approximate series) solution for a cylindrical pore with either one or two diametrically oriented cracks extending outward for treating such pore–crack combinations. Wittman and Saitsev (1972) have also independently used Bowie's model to treat pores in concrete, citing E. V. Panasjuk for an apparently closed form solution. Molnar and Rice (1973) showed that pores having significant anisotropy of shape could be treated as flaws by taking the cross-sectional dimensions of the pore plus the grain size in the observed plane or expected plane of fracture (i.e., in a plane approximately perpendicular to the applied stress) in the region of maximum of stress as the equivalent flaw dimensions.

Rice (submitted for publication) has recently extended this concept of pores as an integral part of flaws by adapting it to pores of essentially all sizes and locations (Fig. 21). This adaptation shows a number of important ramifications of porosity effects, most of which are readily seen by considering Fig. 21 and the equation of Table X. First, porosity and grain size effects on strength become clearly separable as pores approach or become smaller than the grain size. Second, uniform distributions of different types of pores will have similar strength–porosity trends (e.g., will follow the exponential relation), but the b values will depend on pore location, size, and shape. The latter two are important only when the pore causing failure is large in comparison with the grain size, i.e., $R + L \sim R$. For small pores, location is important; i.e., pores at triple points will be substantially more serious than pores elsewhere on grain boundaries or pores within grains,

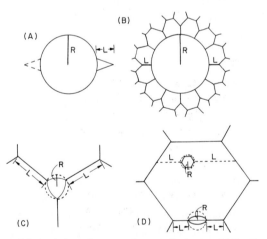

Fig. 21. Flaw model consisting of a pore plus a radiating crack. (A) Sketch of flaw based on cylinder-crack systems originally considered by Bowie. B, C, and D sketch how this concept is applied to pores of different size and location in relation to the grain size being respectively for pores large in comparison with the grain size (B), small pores at triple points, (C) and pores along grain boundaries or within grains (D).

with the latter generally being the least detrimental to strength. Both the potential length of cracks (i.e., involving two grain-boundary facets for triple point pores) and the number of possible associated cracks (i.e., along three possible boundaries, two of which should be at substantial angles to the stress) (Fig. 21) are major reasons for this. Stress concentration effect as suggested by Rice (1972b) may also be a factor in such pore location effects. [Note Spriggs' (1962) modification of the exponential relation to account for open porosity (P_1) and closed porosity (P_2) (e.g., $e^{-b_1 P_1 + b_2 P_2}$) does tend to help strength–porosity and elastic modulus–porosity correlation, because open pores exclude all pores within grains. However, it is very inexact, since closed pores include both grain boundary pores and pores within grains, and the weight of both experimental and theoretical evidence shows that location is the important factor rather than whether pores are open or closed.] The shape of smaller pores will have limited or no significant direct effect on strength unless the pore shape is quite extreme, but the shape of larger pores is more important, since this effects R and $R > L$. Third, this model shows why Young's modulus and strength have similar porosity trends, since the primary porosity dependence of S comes from that of $(E\gamma)^{1/2}$. Thus, since $E/E_0 \sim e^{-bP} \sim \gamma/\gamma_0 \sim e^{-bP}$, $S/S_0 \sim e^{-bP}$. Rice's studies question the applicability of Bowie's cylindrical pore model to spherical pores. He also stresses that the extremes of porosity must be considered; i.e., failure will occur from the largest or most extremely shaped pore or the largest concentration or combination or pores in regions of high stress. Typically, essentially all past models have implicitly or explicitly considered only average porosity, which is clearly a gross oversimplification in many cases. Thus, based on different location, size, and shape of pores, as well as their degrees of concentration, a considerable range of strength behavior can occur for different bodies over the same porosity range.

All of the porosity theories, except Rice's, assume that the strength at zero porosity S_0 is characteristic of a given body, i.e., of a given grain size, impurity distribution, etc. However, S_0 typically depends on surface finish and may depend on porosity. This question has been examined by Rice (to be published), whose model does not present this problem, since no S_0 term is involved because the flaw size is specified in terms of pore and grain size parameters. He finds that the sizes of flaws with small pores as an integral part of them is commonly similar to typical machining flaws, so the strength at zero porosity will be close to the extrapolated value. However, this may not be the case with large pores, with some bodies (e.g., composites), or some machining conditions. Thus, any equation that uses an S_0 term that is not calculable really rests on empirical observations. Clearly more work is needed to delineate such flaw size effects between porous and non-porous bodies.

B. Strength–Porosity Data and Comparison with Theories

1. DEPENDENCE OF TENSILE STRENGTH
   ON AMOUNT OF POROSITY

   Table XII is an extensive, though not completely exhaustive, compilation of room-temperature strength–porosity dependence, and Fig. 22 is an example of a compilation of various data on Si_3N_4. There is also considerable data consisting of a few data points for bodies that can be compared with themselves or other similar bodies showing similar strength–porosity trends, e.g., most of the Al_2O_3 bodies of Duncan et al. (1965) give b values of 3 to 9. As noted in Section I,A, all of the strength–porosity dependence has been expressed in terms of the widely used exponential relation to provide a common basis of comparison, and some values of S_0 and b may differ from those reported by the original authors due to difficulty in reading data from graphs and different curve fitting of data. Further, some previous authors have fit their data simultaneously for both strength and grain size effects, which can alter the porosity corrections.

   While, as with other property–porosity data, there is considerable variation that will be discussed below in terms of the character of the porosity, there is a general trend for the same b values as for Young's modulus. As added support of this agreement, note that fibers also show the same general strength–porosity dependence as bulk materials similar to their Young's modulus behavior (Table VI). Thus, the average of the b values in Table XII is about 4 ± 2, the same as Young's modulus data (Table V). This agreement is in contrast to Davidge and Evans' (1970) suggestion that the b for strength would be half that for Young's modulus, but it is consistent with Rice's model in view of the similar b values for γ.

   It should be noted that, like elastic properties, the porosity dependence of the strength of metals at room temperature is similar to that of ceramics. Thus, for example, Knudsen (1959) analyzed tensile strength data of porous iron and various steels, obtaining b values, respectively, of 5 and 5–7. Eudier's (1962) substantial data on the tensile strength of sintered low alloy steel as a function of porosity gives b values of the order of 3. Gallina and Mannone's (1968) data on flexural and tensile strength and fatigue limits of sintered iron, which give b values from 4 to 7, show the strength values for a given level of porosity increasing as the particle size of the starting powder decreases. Gallina and Mannone attribute this particle size effect to a corresponding pore size effect, i.e., obtaining higher strength with smaller pores, which is basically consistent with Rice's model. Similarly, Rostoker and Liu (1970) present data for sintered brass, having a variety of different pore sizes depending upon sintering conditions, which give b values of the order of 4 with scatter similar to ceramic materials. Finally, Salak et al. (1974) obtained a

TABLE XII

Room Temperature Flexure (or Tensile) Strength–Porosity Corrections

Body character $S = S_0 e^{-bP}$

Material	Fabrication[a]	% P	Porosity shape, spatial distribution, and location[b]	Pore size[c]	G.S. (μm)	Other	S_0 (1000 psi)	b[d]	Source
Borides									
SiB₆	H	2–24	Mostly G.B.	>G and <G	~5.30	Contained other phases	34	8 ± 1	Dutta and Gazza (1973).
TiB₂	H,S	5–15	Many I.G.	<G	Apparently 20		43	4.4 ± 0.2	Mandorf et al. (1963).
ZrB₂	H	0–16	G.B.	<G	6–11		49	3 ± 1.5	Limited data of Clougherty et al. (1968).
Carbides									
B₄C	H						70	4.1 ± 0.3	Unpublished data of King and Wheildon, Norton Co.
Cr₃C₂	H	3–20	Pores mostly at G.B. and ≤G at lower densities and G.S., with more pores <G within grains at higher densities and G		6–10		67	6 ± 2	Hamjian and Lindman (1952, see also Knudsen, 1959).
		6–17			8–12		57	4.5 ± 0.4	
		5–10			8–15		52	2.2 ± 0.8	
		1–10			9–15		47	1.4 ± 0.9	
SiC	H	5–20	I + L,N G.B.	Some >G Some <G	4–15		56	2.4 ± 0.8	Balloffet et al. (1971).
SiC	S	4–16	G.B.	≤G	40–100		26	3.0 ± 0.8	Coppola and Bradt (1972).

TABLE XII (*Continued*)

Material	Fabrication[a]	% P	Porosity shape, spatial distribution, and location[b]	Pore size[c]	G.S. (μm)	Other	S_0 (1000 psi)	b^d	Source
TiC	H	5–6	G.B. and some I.G.	<G	12	Polished	63	4.1 ± 0.6	Harrison (1965).
TiC	H	1–5					84	10 ± 1.0	Glaser and Ivanich (1952).
ZrC	H	3–18	Many I.G.	<G	~15	Ground	35	4.0 ± 1.0	Limited data of Lanin et al. (1968).
ZrC						Electrospark machined	27	4.7 ± 0.2	
Nitrides									
Si₃N₄	H,RS	0–36	Mixed	>G and <G	~0.5–2		65	5.3 ± 0.4	Extensive compilation, see Fig. 22.
UN	S,H	5–40					(10	5)	Diametral compression data of Mönch and Claussen (1968).
Single Oxides									
Al₂O₃ (fibers)	S	6–30	~R mostly I.G.	<G	1–5		113	2.3 ± 0.2	Bailey and Barker (1971).
Al₂O₃ (fibers)	S	6–15	~R mostly I.G.	Some ≫G	2–60		67 / 83	4.1–4 point flexure / 3.9–3 point flexure	Blakelock et al. (1970).
Al₂O₃					0.5–3	Isolated 5–10-μm grains + ~1.3 W/O MgO or CoO	39.8	5.3	Diametral compression data of Gazza et al. (1969).
Al₂O₃	H	0–60			0.5–1		45.0	4.9	
Al₂O₃	S	4–17			1–5 often with large grains	Contained 1%–2% MnO₂ &/or TiO₂	35	1.5 ± 0.5	Cutler (1957).

Body character $S = S_0 e^{-bP}$

Material		Range	Pore location	Pore size	Description	Grain size	No.	Value	Reference
Al$_2$O$_3$	S	2–18	Mostly G.B.	<G	99.5% commercial alumina, round rods	1–5	41	4.0 ± 0.2	Extensive data of Neuber and Wimmer (1971).
	S	2–20				5–10	28	1.6 ± 0.3	
	S	2–9				10–20	27	5.8 ± 0.6	
	S	2–9				20–30	28	6.5 ± 1.0	
	S	5–18				30–70	23	4.7 ± 0.4	
Al$_2$O$_3$	S	2.5–28	Some N.I.G.	Much <G, Apparently <G		1–2	59	3.4 ± 0.3 (3 point flexure)	Steele et al. (1966).
							51	3.5 ± 0.6 (4 point flexure)	
Al$_2$O$_3$	S	5–50	~R mostly G.B.	Most	4 point flexure of round rods	23	30	3.9 ± 0.2	Coble and Kingery (1956).
Al$_2$O$_3$	S	1–7	Mostly G.B.	1000 μ >G and <G	3 point flexure of round and square rods	4–8	70	7 ± 2	Binns and Popper (1966).
	S	2–9		<G		10–20	43	2 ± 4	
Al$_2$O$_3$	S	0.2–8	Much I.G.	Much <G		25–50	44	8 ± 4	Evans and Tappin (1972).
	S	5–50	Most G.B.	Most >G		3	40	3.5 ± 0.8	
Al$_2$O$_3$	S	8–27		~1		~1	36	3.1 ± 0.3	Schofield et al. (1949).
Al$_2$O$_3$	H	0.2–6.5	G.B.	0.06		2.2	92.0	9.35	Passmore et al. (1965).
	H	3–6.5	Much I.G.	5		30	18.5	0.34	
B$_6$O	H	2–19		+	4 point flexure ground bars	~1 + some 5–30 in denser bodies	52	5.0 ± 0.7	Petrak et al. (1974).
					Diametral comp		42.6	4.9 ±	Petrak et al. (1973).
BeO	S	4–14	more intergranular porosity of different size, shape, and distribution in 5 μ G bodies than in larger grain size bodies		4 point flexure of centerless ground UOX + MgO rods having significant axial orientation of c axis	5	30	0.1 ± 0.2	Fryxell and Chandler (1964).
						10	36	1 ± 2	
						20	34	2 ± 1	
						50	24	3 ± 2	
						80	18	2	
					4 point flexure of centerless ground, random grain orientation	5	24	0.7	
						10	36	1 ± 1	
						20	39	3.6 ± 0.3	
						50	20	6 ± 1	
						80	16	1.4	
BeO	S	5–52				13–30	26.5	7.8 ± 0.3	Schofield et al. (1949).
MgO	H	0.7–29	Presumably both mostly at G.B.	<G	As hot pressed and machined	6.5–14	28	4.1 ± 0.6	Sprigs and Vasilos (1963).
		0.8–29			Annealed	4.4–13	38	7 ± 1	

TABLE XII (*Continued*)

Material	Fabrication[a]	% P	Porosity shape, spatial distribution, and location[b]	Pore size[c]	G.S. (μm)	Other	S_0 (1000 psi)	b[d]	Source
						Body character $S = S_0 e^{-bP}$			
ThO$_2$	S-1650	7-32			5-36			5.5	Knudsen (1959).
	S-1725	6-32			5-39			4.7	
	S-1800	6-32			6-43			3.8	
	S-1850	5-31			5-53			3.7	
ThO$_2$	S	1.8-47			4		17.4	2.0 ± 0.6	Curtis and Johnson (1957).
UO$_2$	S,H	0-35			60		19.5	3.5 ± 0.8	Diametral compression data of Mönch and Claussen (1968).
							(7.4)	(6.1)	
UO$_2$	S	8-24			20		16	3.2 ± 0.8	Composite data of Knudsen et al. (1960) and Burdick and Parker (1956).
					35-60		13	1.7 ± 0.7	
Mixed oxides									
Ferrite	S	4-16			1-8		26.5	5.3 ± 0.4	Hoop tensile data of Chen and Weisz (1972).
PZT	H	1-14			3-4		17	3 ± 1	Okazaki and Nagata (1971).
PZT	S	2-11		30-1000	~5		19	11 ± 1	Rice and Pohanka (unpublished)
Silicate glass	S	0-38	S	30-300			12	2.7 ± 0.3	Ali et al. (1967).
Silicate glass	H	2-40	S	60			8.1	2.5 ± 0.6	Hasselman and Fulrath (1967).
Silicate glass	H	2-40	S	20-36			10.2	1.0 ± 0.2	Bertolotti and Fulrath (1967).
				36-44			7.9	1.4 ± 0.2	
				74-105			7.2	1.5 ± 0.2	
				105-186			6.9	1.7 ± 0.2	
Zircon	S						33	6 ± 1	Schoefield et al. (1949).

Other ceramics						
B	H	3–30		50	3.3 ± 0.5	Petrak et al. (1974).
Bone china	S	5–35		18	3.7 ± 0.1	Dinsdale and Wilkinson (1966).
Earthen ware						
coarse flint	S	5–35		17.7	3.9 ± 0.1	Williams et al. (1963).
fine flint	S	5–35		21.1	3.4 ± 0.1	
Porcelain	S	10–19		19.5	3.2 ± 0.2	
Kaolin quartz	S	7–36		18	6 ± 1	Hamano and Lee (1972).
triaxial	S	8–17		28	7.8 ± 0.5	
Sagger clays	S	3–43		42	2.5 ± 0.6	Heindl and Pendergast (1927).
Glassy carbon		29–64		62	4.6 ± 0.2	Compiled from Yamada (1968) and Hucke (1972).
POCO, Graphite		18–32	8	25.8 / 90	8.8 ± 0.9 tension / 10.0 ± 0.9 flexure	POCO Graphite, Inc.
Graphite (Graph-i-tite G)		15–32		7.72	3.3 ± 0.4	With grain flexure data, Carborundum Co.
Graphite infiltrated carbon felt		22–45		10 / 7.2 / 4.1	0.8 ± 0.2 flexure / 1.1 ± 0.4 tensile / 4.0 ± 0.2 interlaminar shear	Kotlensky (1973).
Graphite infiltrated molded graphite		17–28		21 / 15	5.9 ± 0.4 across grain / 5.0 ± 0.2 with grain	Hollenbeck et al. (1974).
Graphite fibers		10–34		690	5 ± 1	Butler et al. (1973).

[a] H = hot pressed, S = sintered, RS = reaction sintered.

[b] I = irregular, L = laminar, S = spherical, N = nonuniform spatial distribution, G.B. = located at grain boundaries, I.G. = located within grain, R = approximately uniform distribution.

[c] Diameter in microns or relative to grain size.

[d] The standard deviation on b is an indication of the fit of the data to the equation $S/S_0 = e^{-bP}$ and that approximate value of b. It cannot be taken as an absolute indication of the value of b since this can be shifted by the accuracy with which data could be read from plots where tabulations were not present. Such data reading errors, should however generally introduce no more than 10% error in b. No standard deviation means that only a curve or equation was given, so no least squares fitting of data could be done.

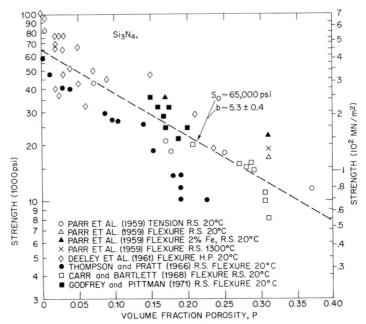

Fig. 22. Strength versus porosity for Si_3N_4. This compilation indicates the similar strength–porosity trend often observed for a given material from various investigators. Variations between investigations and scatter within investigations are due in part to different measuring techniques, sample sizes, and pore character.

value of $b = 4.3$ from tensile strength data on well over a hundred powder iron specimens as a function of porosity ($0.01 \leq P \leq 0.35$) as discussed below.

The limited data on ceramics at elevated temperatures indicate that the porosity dependence of flexure (or tensile) strength does not change significantly at least until $T \geq Tm/2$. Thus, for example, compare data of Table XIII with that of the same investigators data in Table XII. Similarly, Evans and Davidge (1970) found that while the strength of their denser ($\rho = 2.55$ gm/cm^3) reaction sintered Si_3N_4 body slowly decreased to their upper test temperature (1800°C), the strength of their lower-density materials ($\rho = 2.13$ gm/cm^3, tested to only 1400°C) was independent of temperature, showing little change in the porosity dependence of strength.

Neuber and Wimmers' (1968) 99.5% Al_2O_3 strength at 200°C intervals from -200–1000°C (plus -78 and 20°C) indicates that b does not change significantly; i.e., $b \sim 3.0 \pm 1.5$ for $G \sim 10 \pm 5$ μm and $b \sim 4.6 \pm 1.6$ for $G \sim 20 \pm 5$ μm. These higher values of b and their greater scatter in contrast to their $E–P$ data ($b = 2.6 \pm 0.2$) (Section II,B,1) is due substantially to the fewer number of values and more limited porosity range for each of the two grain size ranges.

TABLE XIII

High Temperature Ceramic Tensile Strength–Porosity Dependence

Material				$S = S_0 e^{-bP}$	
Composition (reference)	Fabrication[a]	G (μm)	Test temperature	S_0	b
ZrB_2 (Clougherty et al., 1968)	H	6–11	800	60	0.4 ± 1.8
			1400	40	2.8 ± 0.8
			1800	26	1 ± 1
Al_2O_3 (Passmore et al., 1965)	H	2.2	1200	48.4	9.29
		30	1200	9.7	0.33
BeO (UOX) (some preferred orientation of grains)	S	20	500	40	4 ± 1
(Fryxell and Chandler, 1964)		50	500	27	3 ± 2
		20	1200	34	4 ± 1
		50	1200	31	3.6 ± 0.2
BeO (AOX) (randomly oriented grains)	S	20	500	41	3.7 ± 0.4
(Fryxell and Chandler, 1964)		50	500	26	2.5 ± 5.0
		20	1200	23	2 ± 2
		50	1200	24	2.4 ± 1.5
UO_2 (Knudsen et al., 1960;	S	20	1000	40	6.9 ± 0.6
Burdick and Parker, 1956)		35–60	1000	15	2 ± 2
Sagger clay (Heidl and Pendergast, 1927)	S		750	3.6	1.7 ± 0.5

[a] H = hot pressed, S = sintered.

The approximate average b values ($\pm$ slightly more than 1 S.D.) for all flexural (or tensile) strength data in Table XII, are shown in Fig. 23. Also shown there are curves for theories from Table X that require one (i.e., S_0) or no empirical constants, but may require an assumption on the nature of the porosity dependence of other properties on P. Harvey's theory for simple cubic packing of uniformly sized spherical pores falls particularly high. Bailey and Hill (1970; Bailey, private communication) applied Harvey's model to mixtures of two or more pore sizes by considering the spacing between pores of like size (assuming a simple cubic arrangement of each set) in terms of the spacing if all pores were of the larger size. Then, assuming that failure occurs from the smaller pore, because of higher stress there, and that the maximum separation is the spacing of the smaller pores only, they showed that this can bring Harvey's model down into the range of experimental data, and in fact below it. While this approach indicates an importance of pore size distribution, it is basically wrong, since it places the emphasis on the fine

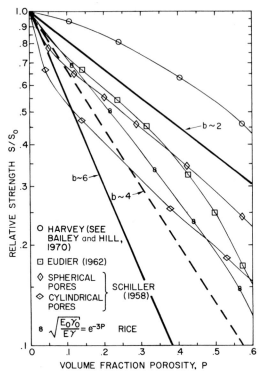

Fig. 23. Relative strength S/S_0 versus volume fraction porosity. The heavy dashed and solid lines represent, respectively, the approximate average and ± 1 standard deviation of the data in Table XII. The four theories not requiring arbitrary assumptions of a critical parameter are plotted, with all but Harvey's showing reasonable agreement with the data. Rice's model is shown assuming that the porosity dependence of Young's modulus and fracture energy are represented by average b values of 3. Use of $b = 4$ for the porosity dependence of E and γ as shown in Tables IV, VIII, and XIX would give near identity with $S = S_0 e^{-4P}$, which was avoided for clarity.

rather than the larger pores. Fracture is clearly seen to originate from large, not small pores in a body having both types of pores and bodies having only fine pores are generally as strong as and often stronger rather than much weaker than an otherwise comparable body with primarily large pores, as discussed later in this section.

Eudier's and Schiller's (spherical pore) theories average about the upper limit of strength data. Rice's model basically follows the porosity dependence of E and γ and hence typically falls within the central region of observed experimental behavior, as does Schiller's (cylindrical pore) theory, except for its initial rapid drop. Salak *et al.* (1974) considered fifteen strength porosity relationships, selecting eight for comparison with their extensive data for powdered iron specimens. Of the three of these that are in Table IX, Balshin and Eudier's equations were among the poorer fitting, while Knudsen's equation, with $b = 4.3$, gave by far the best fit of all equations.

2. Effect of the Character of Porosity on Strength

Table XII illustrates the extremely disappointing lack of characterization of most bodies other than for average porosity. Thus, with the limited exception of work of Fulrath and colleagues on porous glasses, there is generally no qualitative, let along quantitative, stereological data on the porosity or its relation to grain size. Only one or two authors (e.g., Steel *et al.*, 1966) have even commented on any possible porosity inhomogeneity in their bodies, and they made no attempt whatsoever to indicate the extent of this inhomogeneity or its effects on the strength–porosity results. Despite this unfortunate limited characterization, there are trends of data in Tables XII and XIII, as well as additional data that indicate that much of the variation in porosity dependence of strength is due to varying character of the porosity.

The first major trend of Table XII is for samples having more uniformly shaped and spaced pores to have lower strength–porosity dependences consistent with lower $E–P$ trends (Section II,B,1). Thus, for example, all of the silicate glass samples that contained spherical pores that were reasonably homogeneously distributed have low b values. This is not a characteristic of oxide glasses, since, as will be shown later, in Section V, silica glass samples made by processes that clearly must introduce substantial inhomogeneity of spatial distribution and shape of pores lead to very high b values in comparison to processes giving more uniform pores leading to lower b values. Similarly, the lower b values of most of the oxide-based materials listed under "Other ceramics" in Table XII is attributed to their typical liquid phase sintering tending to give more uniformly shaped pores and also limiting to some extent their spatial inhomogeneity. Further, the low to intermediate b value for fibers also supports the concept of inhomogeneous porosity accentuating the effect of porosity, since the small cross-sectional dimension of the fibers and the nature of their processing tend to limit inhomogeneous porosity within a given cross sectional plane (e.g., Fig. 27).

The second major trend is for pores entrapped within grains (e.g., due to grain growth) with higher processing or annealing temperatures to be less detrimental to strength than pores at grain boundaries. Thus, in Table XII, observe a general trend for the porosity dependence of strength (i.e., b values) to decrease as grain size increases. This is indicated to some extent over the limited range of grain sizes in the Cr_3C_2 data, very vividly in the Al_2O_3 data of Passemore *et al.* (1965), who were possibly among the first to note some aspects of this effect, and by Knudsen's ThO_2 and UO_2 data. Similar trends for the latter two materials are also shown in Table XIII. Further, much if not all of the data covering a range of grain sizes that is not fully consistent with this trend, generally appears to deviate for other reasons rather than a violation or nullification of this trend. Thus, for example, Fryxell and Chandler (1964) noted in their extensive study of BeO that the finer grain size bodies, especially the 5 μm grain sized bodies had poorer strength behavior, which they attributed to a considerably different porosity character, i.e., more intergranular pores of different shapes, size, and distribution than in the larger grain bodies. They also specifically noted the tendency for lower b values in their largest (i.e., 80 and 100 μm) grain bodies, which had many of the pores located within the grains. As noted in Section III,C, many fine to medium grain size bodies (e.g., Al_2O_3) often contain grains much larger than the average, which in the absence of other compensating factors will determine strength. Thus, one may often have bodies whose fine grain matrix has most of the porosity at grain boundaries but that is actually failing from isolated larger grains having less porosity, most of it within the grain. This can lead to a compromise effect wherein the porosity within the grain causing failure gives a lower porosity dependence, while porosity around the fine grains gives a higher porosity effect, e.g., because of elastic modulus and/or fracture energy effects. Further, local concentrations of pores alter b values whether the pores are inter- or intragranular or mixed.

Additionally, Rice's (1972a) study of MgO, where bodies of limited, i.e., closed, porosity showed a difference in behavior depending upon whether much of the porosity was within the grains or at grain boundaries. In the latter case, specimens were substantially weaker than specimens without any porosity, while those specimens with many of their pores within grains had substantially the same or greater strength than samples without any porosity. This increased strength due to pores located within grains was attributed to the fact that MgO typically fails by grain boundary crack nucleation due to dislocation motion. Thus, since pores tend to be sources of dislocations, a pore located within a grain and being a source of slip bands that impinge upon a grain boundary has a shorter distance between source and barrier and hence requires a higher stress for failure. The slip bands subsequently emanating from grain boundaries would be blocked by slip bands from pores, i.e., a work-hardening-type effect, reducing the chances of failure from slip bands emanating from grain boundaries. Also as noted earlier, Kessler et al. (1974), using very similar MgO specimens, observed that bodies having limited porosity, much of which was in the grains, have higher fracture energies than samples with no pores. It is also interesting to note that "Mother Nature" has apparently utilized pores located within grains or single crystals to give good strength to softer materials, e.g., where the pores can act as sources of slip or twinning. This is indicated by studies of Weber *et al.* (1969), who found that skeletal components of some sea animals (e.g., spines of sea urchins) consist of single crystals of calcium carbonate containing pores and that they have a higher strength-to-weight ratio than those of calcareous skeletons or rocks that are much denser and are strengthened by their poly-crystalline structure.

It is clear that since fracture is a weak link-type process, a local concentration of the number, size, or shape of pores must be a major factor in failure. Although this is complicated by inhomogeneous pore shape and associated spatial distribution Rice (1974a,b; submitted for publication) has shown a number of clear examples of fracture origins definitely initiating from regions of unusually high porosity. An example is shown in Fig. 24.

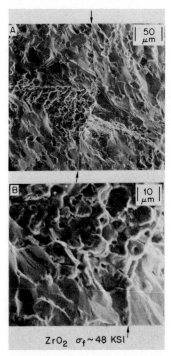

Fig. 24. Fracture initiation from a local region of high porosity in ZrO_2. (A) Lower-magnification SEM photo of the matching and aligned fracture surfaces of each half of a tested sample. Note the top and bottom arrows indicate the mating tensile surfaces. In this dense ZrO_2 body, fracture initiates from a local region of high porosity shown at higher magnification in (B). Note that $S = 48,000$ psi, $\gamma \sim 13$ J/m² (as for similar ZrO_2), and $E \sim 40 \times 10^6$ psi gives a surface half elliptical flaw width of ~ 50 μm for the observed $\sim 2:1$ depth-to-width ratio. In view of the angle between this porous region and the tensile axis, this is in surprisingly good agreement with the width of the porous region, i.e. its dimension parallel with the tensile surface, which should be the dimension of such an elongated flaw dominating strength behavior.

3. Pores as an Integral Part of Flaws

Turning now to other experimental evidence, not appropriate for Table XII, there is substantial support for the model treating pores as an integral part of the flaws themselves. Though not identifying specific fracture origins, the size of large pores in UO_2 (Evans and Davidge, 1969a; Roberts and Ueda, 1972), reaction sintered Si_3N_4 (Evans and Davidge, 1970), SiC (McLaren et al., 1972), Al_2O_3 (Evans and Tappin, 1972), and concrete (Wittmann and Zaitsev, 1972) plus the surrounding grain (or aggregate) size were shown to be generally consistent with flaw sizes predicted using the strengths, fracture energies, and elastic moduli in the Griffith equation. Further, Matthews et al. (1973), though again not identifying specific fracture origins, took a significant step forward in treating pores as an integral part of the flaws by demonstrating that the Weibull distribution of large pores within their SiC bodies matched the Weibull strength distribution.

Subsequently, Molnar and Rice (1973) have specifically shown fracture initiation from large, irregularly shaped pores in a variety of PZT bodies.

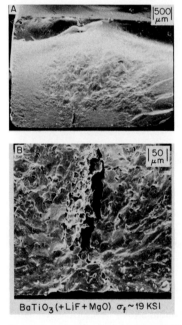

Fig. 25. Failure from a large, irregular, laminar pore in $BaTiO_3$. (A) Low-magnification SEM photo of the complete fracture surface; (B) higher-magnification photo focusing on the large irregular laminar pore from which failure originated. Note use of $E \sim 18 \times 10^6$ psi 6 J/m^2 and $S \sim 19,000$ psi yield a flaw size of ~60 μm, which is in good agreement with the halfwidth of this large pore and the porous region on the left, which should be the dimension dominating failure.

Rice (1974a,b; to be published) has extensively observed such initiation in a wide variety of ceramics, and Bansal *et al.* (1974) has shown this in Pyroceram 9606. Examples of large pores as the direct source of failure are shown in Figs. 25 and 26. In most cases, the sizes of the pores plus the grain size are generally in agreement with flaw sizes, i.e., predicting reasonable fracture energies. However, use of flaw size–pore size factors from Bowie's model for large pores in fine grain bodies yields calculated fracture energies over an order of magnitude too low.

Rice (to be published) has also shown examples of failure from a single void or a few voids smaller than the grain size. Fiber data also supports such failure from voids large or small in comparison to the grain size and further corrobarates the similarity of strength porosity relations for fibers with bulk materials. Thus, for example, Blakelock *et al.* (1970) showed that their weakest polycrystalline Al_2O_3 fibers failed from large pores, e.g., a single large

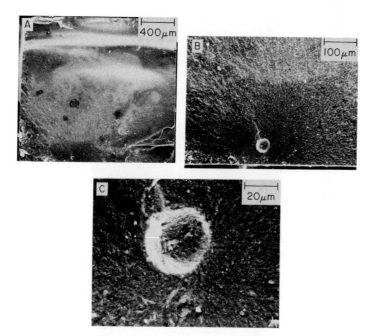

Fig. 26. Failure initiating from a large regular pore in MgF_2. Failure from such pores occurs occasionally in such fine-grain, optical grade, hot-pressed MgF_2. (A) The complete fracture surface, showing the pore approximately centrally located between surrounding symmetrical fracture features approximately radiating from the pore region. (B) An intermediate magnification showing finer fracture features approximately radiating from the void. (C) A high-magnification photo of the void showing fine fracture features eminating directly from the void.

centrally located pore occupying 60% or more of the fiber cross-sectional area at the point of fracture. Intermediate strength samples broke from centrally located pores about one-fourth of the cross-sectional area. Higher strength fracture surfaces did not show any obvious pores. Similarly, many of Simpson's (1971) polycrystalline, mostly Al_2O_3 fibers, failed from one or more pores located in the center of the fibers (e.g., see Fig. 27). The sizes of these pores plus part of the surrounding grains agree with expected flaw sizes (Fig. 27). Similarly, data of Fetterolf's (1970) polycrystalline fibers con-

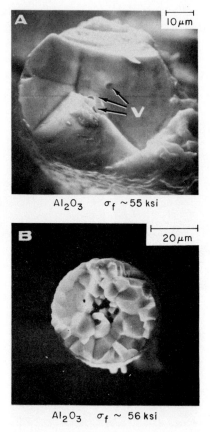

Al_2O_3 $\sigma_f \sim 55$ ksi

Al_2O_3 $\sigma_f \sim 56$ ksi

Fig. 27. Failure of Al_2O_3 fibers (Simpson, 1971). (A) the tensile strength of 55,000 psi, $E = 50 \times 10^6$ psi, and a flaw size of approximately 12 μm as indicated by the region encompassed by the three internal voids (v) yield a fracture energy of about 2 J/cm^2, which appears reasonable in view of the large grain size and intergranular failure. (B) Similarly, the approximately 16 μm, void core indicates a fracture energy of approximately 2–14 J/cm^2, i.e., somewhat less than the single crystal value, depending on how much of the surrounding larger grains were part of the flaw. This appears reasonable in view of the limited, e.g., 1–6 grain boundaries involved.

sisting of mostly alumina and mullite (70–78% Al_2O_3, 20–27% SiO_2) having very fine grain sizes, e.g., 0.05–0.20 μm, show similar results. Thus, if one uses the sizes of the large pores (e.g., 0.05–0.2 μm) plus the expected larger grain sizes, giving flaw sizes of 0.2–0.5 μm and their measured Young's moduli, fracture energies of 1–6 J/m^2 are calculated. Further, their electron microscopy shows that a few pores are often close enough for linking together, especially in bodies with finer pores, thus increasing flaw sizes at least two- to threefold. This raises calculated fracture energies to at least 3–8 J/m^2, in reasonable agreement with 11 J/m^2 for dense mullite ($G \sim 1$ μm) and 18 J/m^2 fine grain ($G \sim 1$ μm dense Al_2O_3) (Figure 15), considering the 10–20% porosity of the fibers and the uncertainties of the composition, pore, and grain sizes at specific fracture origins. Thus, while further work is needed, the concept of pores as an integral part of the flaws causing failure appears to be extremely sound.

4. SUMMARY DISCUSSION OF POROSITY–STRENGTH EFFECTS

While further theoretical development is needed, the concept of pores as an integral part of flaws appears sound. It may have considerable applicability to high-temperature failure where grain boundaries sliding may create and be accentuated by grain boundary pores. Statistical prediction of size and clustering of pores is needed. However, until such models are developed, treating pores as an integral part of the flaws is more of a post failure analytical tool rather than a predictive one. Thus, further development of other models based on the experience from past ones is need for predictive purposes. Also, the question of the dependence of S_0 on various factors needs further attention.

Experimentally, while there is substantial scatter, most data falls in a limited range of b values (4 ± 2). Variation in pore shape and inhomogeniety in spatial distribution are probably the most important factors varying porosity effects.

It is extremely unfortunate that these obvious and dominant factors have been clearly addressed in only a few studies. Pore location, which is relevant only to the important and common case of pores smaller than the grain size, should generally be the second most important aspect of pore character. Triple point pores are worst, other grain boundary pores—intermediate and intragranular pores—the least detrimental to strength. Pore shape is of direct importance when the pores are large in comparison to the grain size, so pore location and shape effects tend to be somewhat mutually exclusive. However, Molnar and Rice (1973) have shown lenticular pores must have aspect ratios of the order of 10 to result in a strength difference of 2. Pore shape always indirectly effects strength through its effect on E and γ.

Finally, grain boundary sliding and separation effects, which should be favored by grain boundary pores at higher temperatures, represent similar failure modes and hence suggest that the pore flaw model might be extended to higher temperatures.

C. The Grain Size Dependence of Tensile Strength of Ceramics

1. BACKGROUND AND THEORIES

The central role that tensile strengths play in many applications of ceramics makes the significant grain size dependence of such strengths very important. Strengths at or near room temperature are reviewed first and most extensively because of their wider study and greater use. The grain size dependence of short-term loading behavior as temperatures increase will be reviewed at the end of this section. Long-term, i.e., creep, behavior will be briefly considered in Section VI. The two basic theories and their application to the interpretation the grain size dependence of tensile or flexural strengths of ceramics, especially at lower temperatures, has been reviewed by Rice (1972b), Bailey and Hill (1970), and Carniglia (1972b). However, major changes in analysis and intrepretation of lower, e.g., room temperature, behavior have occurred in the past few years. These changes will be discussed after first briefly reviewing the two theories and their earlier use.

The oldest theory of the tensile strength of ceramics, derived by Griffith, is for failure from flaws, and commonly has the form

$$S = Y(E\gamma/a)^{1/2} \tag{2}$$

e.g., $Y = 1.12$ for a surface half penny crack and $Y = 1.25$ for an internal penny crack. Griffith originally developed this theory to explain the strength of glass in which microstructure like that of polycrystalline ceramic bodies is typically not present. In order to apply this theory to polycrystalline bodies, authors have typically assumed that there is some relationship between the crack size and the grain size. In the past, this assumption has most commonly taken the form that the flaw size equals the grain size. Later in this section, this will be shown to be frequently quite erroneous.

The second theory is for microplastic failure, i.e., failure from slip or twinning, often without any macroscopic plastic deformation, and has the form

$$S = \sigma_c + BG^{-1/2} \tag{1}$$

This equation, mainly developed by Hall (1951) and Petch (1953) to explain the brittle behavior of metals, is analytically derivable from dislocation mechanics. It reflects the fact that more dislocations in a slip band can pile up at a grain boundary to more readily nucleate a crack as the source–barrier distance increases so that the dislocation source experiences less back stress from the pile-up. Clearly, the maximum and common source–barrier distance is the grain diameter giving the grain size dependence. Twinning also obeys the same form of equation. It should be noted that a few equations have been proposed to explain the grain size dependence of metals that have a G^{-1} dependence. However, as Hirth (1972) points out in his review of grain boundary effects on mechanical properties, the weight of evidence clearly favors the $G^{-1/2}$ dependence.

While it would at first seem simple to determine when strength behavior is controlled by either of the above equations, there are four basic complications:

1. Controlling mechanism can change with test conditions, e.g., temperature, strain rate, and possibly atmosphere, as well as basic material properties, surface finish, and impurities.

2. The effects of microstructure, especially of voids and impurities, must be sorted out. Similarly, effects of grain shape, and of the distribution of grain sizes must be considered, with the latter being particularly important as shown later in this section.

3. Effects of added or inhomogeneous stresses from phase transformations, elastic anisotropy (EA), and thermal expansion anisotropy (TEA) and impurities can be significant.

4. The nature (i.e., size, shape, location) of flaws and the appropriate fracture energies for flaw failure are often uncertain.

An additional complication is that besides competing, the two mechanisms can cooperate; i.e., dislocations can interact with existing subcritical cracks to grow them to critical size (e.g., see Clarke *et al.*, 1962, and Rice, 1972b). While some of these factors also affect the porosity dependence of strength, grain size dependence is more directly affected. Thus, for example, while deformation occurs in grains and is affected more by grain boundaries than by pores, some pore effects are also related to grain size (e.g., see Section IV,A) and elastic and thermal expansion as well as phase transformation stresses are more concentrated at grain boundaries. The above effects will be discussed in this or subsequent sections. However, first previous applications of each of the above mechanisms are briefly reviewed.

Knudsen (1959), as noted in Section I,B, made a valuable contribution to the analysis of the microstructural dependence of ceramic strengths in making the first real quantitative attempts to analyze both grain size and porosity dependence. He generalized Eq. (1) to $S = S'G^{-A}$, where S' is the strength at $G = 1$ μm and A is a constant, based on the fact that most data gives straight lines on a log–log plot, so such strength grain size plots fit his exponential equation. It is interesting to note that Knudsen rejected Eq. (2) despite the fact that five of the six strength–grain size studies he based his analysis on were for brittle metals. Because of the initial success of his results, many investigators used his strength–grain size equation, typically obtaining $A \neq \frac{1}{2}$, usually $A < \frac{1}{2}$, focusing much attention on the meaning of A. Knudsen suggested that $A \neq \frac{1}{2}$ meant that the flaw size was proportional to G^{-2A}, which as will be shown in this section is partially correct. However, Knudsen's suggestion was not seriously investigated, and hence it was not recognized that his equation is not particularly favorable for recognizing the flaw and fracture energy changes with grain size.

Carniglia (1965) was apparently the first to use Eq. (2) in attempting to explain the grain size dependence of ceramic strengths. He reported that Al_2O_3, BeO, and MgO data fitted a two-branched curve on a Petch plot; i.e., the large-grain data fitted Eq. (1) and finer-grain data Eq. (2). He therefore hypothesized that there was an intrinsic change from a flaw to microplastic mechanisms of failure as grain size decreased. Rice (1972b) made a more extensive survey, noting some intercept errors and that negative slopes Carniglia obtained (e.g., for BeO) were probably due to variable porosity and related correction effects. Further, since such branching or change in strength–grain size behavior depends on machining and other factors (e.g., thermal history) it is not an intrinsic change, at least in the range of strengths commonly encountered. Carniglia (1972) has since surveyed much ceramic strength–grain size data, showing that it fits two-branched curves on a Petch plot as per his previous analysis, and he again interpreted this in terms of a flaw mechanism in large-grain bodies changing to a microplastic mechanism in fine-grain bodies. However, the weight of evidence now shows that the fine-grain behavior is generally also due to flaw effects as discussed below.

2. ROOM TEMPERATURE TENSILE STRENGTH–GRAIN SIZE BEHAVIOR OF CERAMICS

Three general advances in understanding flaw failure of ceramics at and near room temperature that significantly expand the strength behavior attributable to flaws have occurred. Fractographic determination and characterization of fracture origins has been a major factor in these advances. These three advances are reviewed, roughly in order of increasing importance, then microplastic failure is reviewed.

The first advance is gaining substantial understanding of the effect of internal stresses from phase transformations and thermal expansion anisotropy. Kirchner and Gruver (1970) suggested that A of Knudsen's equation was determined by elastic and thermal expansion anisotropy. However,

Rice (1970a) showed that while such anisotropies affect A, they do not determine it. He also pointed out that $A \neq \frac{1}{2}$ means there is a nonzero intercept on a Petch plot, which might also have been expected from Knudsen's use of mostly metal data fitting Eq. (1). Rice (1972b) subsequently showed that stresses from thermal expansion anisotropy lower intercept values and can in fact result in negative intercepts. Hence, since Petch intercepts are related to A of Knudsen's equation and stresses from TEA modify the intercept, they affect A, but the primary control of A is the intercept, not its modification. Prochazka and Charles (1973) and apparently Rhodes and Cannon (1974) misinterpreted Rice's discussion of internal stresses modifying positive Petch intercepts, citing such stresses as a cause of positive nonzero intercepts rather than only as a modifier of them. Rice also showed that TEA stresses—estimated as $\Delta \alpha \, \Delta T \, (E/2)$ where ΔT is the difference between the test and fabrication or stress relief temperatures— were commonly in the 10,000–40,000 psi range, in reasonable agreement with the few measured values and differences in Petch plot intercepts between bodies (e.g., of Al_2O_3 and BeO) having random or preferred grain orientations.

More recently, Pohanka *et al.* (1976a,b) have shown that dense $BaTiO_3$ with typical ground or polished surfaces, i.e., flaw sizes of about 10 μm, had strengths about 10,000 psi lower below the Curie temperature (T_c) than above. This approximately 50% reduction in strength was attributed to the development of phase transformation (cubic to tetragonal) stresses below T_c, which are identical in nature to TEA stresses. This reduction in strength is consistent with average transformation stresses estimated from dielectric effects. While the reduction in strength was independant of grain size, Pohanka *et al.* (1976b) subsequently showed that it depended on the flaw size, e.g., going to zero at a flaw size of about 150 μm with a grain size of about 1 μm. This is consistent with the fact that transformation as well as TEA stresses consist of both tensile and compressive stresses that must average to zero over a sufficient number of grains. Therefore, such stresses add to the applied stress to aid failure when flaws are not too large in comparison to the grain size, so a flaw associated with greater than average local tensile stress due to statistical grain orientation effects can more readily cause failure. This flaw size–grain size effect is presumably the reason Fryxell and Chandler (1964) found their oriented (c axis parallel to the tensile and extrusion axes) BeO specimens were not significantly stronger than unoriented ones until the grain size was over 50 μm. Oriented specimens were about 30% stronger than the unoriented ones when grain sizes were 80–100 μm.

The second advance is the recognition of the major role that grain size distribution can play in determining strength. Rice (1972b) first noted that since strength decreases with increasing grain size, the largest grain should represent the weakest link and hence control strength. Davidge and Tappin (1970) also noted in their treatment of large pores plus part of the surrounding grains as flaws in Al_2O_3 that calculated flaw sizes were similar to the largest pore combined with the largest grain size. Rice (1974b,c) has subsequently shown extensive occurrence of failure initiating from single or clustered large grains, e.g., some $BaTiO_3$ and B_4C and nearly all Al_2O_3 bodies. He has pointed out that when flaw sizes are about the size of or smaller than large grains or groups of large grains, they will not only be the source of failure, but they will control strength. This can often occur even if only a few or one such grain or group are present in regions of high stress.

Rhodes *et al.* (1973) also demonstrated that many fractures of their hot-pressed Al_2O_3 samples initiated from single or clustered large grains. The latter often graded from large grains at the center of the cluster out to the normal matrix grain size, apparently due to diffusion of centrally located impurity, such as a silica particle. Large grains in some of their Al_2O_3 powders may also have been the source for some of the large grains in their hot-pressed bodies. Prochazka and Charles (1973) have also observed failure from large, isolated grains in their dense, hot-pressed SiC. Unfortunately, they took the length of these large tabular grains as the flaw size, rather than the width along the fracture surface. The latter is much more appropriate, since for any elliptically shaped flaw, it is the smallest dimension that dominates the strength behavior. Thus, their flaw sizes are an order of magnitude or more high because of the large aspect ratio of their grains. Recognition of this would remove the positive nonzero intercept as well as give a fracture energy close to that expected for single crystals in agreement with the concepts discussed here. More recently, Freiman *et al.* (1975b) have shown preferential failure from larger grains in ZnSe, and Rice (to be published) has observed failure of CaF_2 having flaws 10–50 μm in size in grains several hundred microns in size.

Large grains in the interior of specimens can apparently cause failure only if they are associated with cracks, second phases, or pores, with the latter being very common. Large grains on the surface of specimens will commonly contain machining flaws for failure, but may also have pores etc. associated with them, and these can act independently or cooperatively with machining flaws in the grains. Examples of failures from larger grains are shown in Figs. 28 and 29.

The common occurence of large grains, which are sources of failure, is a major factor for much of the apparent lack of grain size dependence in some studies. Thus, for example, this is probably one of the reasons the Al_2O_3 bodies of Schofield *et al.* (1949) did not show significant decreases in strengths until $G > 100$ μm, since finer grain bodies probably contained larger ex-

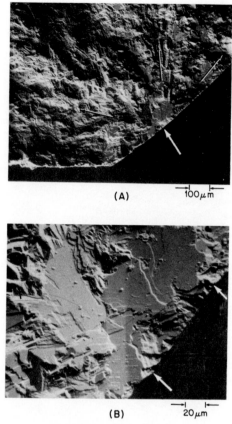

Fig. 28. Fracture initiation from a large grain in CVD ZnSe. (A) the failure initiation site on the rounded corner of a flexure sample is shown by white arrows, and gross fracture marks used to determine the fracture origin site are indicated by black arrows. Note that the flaw is substantially smaller than this larger than average grain, as shown in the higher magnification photo (B).

aggerated grains. Similarly, such grains, which were in many of Cutler's (1957) Al_2O_3 bodies, are probably a major factor in strengths of their 1–5 μm bodies being similar to those with grain sizes to about 400 μm, with strengths decreasing significantly only in bodies with $G > 600$ μm. Decreasing strengths at the very larger grain size is probably due in part to greater strain energy from TEA within each grain as discussed earlier.

The third advance is the recognition of flaw size–grain size and related fracture energy changes. As noted earlier, Rice (1970a) showed that $A \neq \frac{1}{2}$ in Knudsen's equation meant a nonzero intercept on a Petch plot. Initially, he interpreted such nonzero intercepts as being due to microplastic failure,

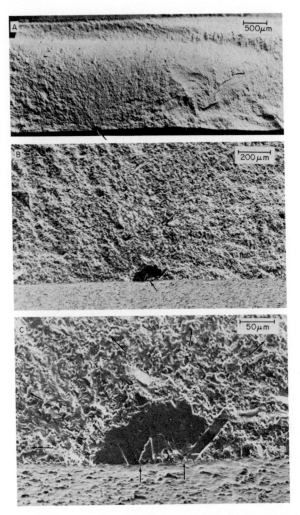

Fig. 29. Failure from large grains on the surface of Al_2O_3. (A) Low-magnification photo showing the large grain at the origin and surrounding fracture features in dense, hot-pressed Al_2O_3. (B) Large grains at the origin of failure at higher magnification. Note that $E \sim 60 \times 10^6$ psi, $S \sim 58{,}000$ psi, and the ~ 35 μm depth of an apparent crack (lower arrows) intersecting the grain boundary between the largest grain and several abutting large grains yields a fracture energy of ~ 11 J/cm^2, which is between the grain boundary and polycrystalline value, but closer to the latter.

as Carniglia had. However, he subsequently recognized that nonzero intercepts on Petch plots also resulted from flaws being larger than the grain size (e.g., Fig. 30). This is readily seen, since if the flaw sizes in a material are larger than, and hence independent of, grain size, a plot of strength versus

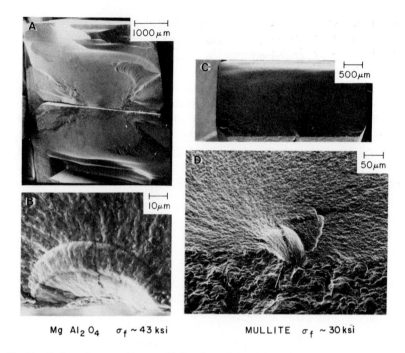

Mg Al$_2$O$_4$ $\sigma_f \sim 43$ ksi   MULLITE $\sigma_f \sim 30$ ksi

Fig. 30. Failure from surface machining flaws larger than the grain size of fine-grain MgAl$_2$O$_4$ and mullite. Complete matching fracture surfaces of a fractured MgAl$_2$O$_4$ sample. The fracture origin shown at higher magnification in (B) is at about the center and focus of the fracture ridges in (A), i.e., near the center of the fracture at the junction of the tensile (mating) surfaces of the two fracture halves. Note that $E \sim 40 \times 10^6$ psi and $\gamma \sim 6$ J/cm^2 (polycrystalline value), and $\sim 43,000$ psi yields a flaw size of ~ 25 μm, in reasonable agreement with the observed flaw size. (C) Lower-magnification photo of a fracture origin (arrow) and surrounding fracture topography approximately radiating from the origin in a fine-grain, dense mullite sample. (D) Higher-magnification photo of the matching fracture half. Note that the origin (arrow) is at a considerable angle to the tensile axis, a case that is not too uncommon. Use of $S \sim 30,000$ psi, $E \sim 32 \times 10^6$ psi, and $\gamma \sim 11$ J/m^2 (polycrystalline value) gives a flaw size of ~ 60 μm, which is in reasonable agreement with the observed flaw size, considering the orientation of the flaw relative to the tensile axis.

$G^{-1/2}$ will be a horizontal line clearly having a nonzero intercept. Thus, such intercepts need not have any relationship whatsoever to microplastic mechanism of failure. Rhodes *et al.* (1972, 1973) also suggested flaws being larger than the grain size as one possible reason for nonzero intercepts on Petch plots of their Al$_2$O$_3$ data.

The author has observed that for a given machining operation flaw sizes remained approximately constant for dense bodies of a given material regardless of grain size. Thus, flaws are smaller than the grain size in larger grain bodies, commonly giving a zero Petch plot intercept, and larger than fine grains, giving a nonzero intercept. The change between these two types of behavior is when the flaw and grain size are about the same. Rice (1974b,c, 1975) has further developed this concept of changing flaw-to-grain size ratios leading to the strength–grain size model sketched in Fig. 31. The major uncertainty in this model is the location of the two boundaries separating the three regions of different fracture energy control of strength. It is clear that flaws within a grain, but sufficiently smaller than the grain, lead to totally catastrophic behavior by the time the crack reaches the grain boundary; i.e., the momentum of the material on either side of the flaw becomes so high that the increase in fracture energy as the crack approaches or passes through the grain boundary cannot arrest crack

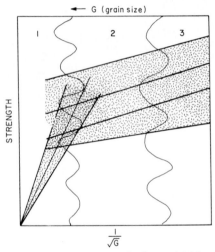

Fig. 31. Strength versus inverse square root grain size model (after Rice, 1974c). Two basic regions of strength versus grain size behavior are indicated. In the large-grain region, strength increases rapidly with decreasing grain size and may show some dependence on surface finishing. At finer grain sizes, strength increases much more slowly or not at all with decreasing grain size, the level and possibly the slope of strength being dependent on surface finishing. The transition between these two regions also depends on surface finishing and should occur when the grain size and flaw size are approximately equal. This need not be an abrupt transition but may in fact be a more gradual transition in contrast to that sketched here. Three regions of different fracture energy control of strength (indicated by wavy lines designating the uncertainty of the exact demarcation) are: (1) strengths controlled by fracture energies at or near those of single crystals or bycrystals, depending upon whether flaws are within grain or along the boundaries of large grains; (3) polycrystalline fracture energies control strength since flaw sizes are substantially larger than the grain size. Region 2 is a transition between regions 1 and 3. Note that if any single large grain or group of larger grains are available in which to contain a flaw, then those strenths should be plotted in region 1 regardless of the average matrix grain size. Published with permission of Brook Hill Publishing Company. Note that some key aspects of the two-branch $S - 1/G^{1/2}$ curves were suggested by B. R. Emrich (review of devitrified ceramics, Air Force Tech. Devel. Rep. No. ML-TDR-64-203).

propagation. One can also see that strengths of large grain bodies are not controlled by polycrystalline fracture energies, but some lower value by the fact that as grain size increases, a point is reached where flaw sizes calculated using polycrystalline fracture energies are smaller than a single grain, which is clearly contradictory. On the other hand, use of single crystal fracture energies gives a flaw size smaller than the grain size, which is reasonable. However, as the flaw sizes increases relative to the grain size, i.e., a/G increases, a point must be reached where the crack can no longer continue to propagate due to increasing fracture energy as it approaches or passes through the first or succeeding grain boundaries. Similarly, a transition from grain boundary fracture energies to polycrystalline fracture energies must depend on the a/G ratio for intergranular flaws. Clearly the uncertainty in the boundary between regions 1 and 2 leads to an uncertainty in the boundary between regions 2 and 3.

There is considerable evidence that strengths of large grain bodies are controlled by fracture energies at or near those for single crystals, an idea apparently first suggested by Wiederhorn (1970). Rice and McDonough (1972a) were apparently the first to observe direct evidence of failure from flaws smaller than the grain size in large grain $MgAl_2O_4$ along with the strength–grain size behavior like that of Fig. 31. This work in fact started the development of the model of Fig. 31. They also interpreted (1972b) similar strength and fracture evidence in ZrO_2 as possibly being due to similar flaw size–grain size–fracture energy effects. Freiman et al. (1974b) also suggested that the strength of many aluminas are controlled by single crystal fracture energies. This was based on the observation that most strengths did not correlate with polycrystalline fracture energies, while single crystal fracture energies gave flaw sizes smaller than the larger grains in many bodies and gave a better correlation with strength.

Rhodes and Cannon (1974) analyzed the strength–grain size behavior of their hot-pressed aluminas that failed from isolated clusters of larger or single large grains, obtaining fracture energies of $6-10$ J/m^2, i.e., in the range expected for single crystals. They thus concluded that the above concept of using single crystal or grain boundary fracture energies for bodies with large grain sizes and polycrystalline fracture energies for bodies with only fine grains with a substantial transition in between was applicable to their data. Freiman et al. (1975b) have extensively shown that fractures originating from larger grains in CVD ZnSe (e.g., Fig. 28), are consistent with single crystal rather than polycrystalline fracture energies. As noted earlier, Prochazka and Charles' (1973) results for SiC samples failing from large grains, upon reevaluation, appear consistent with use of single crystal fracture energies. Also, Tresler et al. (1974) obtained evidence supporting the author's model from their machining studies of Al_2O_3 as discussed by Rice (1975c). Recently, Pohanka and Freiman (to be published) have shown that polycrystalline fracture energies of large-grain $BaTiO_3$ predict flaw sizes smaller than the grains but inconsistent with apparent flaws at fracture origins, while single crystal fracture energies are consistent with the latter.

Rice (1974b) has presented considerable data in support of his model, in particular of the transition in strength versus $G^{-1/2}$ behavior when $a \sim G$. Further, the grain size range (10–50 μm) over which Carniglia's (1972b) curves typically branched agrees with the expected flaw size ranges, providing support for Rice's flaw model in contrast to Carniglia's flaw–microplastic model. Further, data of Table XII supports Rice's model, especially when effects of large grains, as discussed earlier, are noted. Thus, for example, both Neuber and Wimmers' Al_2O_3 data and Fryxell and Chandler's BeO data indicates Petch plot branching at grain sizes in the range of 10–20 μm, which would be in the range of flaw sizes estimated from single crystal fracture energies.

As noted above, major uncertainties are the a/G ratio below which single crystal or grain boundary fracture energies apply and the range of transition to polycrystalline fracture energies. These are intimately related to strength–grain size behavior. Thus, when flaws are controlled by single crystal or grain boundary fracture energies, there should be little or no dependence of strength on grain size unless the flaw is sufficiently large so its associated stress concentration and hence its driving force are perturbed by surrounding grain boundaries, e.g., due to elastic anisotropy. Such effects would presumably not occur until the flaw size was near the grain size, indicating that the transition in fracture energies would begin when $a \sim G$. This would then explain the indicated continued increase in strength with decreasing grain size due to increasing fracture energy. On the other hand, when a/G is some fraction, the flaw may grow but be held up at the grain boundary, so propagation beyond this boundary becomes the strength-controlling step at or near polycrystalline fracture energies. Such a sequence would readily explain the grain size dependence of strengths at large grain sizes, since the grain size becomes the ultimate flaw before failure. However, this sequence makes it less clear what the cause of further strength increases as grain size decrease below the flaw size is. It is also uncertain how rapid the transition between fracture energies is. Thus, Rice (1974b) has also pointed out that if the transition starts at $a \sim G$, then there is a substantial number of grains (e.g., 30–150) that a flaw must extend over in order to complete the crystal or grain boundary to polycrystalline fracture energy control of strength. This is in contrast to Evans' (1973) estimate of about four grains. If this range is related to the flaw size over which internal stress effects go to zero, then Rice's estimate would be correct. Present strength–grain size data is not sufficiently detailed or accurate enough to distinguish between the above mechanisms. Thus, the range of grain sizes investigated is too limited and the scatter too high to clearly show whether there is a continuous rise of strength with decreasing grain sizes in the large grain regime, or at the other extreme, a region of constant strength then a rapid rise in strength with decreasing grain size.

Despite the above uncertainties, several major changes in strength–grain size behavior are established or highly probable. First, nonzero intercepts on Petch plots clearly can arise from flaws larger than the grain size, and need not have any relation to microplastic failure. Second, flaw sizes from which failure initiates in dense bodies are not directly related to the grain size except over a limited range. Third, this limited range of grain sizes where $a \sim G$ appears to be a major transition in strength–grain size behavior and the sizes at which this occurs depends on surface finish. Fourth, there should be an a/G ratio below which single crystal or grain boundary fracture energies control strength and a subsequent range over which the transition to polycrystalline fracture energy occurs. The a/G ratios for these transitions are as yet not well defined and may well depend on the local degree of grain orientation, as well as on the crystallographic dependence of fracture energy in different crystal structures. Thus, crystals having only a few planes of low fracture energy would presumably have lower a/G ratios for the above effects than structures having many low energy fracture planes. In any event, the dependence of fracture energy controlling failure depending on the a/G ratio provides an explanation for the apparent lack of agreement between strength and fracture energy dependences on grain size at large grain sizes.

Turning to microplastic mechanism of failure, MgO and CaO appear to be the only two refractory ceramic materials for which there is substantial evidence for such failure at room temperature. Rice (1968) directly showed microplastic control of MgO strength by finding actual fracture origins from slip bands piled up at grain boundaries. Rice (1972a) has since extensively reviewed MgO strength–grain size data indicating that MgO strength is controlled by microplastic behavior. Subsequently, Sinha et al. (1973) have shown other microscopic evidence for dislocation, crack nuleation control of strength in polycrystalline MgO. Evans and Davidge (1969b) have supported these results and have also shown that the failure of MgO can be changed from microplastic at fine grain sizes to flaw failure at large grain sizes by degrading its surface. More recently, Rice (1972c, 1974c) has indicated that one can have a transition from a microplastic mechanism of failure at larger grain sizes to a brittle mechanism of failure at finer grain sizes due to surface finishing effects, the opposite of what Carniglia originally proposed. Though less extensive, the data on CaO (Rice, 1969, 1972d) in conjunction with the analogous behavior expected between MgO and CaO show that CaO normally will have strength controlled by microplastic effects at room temperature.

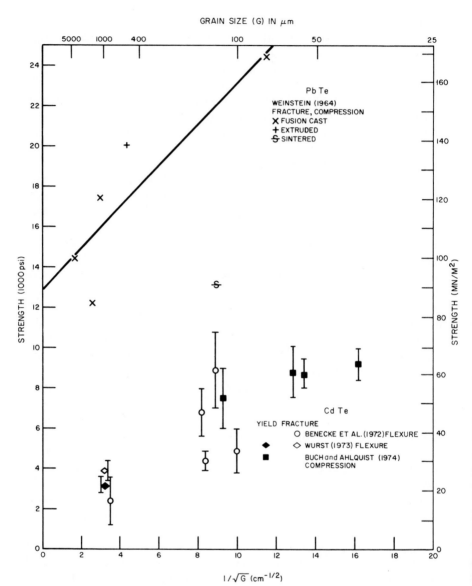

Fig. 32. Strength versus inverse square root of grain size for PbTe and CdTe. The PbTe data is consistent with the Hall–Petch equation (sintered PbTe data was neglected in drawing the least squares line, since it contained porosity and impurities). Observation of yielding in CdTe in both flexure and compression and the general agreement of these values as a function of grain size clearly shows its strenth is affected and in fact probably controlled by dislocation phenomena.

Turning now to other materials, the recognition that nonzero Petch intercepts do not necessarily mean a microplastic mechanism of failure requires that there be other substantial evidence for microplastic failure before concluding that this mechanism is operative. For example, data from single crystals indicating the stress levels at which microplasticity can occur are consistent with the observed Petch intercept would commonly be necessary, e.g., as shown for MgO and CaO. Most of the materials observed by Rice (1972b) to have a nonzero Petch intercept do not meet this criteria. In fact, it appears that microplastic mechanisms of failure are typically confined to materials of the hardness of MgO or substantially less, most commonly the latter. There is some limited evidence that under appropriate conditions $BaTiO_3$ having a hardness of 400–500 kg/mm^2 may exhibit some microplastic type control of strength (Walker et al., 1975), but otherwise, most materials appear to have hardnesses of the order of 100–200 kg/cm^2 or less to commonly exhibit microplastic mechanism of failure at room temperature, CdTe and PbTe (Fig. 32) being recently demonstrated examples. Even ZnSe, having a hardness of only about 150 kg/cm^2, seems to have its room temperature strength predominately, if not exclusively, controlled by flaws, temperature strength/(Freiman et al., 1975b).

3. ELEVATED TEMPERATURE–GRAIN SIZE DEPENDENCE
   OF TENSILE STRENGTHS OF CERAMICS

Increasing test temperatures increases the opportunity for micro- and, eventually, macroplastic mechanisms of failure due both to the reduction in stresses for such flow and to reducing the severity of flaws by rounding and healing. However, there are a number of complications in interpreting when the transitions from flaw to microplastic to macroplastic failure occurs due both to microstructure and to complexities of the flaw and microplastic mechanisms. These complexities are discussed and illustrated, then other limited data on general strength–grain size behavior as a function of temperature is covered.

As an example of the complexity of interpreting flaw to microplastic changes in mechanisms, consider the anomalous minimum in the strength of sapphire and polycrystalline Al_2O_3 at temperatures of a few hundred degrees centigrade reported by Congleton and Petch (1966) and Congleton et al. (1969). Assuming that microplasticity controlled room temperature strength, they suggested that increasing temperature reduces the stresses necessary for crack nucleation due to microplastic flow, lowering strengths until further increased plastic flow becomes extensive enough to blunt cracks, again increasing strengths. Neuber and Wimmers (1968) also observed a minimum and maximum in the flexural strength of $\sim 99.5\%$ Al_2O_3

bodies ($P \sim 0.8$, $G \sim 20$) at about 400 and 600°C, respectively. They observed no such effect in ZrO_2 ($P \sim 0.13$, $G \sim 25$), $MgAl_2O_4$ ($P \sim 0.09$, $G \sim 70$), or MgO ($P \sim 0.39$, $G \sim 65$). Wiederhorn et al. (1973) have verified this minimum in Al_2O_3 and showed that while water does enhance this effect, it is not the basic cause, since the strength minimum occurred in samples baked and tested in vacuum to assure that no water was present. They showed, however, that microplasticity generally does not occur at crack tips in sapphire until temperatures of about 600°C and more. This, combined with other recent evidence of flaw failure in Al_2O_3, rules out the Congleton and Petch explanation, indicating that some other thermally activated mechanism is involved. Becher's (1976) recent observations that interfaces between the matrix and twins introduced in the surface of sapphire by machining are preferred paths for crack formation and growth may be an alternate explanation. Since twins can grow on annealing, the decrease in strength in alumina with increasing temperature might be due to twin associated crack growth along the twin matrix interface, until some microplasticity at the tip of the crack could occur to make it more difficult for the crack to propagate. Since this phenomenon should occur only in bodies having grains large enough to contain twin–flaw combinations large enough to cause failure, it can be tested by seeing if the effect disappears when the grain size is decreased or the flaw size increased sufficiently so no single grain large enough to contain the necessary twin–flaw combination exists.

Grain boundary phenomena, especially when coupled with purity or porosity effects, complicate high-temperature behavior. Impurity effects are discussed in the next section and porosity effects in the last section. While it was noted there that porosity effects did not change significantly until about $T_m/2$, significant increases in porosity effects would be expected at higher temperatures, e.g., as indicated by compressive strength studies (Section V,B) [see also Stokes and Li (1963) and Day and Stokes' (1966) study of MgO]. Another recently discovered effect complicating the analysis of high-temperature strength is that of high-temperature slow crack growth due to grain boundary separation. Apparently the first direct interpretation of results in terms of high-temperature slow crack growth was by Lange (1974b) in hot pressed Si_3N_4. More recently, Kinsman et al. (1974) have elegantly shown such crack growth in a lithium aluminum silicate glass ceramic material. They artificially introduced larger flaws in samples using hardness indents, stressed them for varying periods of time at controlled loads at elevated temperatures, then cooled them to room temperature where they were fractured. Using fractography, they then determined the region of slow crack growth and its test dependence (e.g., Fig. 33). While both of the above examples are related to impurity, e.g., glassy phase effects that accentuate this phenomena. Along with grain-boundary voids, this

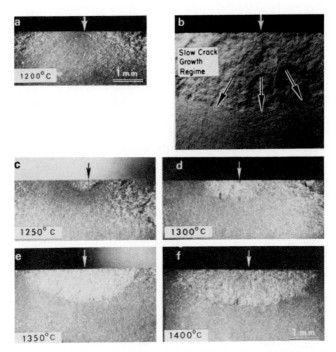

Fig. 33. High-temperature slow crack growth in Si_3N_4. Fracture surfaces of hot-pressed (HS-130) Si_3N_4 specimens that each had a flaw introduced on the tensile surface (indicated by an arrow pointing to the flaw on the tensile, i.e., top surface using a Vickers indenter (4 kg load), then stressed at elevated temperatures, unloaded, and cooled to room temperature where they were fractured. (a) Room-temperature fracture of a specimen previously stressed at 1200°C where no slow crack growth occurred. (b) and (c) Room-temperature fracture surface of specimens stressed at 1250°C where substantial slow crack growth occurred (indicated by arrows in the higher magnification of SEM photo in (b) and the light semicircular region in the optical photo in (c). (d), (e), and (f) are optical photos of specimens stressed at the higher indicated temperatures before fracture at room temperature. Note the increasing semicircular (light colored) regions of slow crack growth. (a), (c), (d), (e), and (f) are to the same scale. Stressing at 1250°C corresponds to the region of rapid decrease of strength of unnotched specimens as a function of temperature, and a peak of strength of cracked (indented) specimens as a function of temperature. The latter peak is due to strengths of cracked specimens initially increasing due to limited crack blunting. Photos courtesy of K. Kinsman, Ford Motor Company.

probably represents a basic failure mechanism in many materials, e.g., those whose plastic deformation of grains is limited. Such separation may be a more severe problem in finer-grain material.

A final complication is the reduction of internal stresses from TEA, i.e., in noncubic materials, as temperature increases. Nonzero intercepts on Petch plots, which will occur in noncubic materials, increase if the flaw size is not too large in comparison to the grain size, can easily be mistaken for the onset of microplastic effects. Thus, for example, Rice (1972b) has shown strengths (and hence intercepts) of BeO (hexagonal) rising with temperature to about 1200°C in contrast to little or no changes in strength of cubic ThO_2 or UO_2 data to 1000°C, only a limited drop in CaO strengths to about 1300°C, and the strength of MgO showing limited decreases by 1315°C (Rice et al., 1968; Rice, 1972a). Further, data on the BeO as well as on an increase in strength of ZrB_2 with initial temperature increases is seen by comparing data of Tables XII and XIII. Also, the rise in strength of Al_2O_3 after the initial decrease discussed earlier, as well as some increases in strength noted later, are most likely due to reduced internal stresses.

Turning now to the general strength–grain size trends with temperature, only a few materials have had significant study of their strength as a function of grain size and temperature. Available data, including unpublished work of the author for Al_2O_3, one of the more extensively studied materials is shown in Fig. 34. Note that while most of this data is not sufficiently detailed

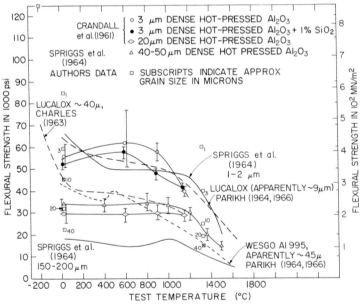

Fig. 34. Flexure strength of various grain-size bodies of Al_2O_3 having limited or zero porosity as a function of temperature. Note that for clarity not all the data of Spriggs et al. (1964) is shown. Their strengths for 10–15 μm grain-size bodies runs just slightly above those shown for Charles; approximately 40 μm Lucalox bodies and their data for 80–100 μm grain-size bodies would fall slightly below that for the approximately 20 μm grain bodies of Crandall et al. Note also that while there is some scatter due to differences in bodies and measurement techniques between different investigators, there is generally consistency of the data.

in the lower temperature region to detect the minimum in strength or rises due to reduced internal stress effects discussed earlier, some (e.g., Crandall's and Parikh's) do show evidence of the latter. In general, strength tends to follow approximately the Young's modulus–temperature trend. Finer-grain bodies, however, though tending to remain stronger than larger grain size bodies, drop more rapidly in strength.

Data on the strength–grain size–temperature trends for MgO are summarized in Fig. 35 from Rice *et al.* (1968) and Rice (1972a), with the latter showing good agreement with flexural data of other investigators. These strengths are also consistent with MgO tensile strength data of Day and Stokes (1966) from 1400–2150°C, which also indicate some decrease of strength with increasing grain size. Again, note that the strength–temperature trend approximately follows that of Young's modulus; i.e., strengths do not begin to decrease significantly until temperatures of about 1100°C. Again,

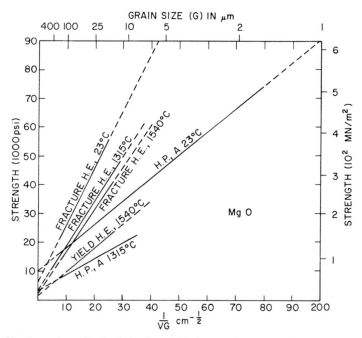

Fig. 35. Strength–grain size behavior of MgO as a function at different temperatures. H.E. means hot extrusion, H.P. means hot pressed, and A refers to annealing at temperatures of approximately 1100°C or more with the initial annealing to approximately 1000°C being slow. Hot-pressed samples did not exhibit any yield, so only fracture data is shown. Note that in general both the slope of the Petch plot and its intercept decrease with increasing temperature. Note also the significant increase of the slope but not the intercept as a result of texture in the material from hot extrusion (samples tested with tensile axis and ⟨100⟩ axial texture parallel).

finer-grain bodies, although stronger, loose strength more rapidly, though this can be interrupted some by macroscopic plastic flow.

UO_2 data of Evans and Davidge (1969) is complicated at lower temperatures by a change from flaw (pore) to microplastic (grain) control of strength at temperatures of about 700°C ($G = 8 \mu m$) and 1100°C ($G = 25 \mu m$). Beyond these temperatures, strengths decrease, with the finer-grain material showing the greater decrease, until there is a limited rise in strength starting at about 1300°C, with both bodies ending with nearly the same strength at 1500°C. Similarly, limited data on SiC (Fig. 18) and Si_3N_4 (Fig. 36) show little drop in strength to temperatures of 800°C or more.

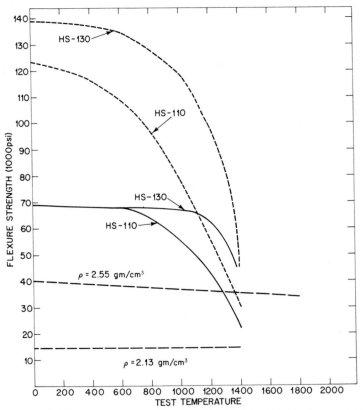

Fig. 36. Strength versus temperature of dense and porous Si_3N_4. Note the marked drop-off of strength of the dense hot-pressed material [(---) Richardson, 1973; (———) Lange, 1974] at higher temperatures in contrast to the reaction sintered material [(— — —) Evans and Davidge, 1970]. This is attributed primarily to the presence of a grain boundary silicate phase or phases in the hot-pressed material. Note differences in hot-pressed strengths due primarily to test differences; but the hot-pressed–reaction sintered differences are real material differences.

D. Effects of Grain Orientation and Shape on Tensile Strength

Elongated (e.g., tabular) grains often appear to affect tensile strength at lower temperatures due to one or more of the following effects. First, elongated grains can increase the area and hence energy of fracture, primarily in samples failing intergranularly. If flaw sizes are bigger than the grains, this will increase strength. Lange (1973a) has, for example, suggested that the higher (e.g., two- to fourfold) strengths and fracture energy (Fig. 10A) of Si_3N_4 hot pressed from α as opposed to β powder because of the more elongated grain structure resulting from the former. Rice and Freiman (to be published) have shown that CVDSiC (cubic), which has oriented, very elongated grains but fails transgranularly, has no significant anisotropy of strength or fracture energy. Thus, the intergranular failure of Si_3N_4, which may be due to its anistropy or its grain boundary phases, appears to be important to this effect (along with the fine grain size so $a > G$). Second, while tabular grains, e.g., from exaggerated grain growth in anisotropic materials such as Al_2O_3, are often associated with lower strengths, this may often be due to the greater size and not the shape of such grains. In fact, the greater surface area for a given volume of an elongated versus an equiaxed grain may reduce microcracking, e.g., from TEA, and hence give a higher strength for equivalent grain volume. On the other hand, if elongated faces of tabular grains result in increased contact between anisotropic grains having high differences in properties, e.g., thermal expansion, then strengths may be lowered.

Grain orientation can affect, usually increase, strengths whether failure is from flaws or microplasticity. The main effects of orientation on flaw failure stem from its effects on internal, e.g., TEA, stresses in anisotropic materials and orientation of elongated grains or weak fracture planes. As discussed earlier (Section III,C), Lange (1973a) has attributed an anisotropy of about 20% in fracture energy and strength of hot-pressed Si_3N_4 to orientation of elongated grains, with values being higher for fracture parallel with the hot-pressing direction, i.e., normal to the orientation of the elongated grains. More recently, Meyer and Zimmer (1974) have discussed the failure of graphite in terms of the probability of having several grains adjacent to one another, one or more pores, or both such that the preferred planes of weakness or existing microcracks are favorably oriented to constitute a larger flaw. This is an important concept, since it not only can increase the flaw size as Meyer and Zimmer indicate, but it can also lower the fracture energy to that for grain boundaries or single crystals. Freiman et al. (1975b) for example noted failure frequently associated with two grains of limited misorientation in CVD ZnSe. Failure from the combination of microcracks and possibly grain boundaries or cleavage planes giving the minimum stress intensity is probably the reason why the strength of highly anisotropic

$MgTi_2O_5$ drops continuously with increasing grain size (Fig. 12) despite initially increasing fracture energy.

Grain orientation clearly affects internal, especially TEA, stresses. If such orientation is not uniform, then oriented patches may act as large grains, which can lower strengths. On the other hand, uniform crystallographic orientation of grains will clearly reduce internal, e.g., TEA, stresses, and hence increase strengths if flaw sizes are not so large that they are uneffected by such stresses. Thus for example, Rice (1972b) has noted lower Petch plot slopes but higher intercepts due to orientation of Al_2O_3 and BeO specimens in contrast to unoriented specimens. The increased Petch plot intercepts (now attributed to flaws larger than the grains rather than microplasticity) correlated with estimated TEA stresses. Thus, for example, the intercept of hot-pressed Al_2O_3 was about 30,000 psi lower than press forged, i.e. oriented, Al_2O_3 at room temperature and about 10,000 psi lower at 1200°C, with the net values being lower at 1200°C as expected.

The indicated lower Petch plot slopes are less well understood but may reflect the combination of higher strengths at larger grain sizes and lower strengths at finer grain sizes of textured bodies relative to bodies of randomly oriented grains. Thus, in large grain bodies with flaws within grains or along grain boundaries, polycrystalline fracture energies do not control strength, so lowering of such energies due to orientation is not important. However, internal stress will directly aid failure, giving higher strengths in bodies with grain orientation due to lower internal stresses. In fine grain bodies with flaws much larger than the grains, internal stress will not effect strength, but polycrystalline fracture energies will be lowered by orientation, giving the lower strengths relative to bodies of randomly oriented grains.

The mechanisms and extent to which polycrystalline fracture energies might be lowered due to grain orientation is not clear. Thus, for example, the energy for basal plane fracture Al_2O_3 is at least four to six times that of most other planes (e.g., Becher, 1975), forcing fracture on other planes, increasing fracture roughness and energy over that for fracture on the lower energy planes. Since Al_2O_3 basal planes are oriented approximately perpendicular to the forging direction and only thinner forgings are made, fracture of forged specimens would thus involve almost no attempted basal plane fracture of grains. On the other hand, oriented BeO specimens have generally been fractured normal to the C axis texture rather than parallel with it as Al_2O_3 has. However, BeO (wurtzite structure) has the opposite thermal expansion anisotropy (i.e., $\alpha \| C < \alpha \perp C$) than Al_2O_3 (hexagonal), so it may also have different fracture energy anisotropy. Yet, Bentle and Miller (1967) report that cleavage does occur on the basal plane of BeO, but that prismatic cleavage is favored, so clearly more study and analysis of the indicated lower Petch plot slopes from orientation is needed.

Turning now to the effects of grain orientation on microplastic failure mechanisms, a sufficient degree or orientation could reduce the ease of dislocation or twin crack nucleation, and hence increase strength. Of course, extreme orientation, i.e., approaching a single crystal, could lower strengths due to reduction of grain boundaries inhibiting dislocations or twins. Rice *et al.* (1968; see also Rice, 1968, 1970b) showed the room temperature strengths of dense MgO with a $\langle 100 \rangle$ axial texture from hot extrusion were increased when the texture was parallel with the applied stress (type A) (Fig. 35) or normal to both the applied stress and the specimen tensile surface (type B). The third orientation, i.e., with the texture normal to the applied stress but parallel to the specimen tensile surface (type C), showed no increase in strength; i.e., the same as dense, unoriented MgO bodies. The increased strength of type A samples is clearly consistent with the fact that grain boundaries of lower misorientation, e.g., 5–15°, effectively block slip bands but are less prone, i.e., require higher stresses, to nucleate cracks from such blocked slip bands, since the slope but not the intercept of the Petch plots increased. The reason for the higher strengths of type B but not type C samples was not resolved, but it was suggested to be due to the orientation of high-angle tilt boundaries relative to the flexure axis, a difference that would not occur in true tension.

At higher temperatures, internal, e.g., TEA, stresses will be reduced, hence limiting the effect of orientation in further reducing such stresses. Similarly, the effect of orientation or reducing crack nucleation due to slip or twinning will be reduced, though this may require higher temperatures. However, at high temperatures, grain orientation may reduce the tendency for grain boundary sliding and hence improve strengths. Thus, the effects of orientation on strength may stay about the same, decrease some, then increase again, or increase steadily with increasing temperature. Thus, Al_2O_3 data indicate a reduced effect of orientation, while BeO may be about the same or show some reduction of orientation effects at 1200°C (see Rice, 1972b). MgO data shows no obvious change in the effect of orientation on strength to about 1300°C, the limit of comparative data (Fig. 35).

E. *Effects of Binders, Additives, and Impurities on Tensile Strength of Ceramics*

1 BACKGROUND AND THEORIES

The effect of additives or impurities on the tensile strength of ceramics depends on both the mechanism of failure and the distribution of the additive or impurity. Impurities or additives in solid solution in materials failing due to plastic processes (i.e., slip, twinning, or diffusion) can significantly effect strength, usually increasing it. On the other hand, solid solution effects on strength in materials falling in a brittle fashion from flaws will be realized only through possible solution effects on Young's modulus and fracture energy, or changes in hardness that may affect flaw sizes induced as a result of machining or other handling. These effects will typically be rather limited unless substantial solution content is present or complicated by other factors.

Three complications can occur with solution effects on either microplastic mechanisms of failure or brittle mechanisms of failure. First, inhomogeneous solution can result in stress concentrations, e.g., at grain boundaries if there is a boundary excess or depletion of the additive or impurity, which can often lower strengths. Second, species in solution may affect grain size and porosity, e.g., through effects on sintering. Third, solid solutions can lower (or increase) internal stresses from EA and TEA; e.g., Kirchner (1969) has significantly reduced or eliminated TEA of TiO_2, Al_2O_3, and SnO_2, usually with 20% or more solute, and from phase transformations, e.g., by solution of stabilization of ZrO_2. Whether reduction of these internal stresses will affect the strength of the sample can depend on the grain size and shape and particularly on the ratio of the flaw size to grain size, as discussed in Section IV,C.

The effect of second phases, e.g., inclusions, dispersions, precipitates, etc., in a body depends upon the shape, size, and particularly the distribution (including spacing and location) within the body. Second phases located within glasses, single crystals or a single grain, can inhibit crack propagation in all of these or in the latter two (see Section III) or hinder the motion of dislocations or twins to increase strengths. Second phases at grain boundaries as precipitates, dendrites, or films can be very significant. They commonly lower strengths, particularly as films, by weakening grain boundaries, especially for anion impurities such as fluorine, hydroxyl or carbonate ions. However, grain boundary phases can also increase strengths by relaxation of EA, TEA, or phase transformation stresses, increasing fracture energy, e.g., as with Fe or Zr metal grain boundary phases, respectively, in wustite and ZrO_2 (Section III,D), or possibly from enhanced boundary microcracking under stress. Rice (1974b) and Rice and Freiman (to be published) have suggested for example, that the latter may be an important reason for the high fracture energy of some composite ceramics such as hot-pressed Si_3N_4, which appears to have an additive derived grain boundary phase, e.g., a magnesium silicate.

Bache (1970) is one of the few investigators to analytically address the problem of grain boundary phases. As noted in Section II,A, his model based on spherical particles bonded together by a brittle phase can be expressed in terms of the volume of bonding layer as follows:

$$S = k(V_b/V_p)^{3/4}(\gamma E/G)^{1/2} \tag{4}$$

where k = a proportionally factor, V_b = the volume of the boundary phase (assumed to be spread uniformly over the spherical particles), V_p = the volume of the particles (or grains), E = Young's modulus of the particles, G = the particle (grain) diameter, and γ = the fracture energy of the boundary phase. Unfortunately, this model is based on limited contact between particles, e.g., high porosity, and assumes compatibility between the bonding phase and the matrix, i.e., no mismatch stresses, but it does provide a starting point. The effects of boundary phases can be complex, since several changes can occur at the same time. Thus, for example, a grain boundary phase may reduce internal stresses but provide a lower fracture energy path, or they may add internal stresses but increase fracture energy due to multiple cracking. Such phase may also be inhomogeneously distributed, leading in the extreme to local concentrations acting similar to increasing.

Bache also derived a model for the same conditions as above, only for a ductile bonding phase. This gives S equal to a constant times V_b/V_p. While this equation might be considered for poorly bonded cements, both it and his above equation could be used, respectively, for ductile and brittle binders for green bodies. Onada (1976) has specifically considered the binder problem. His solutions, which appear applicable only to ductile (i.e., nonbrittle) binders have the same form or Baches for completely coated particles. When particles are bonded only near their contact point, Onada obtains the strength as a function of $(V_b/V_p)^{1/2}$.

The effect of spacing of second-phase particles etc. on crack propagation have not been appropriately recognized. First, it is important to note that since mechanical failure is a weak-link process, a flaw in a region of maximum particle separation will represent the weakest link, i.e., the composite will preferentially fail from a region representative of the mechanical behavior of the matrix, not that of the composite. Commonly, the extreme separation of particles or fibers will be two or three times the average and can be substantially greater in randomly oriented fiber systems. Second, if the flaw size is smaller than or even comparable to the smallest spacing between particles or fibers, its propagation can become completely catastrophic before it ever has an opportunity to be significantly effected by the particles or fibers. This results from (1) the increasing excess strain energy for further crack propagation as crack growth occurs and (2) the more limited effect particles have on propagation of cracks with small radii of curvature due to the more limited changes in radii of curvature and hence line tension and fracture energy for small crack–second phase interactions (see Rice and Freiman, to be published; Lange, 1970) (Eq. 3, Section III,D). When inclusions or precipitates within grains are inhibiting plastic processes, then the size and spacing of the particles controls properties in a fashion similar to that found in metals (e.g., see Bansal and Heuer, 1973).

As second-phase particles become larger, they can begin to act as sources of cracks, which may either be a part or all of the flaw causing failure, or interact with cracks causing failure to either raise or lower fracture energies. The occurrence of cracks around inclusions only above certain sizes depending upon the nature of the inclusion and the matrix (e.g., see Lange, 1974a,b) is not predicted by normal stress considerations, since, for example, the stress due to difference in thermal expansion between the matrix and the inclusion is independent of the inclusion size. Davidge and Green (1968) successfully explained the dependence of cracking on inclusion size by an energy balance criterion similar to Griffiths; that is, they assumed cracking would occur around an inclusion when the total strain energy stored within the inclusion and the matrix was equal to or greater than that for cracking. Subsequently, Lange (1974a,b) has indicated that a crack around an inclusion will not propagate unless the product of (1) the square of the stress on the crack that includes both the applied stress and the local stress and (2) the particle size is greater than a constant value determined by parameters of the system. Evans (1974b) has shown that the inclusion size can be taken as the flaw size for a conservative prediction of failure if the shear modulus and thermal expansion of the inclusion are larger than the values of these parameters for the matrix. If the shear modulus and thermal expansion of the inclusion are smaller than that of the matrix, extensive cracking can develop at the inclusion and may lead to premature failure. Thus, for example, Evans notes that of the typical oxide, carbide, and nitride inclusions that may occur in Si_3N_4, only the silicate base inclusions are likely to be a problem. Similarly, he suggests that the use of B-based additions in fabricating SiC could present problems if they form B_4C inclusions.

Another important aspect of the effect of impurities or additives on the strength of ceramics is their effect on microstructure. Impurities or additives can either aid or retard densification and hence influence the amount of porosity present. They may also influence the shape and size of pores present, e.g., as noted earlier in the elastic property section, liquid phase sintering may often result in more uniformly shaped pores as well as possibly more uniformly spaced pores. Similarly, additives or impurities can either enhance or retard grain growth; e.g., MgO is used to limit grain growth, while sodium and apparently fluorine impurities enhance exaggerated grain growth in Al_2O_3. Thus, it should be recognized that there are a variety of complications in strengthening materials with additives or using the effects of impurities or additives as a guide to determining the mechanism of failure. For example, in order to determine whether there is an intrinsic effect of an additive or impurity on the mechanical properties of a ceramic, one must separate out its possible effects on microstructure. One can use effects of additives or impurities to distinguish between brittle and microplastic mechanisms of failure, but a number of parameters must be clearly verified, e.g., the homogeneity of the distribution, the nature of the distribution, and whether anisotropy effects have been altered and whether precipitates may be either inhibiting microplastic processes or simply increasing fracture energies. It should be noted that effects of additives on microstructure, especially in limiting grain growth can be an additional advantage. To be most effective, it is often useful to use two immisicible phases. Thus, for example, several oxides with SiC, especially (BeO–40 wt. % SiC), ThO_2–5 wt. % BeO, and Al_2O_3–TiC composites look promising for a variety of applications, (J. Rubin, private communication) (also see Section V,C).

2. Room Temperature Experimental Observations

Only a limited review of the effect of experimentally observed effects of impurities and additives on the room temperature on the tensile strength of at or near room temperature is presented here, first, because the literature is very voluminous and, second, because a great deal of the work does not present sufficient information, i.e., characterization of the body to determine unequivocally the validity of the effects claimed or whether or not the observed effects are intrinsic or not.

Consider first the effect of solid solution additives or impurities on strength. Rice (1969b) reported a greater increase in strength of CaO at finer grain sizes due to possible changes in impurities effecting dislocations upon annealing. Rasmussen et al. (1965) added up to 50% CoO or NiO to MgO and, after correcting for both porosity and grain size effects, concluded that both did indeed solid solution strengthen MgO as would be expected from anticipated dislocation crack nucleation strength control in MgO. Most of the strength increase occurred by about 10% addition. Similarly, these authors investigated the effect of up to 50% of Cr_2O_3, 20 wt. % Fe_2O_3, or 2 wt. % TiO_2 additions to Al_2O_3. Again, correcting for porosity and grain size effects, they found no intrinsic effect of any of these additives within the solid solution range consistent with brittle failure. Also, no effect of the Fe_2O_3 was found even when it formed a second phase. However, when a second phase was formed with TiO_2 additions, there was an intrinsic decrease in strength, i.e., over and above that caused by the common increase in grain size with TiO_2 additions and its effect on densification. The appearance of this second phase and the intrinsic loss of strength, e.g., approximately 20%, depended not only on the level of TiO_2 addition but also on the firing atmosphere, since precipitation was retarded in reducing atmospheres. It should be noted that while Rasmussen et al. did not evaluate possible effects of their solid solutions on other aspects of mechanical behavior such as the internal stress, one would not expect significant effects by changing thermal expansion anisotropy or elastic anisotropy based on their low levels of additions and the work of Kirchner. Also as noted earlier (Section IV,B), no obvious effect of any of the additives was found on the porosity dependence of the strength of either MgO or Al_2O_3.

The indicated strength reduction of Al_2O_3 with a second phase from additions of TiO_2 and possibly its close chemical relative, MnO_2, is also indicated by the fact that two of the three types of alumina studied by Binns and Popper (1966) that were intrinsically low in strength, i.e., after correction for grain size and porosity effects, contained either or both MnO_2 and TiO_2 (Binns, private communication). This effect of TiO_2 on reducing strength of Al_2O_3 has been further studied and corroborated by Clarke et al. (1970), who fabricated polycrystalline Al_2O_3 with various TiO_2 additions under a variety of firing or annealing temperatures and atmospheric conditions. Again, more oxidizing atmospheres enhanced precipitation, found to be Al_2TiO_5, usually preferentially at grain boundaries, especially on triple lines. The size and shape of precipitates depend on a variety of processing conditions, e.g., temperature, time, and atmosphere. Cracks between the Al_2TiO_5 precipitates in the matrix were frequently observed, and samples containing 1.4% or more Ti additions disintegrated to powder after they were aged in air at 1500°C for 24 hours due to cracking from the precipitate. This effect of the Al_2TiO_5 on the strength of alumina appears to be basically consistent with the predictions of Evans in that this material has an extremely anisotropic thermal expansion such that at least one expansion coefficient will be significantly higher than that of alumina and the sizes of some of the precipitates cited; e.g., the studies of Clarke et al. would be sufficiently large to act as critical flaws in the material.

Crandall et al. (1961) report Al_2O_3 hot pressed to essentially zero porosity with 1% SiO_2 averaged about 10% lower in strength at room temperature than undoped samples of comparable density and fine grain size. Larger additions would presumably make greater reductions in strength, e.g., as suggested by their data on elastic moduli showing about a 9% reduction for 1% silica additions but about 16% reduction for 5% SiO_2. The author has estimated the effect of variable SiO_2 content in the different Al_2O_3 bodies studied by Binns and Popper (1966) by estimating the true porosity and the effects of porosity on strength by comparing strengths corrected to zero porosity with those of Lucalox. This shows an increasing effect of increasing SiO_2 content as indicated in Table XIV.

The effect of anion impurities can be equal to and often substantially greater than cation impurities, but they are often neglected, in part because of greater difficulty in analysis. However, Rice (1972a) and Rice and Mc-Donough (1972b) have shown that residues from the addition of LiF to MgO or $MgAl_2O_4$ for fabrication purposes significantly lower strength. Auger analysis by Johnson et al. (1974) shows primarily fluorine remaining

TABLE XIV

Estimated Effects of SiO_2 on Room Temperature Strength of Al_2O_3

Percent SiO_2 content	Approximate[a] true theoretical density (gm/cm³)	Percent porosity		Estimated[b] percent of zero porosity, zero impurity strength due to:		
		Calculated neglecting SiO_2	Using approximate true theoretical density	Porosity and SiO_2	True porosity	SiO_2
1	3.97	2.6	2.0	84	87	97
2	3.95	4.3	3.3	74	79	94
4	3.89	6.8	4.4	62	74	84
8	3.79	10.0	5.3	50	69	73
12	3.70	13.8	7.0	38	61	62

[a] Using data of Binns and Popper (1966).
[b] Calculated using corrected Lucalox data and $b = 7$.

at the grain boundaries of these bodies, so this is primarily an anion effect. Similarly, Rice has shown that residual hydroxide and especially carbonate content, as indicated by IR analysis, correlate with significant reduction in strengths of hot-pressed MgO, e.g., 50%. This may result from grain boundary weakening, since these ions, especially the carbonate ions would be too large for solution in the lattice and would therefore be expected to be at the grain boundaries, and more intergranular failure is found in fine-grained hot-pressed material that typically contains these impurities. However, auger analysis by Johnson *et al.* did not find carbon at the boundary of such factured samples.

Much of the information on effects of dispersed particles and fibers on tensile strength of ceramics, has been presented in conjunction with the section on fracture energy (Section III,C) (see also Lange, 1974b). As noted there, strengths generally did not increase; in fact they often decreased as fracture energy was increased by the addition of various particles or fiber contents to the matrix. The majority of these results as discussed earlier are explained by two factors, namely the flaw size to the particle spacing and the particles or fibers acting as flaws, as reviewed by Rice and Freiman (to be published).

Analysis shows that many composites have average particle or fiber spacing greater than expected matrix flaw sizes (Table XV). Furthermore, that spacings between some particles or fibers can be severalfold more than the average due to statistical variations in particle or fiber spacings is a major reason why most composites have not had higher strengths than the matrix alone. Thus, for example, it is seen that the spacing of partially aligned carbon fibers in the samples of Sambell *et al.* (1972a) is not sufficiently small at 10 vol. % so that there is no strengthening, while with greater additions, the spacing should be sufficiently close to give strengthening as observed. This effect of spacing was generally shown by Hasselman and Fulrath (1966) in their study of Al_2O_3 particle additions to glass samples resulting in strengthening only when the mean interparticle spacing that they directly measured was 40 μm or less, regardless which of the eight different sizes of alumina particles were added to the glass samples. Estimates of the flaw size and particle spacing in Table XV are consistent with their results; e.g., strength of samples with the largest (60 μm alumina) additions increased only as the volume fraction of alumina approached about 40%, while the smallest (15 μm alumina) additions increased strengths when their volume fraction was slightly over 10%. This indicates an idealized average particle somewhat less than half the flaw size; i.e., that the extreme separations of particles were between two and three times the average. Unfortunately, these results have not received sufficient attention by subsequent investigators studying ceramic composites. Bansal and Heuer (1974) have reported

TABLE XV

FLAW SIZE PARTICLE OR FIBER SPACINGS, AND STRENGTHS OF CERAMIC COMPOSITES

Material	Flaw size[a] (μm)	Particle or Fiber Material	Diameter (μm)	Spacing (μm) for volume fraction 10%	20%	30%	References
Al$_2$O$_3$	65	Random carbon fibers	8	32	21	14	Sambell et al. (1972a)
Glass	23	Al$_2$O$_3$ particles	60	45	23	12	Hasselman and Fulrath (1966)
			15	11	5.7	3.1	
Glass	30	Aligned carbon fibers	8	14	8	5	Sambell et al. (1972a,b)
Glass	30	Random carbon fibers	8	32	21	14	Sambell et al. (1972a)
Glass	72	Al$_2$O$_3$ particles	3.5	2.6	1.4	0.7	Lange (1971)
			11	8	4.2	2.2	
			44	33	17	9	
MgO	82[c,d]	Random Ni fibers	89	360	240	160	Hing and Groves (1972a,b)
	G[d]	Co, Fe, or Ni particles	~40	30	15	8	
Si$_3$N$_4$	62[e]	SiC	5	3.7	1.9	1.0	Lange (1973)
			9	6.7	3.4	1.8	
			32	24	12	6.6	

[a] Flaw size calculated from S, E, and γ of matrix.

[b] Assuming uniform spacing of equal spherical or cylindrical particles, for random particles, only one-third were assumed to be oriented to significantly effect crack propogation.

[c] Flaw size calculate by authors.

[d] Grain size (G) 10 μm or less and $0.03 < P < 0.07$.

[e] Lange obtained a value of 160, but this is based on an unusually high value of the geometrical (Y) parameter in the Griffith equation.

increases of about 50% in both the strength and stress intensity (i.e., nearly a 100% increase in the fracture energy) of Al_2O_3-rich spinel by precipitation. Porter and Heuer (1975) have also indicated similar phenomena in partially stabilized ZrO_2. Clearly, the spacing of such precipitates can be much finer than the flaw sizes and can hence strengthen materials, provided grain boundary precipitation or other effects do not occur.

The second factor limiting composite strengths is the particles or fibers themselves acting as flaws as discussed earlier. This can be especially serious when the spacing between particles or fibers is of the order of or greater than the size of the particle and the associated cracks so that the resultant flaw propagates catastrophically with little or no effect of the surrounding particles. This appears to explain the initial drop in strength of Lange's (1971) glass samples reinforced with the order of 10 vol. % alumina particles approximately 44 μm in size. Again, statistical variations will accentuate this, since, for example, a significant percentage of particles twice the average size will commonly exist in many particular dispersed systems. Similarly, this may explain the drop in strength of Lange's (1973b) Si_3N_4 bodies with 32 μm average size particles of SiC, especially since he found some of these SiC particles were fractured in the Si_3N_4 matrix. Hing and Groves (1972b) observed that computed flaw sizes for their fiber-reinforced MgO bodies are of the order of the fiber size and that poor bonding of the fiber to the matrix gave flaws of that size.

Finally, it should be noted that second-phase effects can be complicated by their possible multiple effects. Thus, for example, stress or cracking effects from second phases tending to lower strengths may be counteracted in part by their giving finer grain sizes, which may, for example, limit crack sizes. This is suggested by results of Hill et al. (1967) for BeO + 8 wt. % SiC. While strengths of the composite were lower than for BeO at comparable grain size, sufficiently finer grain sizes were obtained with the SiC to give higher strengths than they obtained with BeO only. Also, composite strengths will depend on the composite modulus, which may often be lowered by the reinforcing phase (see Section II,C).

3. HIGH TEMPERATURE EFFECTS OF SECOND PHASES

As test temperatures are increased, the same basic mechanisms of failure are applicable, but there are three shifts in the effects of impurities and additives. First, because deformation processes due to slip and twinning and grain boundary sliding become more important, and effects that alter such deformation become more important. Second, the properties of the bulk and added phase often become more widely separated accentuating effects. Also, some solution or redistribution of second phases may occur.

Third, internal stresses due to thermal expansion differences between the matrix and the second phase will generally decrease.

Consider the strength-reducing effect of impurities at high temperatures. Thus, for example, silica-based grain bounbary phases typically cause a rapid drop in high-temperature strengths as the silicate phase softens. This is often preceeded by a rise to a strength maximum over a range of strain rates, due to crack blunting or branching effects. Both strength maxima and subsequent rapid drops in strength have been reviewed by Chu (1966) in a number of materials, e.g., superduty fire clay brick and electrical porcelain. (Note the maxima of the latter are dependent on heat treatment. Also, its strength rises again at higher temperatures, apparently due to reduction of internal stresses and crack healing). Similarly, Padgett and Clements (1970) have observed maxima and rapid drops of strength in four different mullite refractories, and Davidge and Tappin (1970) noted such effects in debased Al_2O_3. Though not as detailed, some of the Al_2O_3 data in Fig. 34 also indicate decreased strength with SiO_2 content, e.g., data of Crandall *et al.*, and Wesgo Al_2O_3 in contrast to Lucalox.

Similarly, it is generally accepted that the high-temperature strength of hot-pressed Si_3N_4 drops rapidly due to the silicate grain boundary phase believed to be present in contrast to reaction-sintered material, e.g., Fig. 36. Further, impurities, e.g., of larger ionic size that will not go into solution and hence will be in the grain boundary silicate phase, can further lower high-temperature strengths by lowering the silicate grain boundary phase viscosity (e.g., see Richardson, 1973). Reduction in impurities has been the reason for improved high-temperature strength of commercial Si_3N_4 (Fig. 36). Lange and Iskoe (1973) have shown that Ca is especially detrimental to high-temperature strength, which is important, since Ca is a common impurity in Si_3N_4 (due to the common production of Si by reduction of SiO_2 with CaC_2). More recently, investigators have recognized that the low (about 1500°C) melting temperature of magnesium silicate grain boundary phases likely to form from the common use of MgO as a hot-pressing aid for Si_3N_4 will inherently limit the strength of material so fabricated. Thus, investigators have sought additives that would densify Si_3N_4 and give more refractory silicate phases. Gazza (1973) and Tsuge *et al.* (1974) successfully used Y_2O_3 additives, Mazdiyasni and Cooke (1974) used rare earth oxides, especially CeO_2, and Rice and McDonough (1975) used ZrO_2 or other zirconium compounds. All show improved high-temperature behavior.

While silicate grain boundary phases are a particularly common source of high-temperature strength reduction, other phases can also cause this. Thus, for example, Rice (1969b) noted that CaO containing quite limited ($<1\%$) fluoride residues from hot pressing with LiF showed pseudoplastic behavior with low strengths, total intergranular failure, and no obvious grain size dependence of strength at about 1300°C. This was in contrast to an order of magnitude higher strength, no apparent plastic flow, decreasing strength with increasing grain size, and considerable transgranular failure in CaO hot pressed without LiF. Similarly, McLaren *et al.* (1972) report a precipitous drop in strength of SiC containing free Si near the melting point (1410°C) of Si. Since the Si occurs as particles, strengths drop to the constant level of the same material from which Si has been removed by etching.

Turning now to second phases strengthening ceramics at high temperatures. Lange's (1973b) addition of various sized SiC particles to Si_3N_4, while not raising room-temperature strengths, approximately doubled flexural strength at 1300° and 1400°C. Similarly, Rubin (private communication) reports substantially improved high-temperature strength of ThO_2 with BeO additions. Further, other composits, e.g., directionally solidified eutectics (Fig. 19), can retain high strengths at elevated temperatures much better than single-phase materials. This is to be expected, since any plastic or viscous flow process that is typically involved in the high-temperature failure should be inhibited by the presence of more refractory, less deformable particles, fibers, etc. It is clearly expected from extensive studies of metals containing refractory precipitates or dispersed particles as well as from studies of the mechanical behavior of ice (e.g., Kingery, 1961) showing that addition of kaolinite, while not improving short-time strength, did substantially reduce the rate of deformation under a fixed stress, while addition of wood pulp fibers or wood sawdust to ice could substantially increase short-time strength as well as increase both the resistance to deformation and impact or shock loading damage.

F. Effect of Microstructure on the Environmental and Loading Dependence of Tensile Strength and on Electrical Failure

1. Interactions between the Effect of Environment, Cyclic Loading, and the Microstructural Dependence of Tensile Strength

The effect of microstructure on the environmental strength dependence of ceramic materials has received almost no study, in part, because many apparently feel there will be no significant effect of most microstructural variables. However, impurities are clearly an important exception to this, since it is generally recognized that impurity concentration at grain boundaries can lead to significant changes in the environmental effects, e.g., stress corrosion of a material. Thus, for example, a silica or silicate grain boundary phase in a body whose grain composition was not subject to stress corrosion could well lead to a stress corrosion phenomenon along the boundaries in the resultant contaminated body. Rhodes *et al.* (1974) observed stress corrosion in polycrystalline MgO in which failure appeared to be exclusively intergranular, indicating either an intrinsic or, more likely, an impurity grain boundary effect, since Shockey and Groves (1969) showed that water actually increases the toughness of MgO single crystals. Impurity effects cannot be discounted, since polycrystalline MgO, even of quite high purity, typically tends to have some silica and/or calcia at grain boundaries. However, recent studies by Freiman *et al.* (to be published) show delayed failure of extremely high purity (laser window grade) CVD ZnSe preferentially initiates from grain boundary, whereas normal failure commonly initiates transgranularly. In both cases, the overall mode of crack propagation was transgranular; only the origins differed in fracture mode. Thus, delayed failure may represent a true grain boundary effect in some, and possibly all, polycrystalline ceramics.

Whether flaws are comparable to or contained within the grains (instead of being substantially larger than the grains) so strengths are controlled by fracture energies at or near those of single crystals instead of polycrystalline fracture energies can effect environmental dependence of strength. Thus, for example, Freiman *et al.* (1975b) show that single crystal fracture energy control of strength leads to a predicted life approximately twentyfold shorter than one would obtain using polycrystalline fracture energies. However, since the proof stress-to-service stress ratio in proof stress theory is typically not sensitive to the lifetime, one need only increase this ratio by about 10% to achieve the same lifetime as predicted by polycrystalline fracture energies. On the other hand, this higher proof stress-to-service stress ratio can mean a significant increase in the number of rejected, i.e., failed components in the proof test.

While no direct effect of porosity on environmental effects of strength is known, there could be some. Thus, for example, if during the fabrication process pores close entrapping firing atmosphere species that may lead to stress corrosion, then such porous bodies may exhibit stress corrosion in the absence of this species in the external test environment. This, for example, is indicated in studies by Freiman *et al.* (1975c) of slow crack growth in graphite materials.

The increasing interest in mechanical applications of ceramics has, of course, increased the interest in their dynamic fatigue. Studies do show definite dynamic fatigue effects at room and elevated temperatures, but microstructural and mechanistic data are limited, particularly at room temperature. Thus, for example, Williams (1961), Sarkar and Glinn (1970), Krohn and Hasselman (1972), and Acquavia and Chait (1972) have all reported dynamic fatigue in Al_2O_3 at room temperature and the latter authors in B_4C as well. It has been recognized that static fatigue, i.e., stress corrosion, is an important factor, but authors have generally concluded that there are additional dynamic effects. Previously, these have been tentatively attributed by some to possible microplastic effects, e.g., as possible extrapolations of behavior of softer materials such as NaCl and LiF. However, the increased understanding of flaw effects strongly suggest that any additional dynamic effects are due to flaws. Coalescence or opening of cracks caused or aided by TEA stresses is a possibility as indicated by Seaton and Katz's (1973) kilohertz sonic fatigue tests. They found no fatigue in Si_3N_4, a limited amount in Al_2O_3, and a substantial amount in TiB_2. Also, in Al_2O_3 and other materials that may have twins introduced with associated flaws by machining or other surface contact, as discussed in Section IV,C, such dynamic effects could increase the severity of the flaw, possibility aided by TEA stresses. However, much work is needed to define whether true dynamic effects exist in the absence of or in conjunction with environmental effects and what mechanisms and parameters are involved.

Some work on elevated temperature fatigue has also been conducted, where plastic deformation clearly plays a role. Thus, for example, Kossowsky (1974) concluded that above 1200°C, plastic deformation, most likely by grain boundary sliding, was rate controlling in hot-pressed Si_3N_4. Similarly, White and Ashbee (1974) report that the high-temperature fatigue of a $Li_2O–ZnO–SiO_2$ glass ceramic is determined by the remaining glass phase in which voids nucleate, grow, and coalesce to cause failure. It is also important to note that the much higher compressive than tensile yield stress of these glass ceramics lead to a dimensional instability.

2. The Effect of Microstructure on the Strain Rate Dependence and Impact Strength of Ceramics

No extensive studies have been made on the strain rate dependence of strength of ceramics as a function of their microstructure. However, as discussed earlier, Freiman *et al.* (1975) showed a significant, e.g., twentyfold difference in expected life at a given service stress for ZnSe specimens, depending on whether the flaw size was smaller than the grain size or substantially larger than the grain size. This could result in different strengths with different strain rates as a function of flaw-to-grain size ratio. More work is clearly needed to explore this area.

Dinsdale *et al.* (1962) carried out one of the earlier, more comprehensive studies of impact strength as measured by typical Charpe or Izod tests methods on brittle ceramics. They showed a reasonably good correlation of impact strength with the strain energy density (i.e., $2^2/SE$) and hence the stored elastic energy at static failure. More recently, Davidge and Phillips (1972) found that in strong brittle ceramic materials their impact strength generally also correlated with stored elastic energy. However, they noted that at the other extreme, i.e., weaker, tougher materials like graphite and fiber-reinforced plastics, that impact strengths may be controlled more by the work of fracture of the samples than the elastic energy at static failure. Bertolotti (1974) also found in his study of instrumented impact failure of Al_2O_3 samples that impact strength again correlated with strain energy density and hence stored elastic energy in the samples. This was somewhat higher than the strain energy density of samples in conventional slow-bend tests; however, the higher value in the impact test was consistent with estimates of slow crack growth effects present in the conventional bend test. More recently, Kirchner et al. (1975) found that their impact strengths of two commercial aluminas, dense Si_3N_4, and SiC, as well as steatite and flint glass all generally increased with increasing strain energy density at static failure and that normalizing the impact energy for a cross-sectional area gave a good correlation between his results and those of Dinsdale et al.

Correlation of impact strength with the strain energy density at static failure, i.e., $S^2/2E$, indicates the general microstructural dependence of impact strength, but details are uncertain. Thus, for example, since the porosity dependence of strength and Young's modulus are very similar, one would expect a porosity dependence of impact strength similar to that of either (note that this is a typical example of the utility of simple empirical expressions, e.g., the exponential relation, that give a good fit to strength— and elastic modulus—porosity data allowing ready calculation of the expected porosity dependence of other mechanical properties). Since Young's modulus is essentially independent of grain size while strength typically increases with decreasing grain size, one would expect impact strength to increase substantially faster with decreasing grain size than static strength. On the other hand, $S^2/2E$ is proportional to $\gamma/2a$, indicating dependence on porosity through γ and possibly a, as well as reduced dependence on grain size (through γ and possibly a). The porosity dependence of γ and E are also similar, so either porosity prediction may be similar, depending on whether a depends on P.

Data on the microstructure dependence of impact strength is extremely limited, especially where one investigator has determined the microstructural dependence of both static and impact strengths as well as Young's modulus. Hamano and Lee's (1972) impact data is less extensive than the fluctural data, but it clearly indicates a lower porosity dependence than their flexural strengths. Impact strengths of their kaolin and quartz body gave $b \sim 1$ in comparison to $b \sim 6$ for flexure strength. Similarly, impact strengths of their triaxial body gave $b \sim 3$ in comparison with $b \sim 8$ for the flexural strength of this material. This lower impact porosity dependence may be in agreement with the microstructural dependence expected from $\gamma/2a$ if a also depends on porosity, e.g., pore size due to pores acting as flaws.

More recently there has been a substantial increase and interest in impact and erosion from particles (e.g., from rain, hail, and dust) on a variety of materials ranging from those used for irdomes or radomes to those used for turbine blades. Adler (1974) has made significant experimental studies of such particulate impact and erosion, but as yet no significant micro-structural effects have been studied. Evans (1973b) has conducted some analytical and experimental studies of Hertzian crack formation from impact of a spherical particle in terms of the particle momentum and the elastic properties of projectile. He shows that a critical momentum must be exceeded before the size of resultant cone cracks exceed the size of the preexisting cracks in the material, which are of course dependent upon the surface condition or processing of the material. Since both the extent of the preexisting flaws and the cone cracks resulting from impact should have a significant microstructural dependence, one would clearly expect such impact phenomena to show a significant variation with microstructure. However, detailed studies have not yet been done. It should be noted, however, that a number of investigations are currently underway to improve the particulate impact resistance of ceramic materials, e.g., through use of a porous, crushable surface layer.

Higher-velocity impacts such as some of the above impact and erosion studies, and especially ballistic impact, begin to involve high compressive as well as tensile loads, but they are briefly noted here for continuity. Ferguson and Rice (1971) showed that the ballistic limit velocity, i.e., the velocity of projectile just stopped, of Al_2O_3 bodies showed a limited, but definite Petch-type grain size dependence. They also showed that reducing the Al_2O_3 content, i.e., due to additives or impurities, lowered the ballistic limit over the grain size range investigated (1 μm to ∞).

3. MICROSTRUCTURAL DEPENDENCE OF THERMAL STRESS
   (AND THERMAL SHOCK) FAILURE OF CERAMICS

The resistance of ceramic materials to failure as a result of thermal stress conditions, especially those resulting from a rapid change in temperature difference, typically referred to as thermal shock, are extremely important. This is a complex and extensive subject in itself, and hence this review must be limited some in scope and especially in depth. However, since much of the past literature is based on relative empirical tests with questionable comparability to one another, it is still possible to discuss the major micro-structural dependence of thermal stress and thermal shock failure in a limited review.

There are two major concepts of thermal stress fracture of brittle material.

The first and oldest is basically concerned with whether a material survives a given thermal stress environment or change in thermal stress environment without any degradation in mechanical properties. This typically indicates that the resistance to thermal stress or shock damage increases as the ratio of the fracture stress divided by the product of the thermal expansion and Young's modulus ($S/\alpha E$) increases. The microstructural dependence of thermal stress resistance is thus indicated by the dependence of fracture strength and the porosity dependence of Young's modulus. This basic thermal stress or shock resistance parameter is often modified, particularly dependent upon the nature of the heating conditions, as well as the nature of the stresses. The latter often adds a $1-v$ factor, usually in the numerator, hence adding a small amount of additional porosity dependence. Fracture energy also frequently occurs in the numerator in a number of formulations, adding its microstructural dependence. Many formulations also involve coefficients determined by the mode and extent of heat transfer to or within the material. These coefficients can appear in either the numerator or denominator, and of course, those dependent upon thermal conductivity of the material will depend on porosity and second-phased inclusions. Hasselman (1970b) has presented a summary of thermal stress parameters and considerable discussion of them is also given by Nakayama (1974). Some of these, as well as a fair amount of experimental data, are found in the paper by Walton and Bowen (1961).

More recently, a number of investigators have recognized that to prevent any damage to samples in extreme thermal stress environment requires extremely high strength, e.g., approaching 1,000,000 psi, and therefore that significant attention must be paid to minimizing the amount of damage that occurs. Hasselman (1969b, 1971) has been a leader in the analytical development determining the parameters necessary to maintain the best level of strength after thermal stress or shock damage occurs as a function of the thermal stress environment. Basically, the analytical approach uses fracture mechanics considerations of crack propagation, which can be stable rather than catastrophic depending on material and especially crack parameters. This theory emphasizes that limiting strength loss after damaging thermal stress exposure will be favored by long initial crack lengths and higher crack densities and related lower values of strength. This model predicts that there is a critical temperature difference, dependent upon material parameters, over which no change in strength will occur. Immediately beyond this temperature difference, a precipitous drop in strength occurs to a lower level, again determined by material parameters. As the temperature difference is further increased, strengths will again be stable at this lower level and finally begin to decrease again. Manning and Lineback (1974) have more recently pursued a similar course of development. Clearly the importance of flaw size and flaw density in determining the residual strength after thermal stress damage depends both on the microstructure as well as the processing history of the material, with the microstructural dependence of crack size and density interacting with the various processing parameters not being particularly well understood.

More recently, Hasselman has shown that in glass samples, slow crack growth during thermal shock can play a significant role in the thermal stress failure of glasses (e.g., Badaliance *et al.*, 1974). How significant this effect will be in polycrystalline materials, which generally show significantly less slow crack growth than glasses and which may also have significantly microstructural effects on slow crack growth, e.g., whether cracks are smaller or larger than the grain size, remains to be determined.

Finally, it should be noted that more recent work by Mecholsky *et al.* (to be published) shows that when one is dealing with a very high-intensity heat source, such as a high-intensity laser, significant changes in thermal stress behavior can occur. They developed a model that shows that the time that it takes a specimen to failure in such an extreme thermal stress environment is directly proportional to two sets of terms. The first term is essentially the specimen thickness squared divided by the thermal diffusivity, and the second term is determined basically by the strength divided by the product of thermal expansion and elastic modulus, i.e., the normal thermal stress parameter. However, the second term also has the power density coupled into the sample in the denominator. Thus when this power density becomes sufficiently high, the second term becomes negligible and the survival in the extreme thermal environment is determined primarily by the specimen thickness and thermal diffusivity. Since one can significantly decrease the thermal diffusivity as a result of increasing porosity or second phase, this then shows that significantly increased lifetimes can be achieved by modifying microstructures in this fashion.

Hasselman's model has been extensively verified for Al_2O_3 (Hasselman, 1970), ZnO (Gupta, 1972), SiC (Coppola and Bradt, 1973), and B_4C (Seaton and Dutta, 1974). The last specifically addressed the effect of grain size, showing that the decrease in strength is less for larger grain bodies as predicted by Hasselman's theory, since strength typically increases with decreasing grain size, and the strength after thermal shocking of larger-grain B_4C bodies was comparable or greater than that of fine-grain bodies, which initially had higher strength. Gebauer *et al.* (1972) have shown that one can increase the residual strength after thermal shock by increasing the initial strength through surface compression in contrast to lowering the residual strength as a result of increasing strength of the bulk material. Gupta (1973) has also shown the reduction of strength as a function of a given thermal stress environment with increasing grain size of Al_2O_3 up to a critical grain size in the range of 70–200 μm, where failure became total. Ainsworth and Moore (1969) present Al_2O_3 data showing residual strength trends after thermal stress damage similar to those predicted by Hasselman dependent on initial crack density. Also thermal shock testing of Al_2O_3 by Duncan et al. (1965) shows the trend subsequently predicted by Hasselman. Larson et al. (1974) have extensively shown the applicability of Hasselman's theory for predicting the spall resistance of high-alumina refractories.

Rossi (1969b, 1971) recognized the importance of low strength and Young's modulus as a result of weak interfaces and has developed highly thermal-stress-resistant ceramic composites by providing extensive networks of weak interfaces, e.g., between graphite particles and surrounding carbide matrices or tungsten and MgO to inhibit the extent of crack propagation. More recently, El-Shiekh and Nicholson (1974) have shown some development of ZrO_2 partially stabilized with CaO so that a fairly extensive microcrack network is developed as a result of thermal stress, giving the material fairly good thermal shock resistence.

4. Effect of Microstructure on the Electrical Failure of Ceramics

High electrical fields, as applied, for example, to some ceramic insulators and to some piezoelectric ceramics, can develop sufficient tensile stresses to lead to mechanical failure. Davisson and Vaughn (1969) have extensively reviewed such electrically induced fractures in single crystal and glass materials, i.e., fully dense materials, which provides a baseline against which to compare results of bodies containing pores, grains, and/or second phases. Of these parameters, only the effect of porosity has been studied to any significant extent. A brief summary is given of the limited work studying these variables, primarily porosity, on electrical failure of ceramics.

Bratschun's (1965) study of lead zirconate titanate materials clearly showed tensile failures specifically originating from large pores as a result of electrical breakdown. Gerson and Marshall (1959) developed a model for the dielectric breakdown of porous ceramics. This model is particularly interesting among models for porosity effects on mechanical properties in that it considers the probability of particular inhomogenities in the spatial distribution of pores favorable to breakdown, i.e., the probability of a number of pores being contained within a thin column of material so that breakdown would preferentially occur from pore to pore along this column. This model assumes spherical pores of uniform size, although, it should not be affected significantly by variation in pore shape, since the critical parameter of the pore is the dimension parallel with the electrical field. In other words, this model basically assumes that the void dimension parallel with the applied field is the only significant parameter and that the void position through the thickness is not important, which is consistent with experimental observations of Hall and Russek (1953). However, the model would be seriously affected by significant variations in both the volume and size of pores within the body. Gerson and Marshal show that the logarithm of the dielectric breakdown strength is a linear function of porosity over much of the range of porosity. There is a slight upward curvature as zero porosity is approached and a much more rapid decrease of dielectric strength at higher porosities with this latter breakover occurring at higher porosities as the size of the voids decreases. It is interesting to note that all of these effects, except the pore size effects, are the same as Knudsen's semianalytical development of the exponential relationship, which can therefore be applied to Gerson and Marshall's curves. These yield b values ranging from about 3.3 to about 7.9, respectively, for voids 50 μm and 2500 μm in diameter. Gerson and Marshall experimentally tested their theory by fabricating lead zirconate titanate bodies with voids approximately 1200 μm in diameter and found that their theory for this size voids fit their experimental data, which gives $b \sim 5$, quite well. More recently, Johnston and Tibbetts (1974) demonstrated that corona breakdown within pores, which should be a precursor to dielectric breakdown, could be used for determining their internal location within 2 mm in lead zirconate titanate body.

Morse and Hill's (1970) data on the electrical strength of 99% pure hot-pressed Al_2O_3 having about 2–16% porosity fits the exponential relationship well, giving a b value of 6.9 $\pm$ 0.8 and extrapolates very closely, i.e., within about 2% of the weighted average of the breakdown field, for various single-crystal orientations. These authors also observed apparent melting along grain boundary breakdown paths, which they felt was possibly due to prebreakdown effects but could not unequivocally demonstrate that it was not due to postbreakdown conduction.

Morse and Hill also showed that the electrical breakdown field of their hot-pressed as well as some commercial aluminas decreases as the specimen thickness increases, e.g., the field was proportional to the -0.3 power of thickness. Gerson and Marshall observed a very similar dependency for dielectric breakdown in their lead zirconate titanate samples. The cause of this thickness effect, which may well be due to surface microstructure, is not known, but it clearly complicates the microstructural analysis of electrical breakdown effects. However, from the above results, it is clear that porosity has a significant effect on dielectric breakdown, and such porosity effects appear to be quite similar to those for tensile failure.

Binns et al. (1970) investigated the effects of temperature on the porosity dependence of electrical breakdown of a ball and a china clay body, both containing 40% molochite grog, from 100° to 1200°C in 100°C intervals.

At temperatures to about 300°C, both bodies showed decreasing levels of electrical breakdown with increasing porosity, giving b values of 2–6 for the exponential relation. At higher temperatures, e.g., in the 400–600°C range, the porosity dependence of electrical breakdown decreased, and in fact reversed, indicating that the pores became better insulators than some of the solid phase.

G. Microstructural Dependence of Fracture Path

General effects of microstructure on the fracture path, e.g., whether the crack propagates transgranularly or intergranularly are generally well known and have been reviewed by Rice (1974a). Thus, it is widely recognized that increasing porosity, because it usually occurs mostly at grain boundaries, increases intergranular failure, with impurities often having a similar effect. Both of these effects are accentuated as test temperatures increase, e.g., see discussion of CaO in Section IV,B. It has also been generally recognized that transgranular failure tends to increase with increasing grain size but decreases with increasing test temperature, however, there is little quantitative data on these effects and considerable uncertainty as to whether or not the grain size effect is intrinsic or extrinsic. Further, only limited attempts have been made to differentiate the fracture topography from which failure initiates in contrast to fracture topography of the resulting crack propagation, which may have significant bearing on failure mechanisms. Also, as noted in Section IV,F, environmental effects can alter the mode of fracture, at least at the fracture origin.

Rice (1974a, 1972a) is one of the few researchers to have obtained quantitative data on the change of fracture topography as a function of grain size and temperature in relatively pure, nearly theoretically dense materials. His data for hot-pressed MgO is shown in Fig. 37, indicating the trend noted above for grain size and temperature. The high-temperature data do not necessarily represent intrinsic behavior. Day and Stokes (1966) have shown that fully

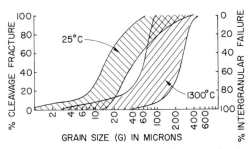

Fig. 37. Percentage cleavage, i.e., transgranular and intergranular fracture, in MgO at 23 and 1300°C as a function of grain size (after Rice, 1972a).

dense MgO with very little grain boundary impurity has changed to intergranular failure at test temperatures below 1600°C, but then true ductile
failures occured beyond 1700–1800°C similar to metal failures. On the
other hand, dense hot-pressed samples having trace porosity and some grain
boundary impurities similar to those of Rice continued to exhibit intergranular failure beyond 1700–1800°C. Similarly recrystallized MgO and
CaO exhibit predominately transgranular fracture to over 1500°C, as
noted later.

Hot-pressed Al_2O_3 bodies of similar purity and density do not show this
grain size trend; typically, averaging of the order of $40 \pm 30\%$ transgranular
failure over much of the grain size range, then tending to drop off at large
grain sizes (Fig. 38). [NOTE: In Rice (1974a), a typographical error incorrectly
indicates that the transgranular failure of alumina increases with increasing
grain size.] Heuer (1969) also observed that dense, hot-pressed Al_2O_3 with
$\sim 1~\mu m$ grain size had high percentages of transgranular failure (often over
50%). This trend for a decrease in transgranular failure at larger grain sizes
as well as the resultant spontaneous intergranular fracture of large-grained
Al_2O_3 can generally be explained on the basis of Davidge and Green's
(1968) model developed for composites, where they relate strain energy in
an inclusion to the energy for fracture around or through the inclusion.
In view of the concentration of stresses from thermal expansion anisotropy,
intergranular failure is expected, as generally observed. Rice (1974a) used this
concept to consider the spontaneous failure of grains of anisotropic materials
where the strain energy is determined by the anisotropy of thermal expansion.
This approach gives reasonable approximations for the grain size range over
which spontaneous failure is observed in Al_2O_3, $PbTiO_3$, and $BaTiO_3$.
Clearly, since this model indicates that strain energy for spontaneous

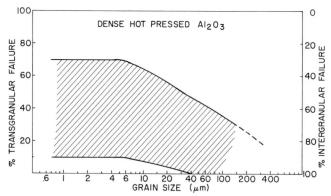

Fig. 38. Percent transgranular and intergranular failure versus grain size of Al_2O_3 at
approximately 23°C.

fracture builds up as grain size increases, it also explains the dropoff in transgranular failure with increasing grain size in Al_2O_3.

Many have felt that the trend to more intergranular failure at finer grain sizes is intrinsic (e.g., Spriggs et al., 1966; Spriggs and Vasilos, 1966). However, Rice (1972a, 1974a) has pointed out that this is not necessarily so, since many of these observations may be clouded by the effects of trace porosity and in many cases, impurities at grain boundaries, e.g., anion impurities in bodies such as MgO. Rice (1974a) also sites examples of transgranular failure in quite fine grain bodies, e.g., dense ZrO_2 having average grain sizes of approximately 0.4 μm. Thus, there is extreme doubt about an intrinsic trend for increasing intergranular failure as grain size decreases, at least over grain size range investigated, e.g., $\sim$0.5 μm or greater.

Turning next to the effect of porosity on fracture as noted earlier, grain boundary porosity typically enhances intergranular failure. Less well known, although certainly not unexpected, is the fact that pores within grains tend to enhance transgranular failure (e.g., Rice, 1972a, 1974a). Another important aspect of porosity related to fracture is the fracture steps that are left on the side of a pore opposite from which the crack approaches it. Excellent samples of this have been shown in the past (e.g., Passmore et al., 1965), and this has been discussed with a number of examples shown by Rice (1974a). This phenomenon basically results from the two sides of the crack becoming separated as they pass around the pore. Minor perturbations lead to these two halves almost never completing their passage around the pore on exactly the same fracture plane. Thus, on the opposite side of the pore from which the crack approached it, a step of extra fracture will form until the two slightly separated fracture planes can again pull together to a single plane. Examples of this phenomenon shown in Fig. 39 are extremely valuable in identifying crack propagation directions, for example, to locate fracture origins. Such steps always occur with pores within grains, they also typically

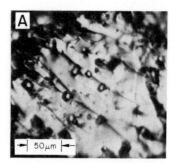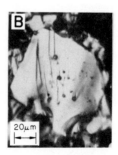

Fig. 39. Fracture tails from pores. (A) $BaTiO_3$ (fracture direction: upper left to lower right); (B) $MgAl_2O_4$ (fracture direction: bottom to top). (After Rice, 1974a, published with permission of Plenum Press.)

occur with pores large in comparison with the grain size. They typically do not occur with pores at the grain boundary unless the fracture is changing from intergranular to transgranular at the point it passes through the pores, as discussed by Rice (1974a). Lange (1970) has investigated in more detail effects in crack interaction with pores in MgO single crystals, showing that significant increase in cleavage step density can occur on the side that the crack front approaches the pore in addition to the fracture steps noted earlier. These additional fracture steps sometimes cloud the fracture tail pattern.

Turning now to the question of fracture topography at fracture origins as opposed to surrounding crack propagation areas, Rice *et al.* (1968) and Rice (1972d) have shown that fully dense high-purity polycrystalline samples of MgO and CaO made by recrystallization of single crystals exhibit almost exclusive transgranular failure over a wide range of test temperatures (e.g., to over 1500°C) in contrast to less pure, slightly less dense samples showing considerable intergranular failure, especially as temperature increases. The texture (i.e., preferred crystal orientation) that many of these recrystalized samples exhibit may be a factor in some of these effects. However, similar recrystallized samples containing limited grain boundary impurities and porosity comparable to normal high-purity, high-density hot-pressed MgO show the normal trends, i.e., significant intergranular failure increasing substantially with increasing test temperature. In the recrystallized crystals, fracture origins were occasionally observed to occur from grain boundary surfaces at room temperature, e.g., in the MgO. However, as the test temperature increases, there was an increasing tendency for failure to initiate at sections of grain boundary surfaces, e.g., at triple points, though at test temperatures of the order of 1200°C, this was still not necessarily the dominant source of failure. Since these materials still contain impurities, some of which should be at the grain boundaries, this still does not represent intrinsic behavior, indicating that substantially more work is needed to understand the causes of fracture topography particularly at fracture origins. This is consistent with Kingery's (1974) observation that intrinsic grain boundary behavior has generally not been observed.

V. Hardness, Compressive Strength, and Wear of Ceramics

A. Hardness

1. BACKGROUND

Hardness is extremely important, because it correlates with many other properties as reflected in compressive strength, ballistic performance of armor, machining behavior, and friction and wear phenomena. Its value for such correlations is further enhanced by the simplicity of the tests and the small sample size that can be used. There are a variety of hardness test available based on an object penetrating the surface either with or without motion parallel with the surface. The former are the less used scratch hardness tests, which were developed first on a relative basis, e.g., Mohs' scale. The more recently developed tests, which use an indentor penetrating only normal to the surface with no transverse motion use a variety of indentor shapes, e.g., balls and various cones, but pyramids, particularly the Vickers or Knoop pyramid indentors, are by far the most commonly used shapes (e.g., see Shaw, 1973).

There has been substantial confusion and controversy as to the meaning of hardness, due to a lack of understanding of mechanisms involved in the indenting process, compounded by the frequent variability of results from different test and material parameters. The controversy of mechanisms controlling hardness of ceramic materials has centered primarily around the relative rolls of cracking and plastic flow in determining hardness. Even removing the question of the mechanisms by which plastic flow might occur in noncrystalline materials, i.e., glasses, many have doubted the existence of plastic flow in very hard crystalline ceramic materials at least until high temperatures are reached.

Focusing on crystalline ceramic materials for which the microstructural dependence of properties is most important and most common, it now seems clear that significant microplastic flow accompanies indentation of these materials. This is shown in recent surveys by Rice (1971) and Hockey (1973), as well as data presented later in this section. Though further work is clearly needed in examining the role of fracture in forming indents, the clear occurrence of microplasticity indicates that it is the fundamental and controlling process of such indentations, at least at lower loads, e.g., 100 gm, where cracking does not occur or is limited. Recognizing this importance of plastic deformation allows much of its variation to be understood at least qualitatively, e.g., effects of stoichiometry due to their effects on plastic flow. Also, as shown by Becher (1974) for sapphire and rutile and by Rice (1973a) for MgO, surface work hardening, which varies with surface preparation and possibly microstructure, can increase hardness values 50%. Such work hardening is important in the load dependence of hardness, since more of the greater depth of indentation at higher loads takes place below the work-hardened zone, giving lower hardness values.

It is clear that plastic deformation, and probably cracking should give a microstructural dependence, e.g., porosity and grain size dependence of hardness. However, adequate microstructural characterization is generally not given with hardness data, so apparent variations in hardness from investigator to investigator is often due to differences in microstructure. There are, however, enough hardness–microstructure data to indicate the major microstructural trends.

2. The Porosity Dependence of Room-Temperature Hardness

Available hardness values systematically decrease with increasing porosity, fitting well the exponential relation used to put all data on a common basis (Fig. 40). Note that the glassy carbon data follows the relation shown to the upper extent of its porosity ($P = 0.7$) as does gypsum (discussed below). Note also Neuber and Wimmer's (1971) ~ 6 μm grain Al_2O_3 data ($H_v \sim$ 100 μm off the scale of Fig. 40) gives $b = 5.5 \pm 1.4$ with $H_0 \sim 2900$ kg/mm^2,

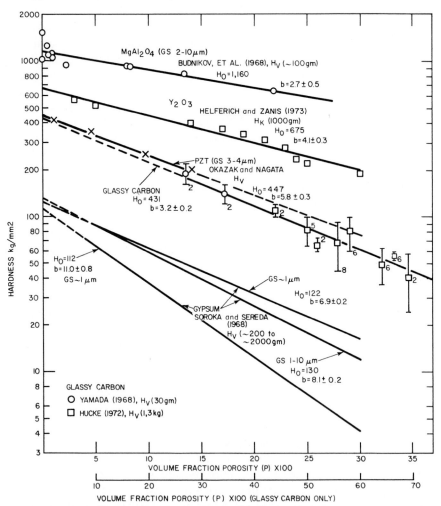

Fig. 40. Hardness of various ceramic materials versus volume fraction porosity. Note that the lower scale is for glassy carbon only. Sub- or superscripts represent the number of tests averaged for a given data point, and bars indicate the standard deviation.

or if one particularly low point is discarded, $b = 6 \pm 1$ and $H_0 = 3150$ kg/mm^2. These b values, particularly the latter, are consistent with the average b value from Fig. 40 of 6 ± 3.

As with other mechanical properties, inhomogeneous porosity can lead to variability within a given set of data and, more commonly, between different sets of data. This is probably the major reason why two sets of Soroka and Sereta's gypsum data at higher porosity ($P = 0.18–0.42$ and $0.5–0.7$) have similar slopes ($b \sim 10.4$ and 12.8, respectively) to their other data but are offset so they give higher zero intercepts ($H_0 \sim 396$ and 2120 kg/mm^2, respectively). (Note also that Soroka and Sereda increased the indent load as hardness decreased to maintain the indent depth at 35–50 μm, which may increase the apparent porosity dependence some.) Thus, for example, indents distorted by being near a large pore in a body having part or all of its porosity as large pores are likely to be discarded, while values further from pores will be higher, leading to higher hardness values than obtained in a body of the same porosity but having all fine pores. Also, different grain sizes may be a factor in different hardness values as discussed below.

3. THE GRAIN SIZE DEPENDENCE OF ROOM-TEMPERATURE HARDNESS

The available data for materials having hardness less than approximately 1000 km/m^2 are shown as a function of grain size in Fig. 41. MgF$_2$ and ZnS data of Huffadine et al. (1969) agree very well with that of Carnall, since their samples had essentially the same grain sizes (Huffadine, private communication). Note that the ZnS single crystal data were obtained with a Vickers indentor, which often gives higher, e.g., 20%, hardness values, so there may be more grain size dependence than indicated. The materials of Fig. 41, most of which are for optical purposes, have generally been better characterized and are of very high quality, in part because their lower processing temperatures are amenable to higher densities and often higher purities, as well as because their applications require high quality. All of these materials show an increase of hardness with decreasing grain size in accordance with a Petch-type relationship. Recent hardness (K, 50 gm) data of Swanson and Pappis (1975) data on CVD ZnSe having $G = 30–100$ μm, fitted the equation $H = 77 + 169G^{-1/2}$. Similarly, Rice (1971) has compiled data of Brace (1960, 1961) showing a similar Petch-type dependence of hardness of limestone, quartizite, and basalt on grain size. Further, the fact that the Y$_2$O$_3$ data of Helferich and Zanis (1973) (Fig. 40) extrapolates to 675 (a higher value than that reported by Anderson (1970) for Yttralox of 600 kg/m^2) also indicates hardness increases with decreasing grain size. While Helferich and Zanis do not give grain size data, their use of powders having particle sizes less than 2 μm, moderate sintering temperatures, and

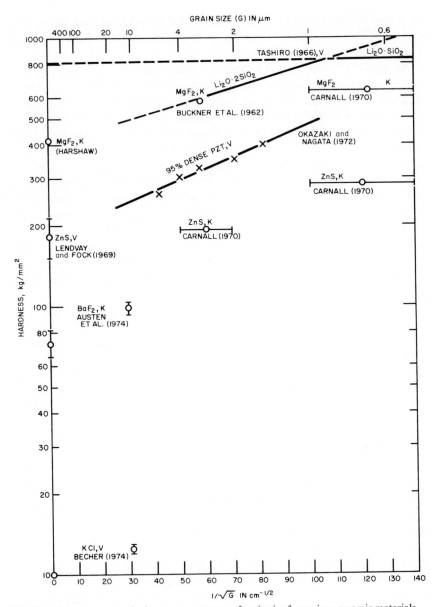

Fig. 41. Hardness versus the inverse square root of grain size for various ceramic materials.

substantial final porosity indicates that grain sizes in their samples were substantially less than the 40 μm in Yttralox. This is corroborated by Dutta and Gazzas' (1969) hardness (K, 100 gm) on dense hot-pressed Y_2O_3 ($G \sim$ 1 μm) of 900 kg/mm^2. Similarly, Soroka and Sereta's (1968) intercept values of 110–130 kg/mm^2 for gypsum, in contrast to the typical Vickers hardness of 50–60 kg/mm for gypsum crystals, probably represent effects of their fine grain size (e.g., 10 μm or less).

Data for substantially harder materials are generally incomplete because of limited data from any investigator and variable character and quality of specimens, especially between different investigators. However, Floyd's (1965), Rockwell (45 N scale) data for 96% Al_2O_3 bodies having average grain sizes of 6–25 μm follow a Petch-type relationship, $H = 57.7 + 0.47G^{-1/2}$ (G in centimeters), very well. Neuber and Wimmer (1971) showed random Vickers hardness (100 gm) measurements on various bodies of a commercial (99.5%) Al_2O_3 ($P \sim 0.085$) increased from ~ 2000 kg/mm^2 to ~ 3200 kg/mm^2 as grain size increased from ~ 5 to ~ 15 μm then decreased to ~ 2700 kg/mm^2 for larger grain sizes (~ 25 μm in contrast to typical single crystal values, ~ 2500 kg/mm^2). Though obtaining lower values, they found the same trend for measurements within different grains of a given specimen, i.e., ~ 1500 kg/mm^2 at $G \sim 20$ μm then falling to ~ 1360 kg/mm^2 at $G \sim 50$ μm, again indicating a grain size effect. The lower values of these single grain tests, and the lower hardness at finer grain size is probably due substantially to the amount and location of porosity. Kalish et al. (1966) did not directly determine the dependence of Knoop hardness (200 gm) of B_4C on grain size but noted that hardness passed through a maximum with the boron to carbon ratio. Analysis of their data ($G = 2-8$ μm) indicates the following hardness–grain size relationship: $H = 3000$ kg/mm$^2 + 2.3G^{-1/2}$ (G in microns), i.e., that hardness, though not extremely sensitive to grain size, does increase with decreasing grain size.

An exception to the above grain size trend has been reported by Armstrong et al. (1970) for commercial BeO samples using Vickers indentation at 1 and 2 kg loads. They report hardness increasing with increasing grain size (5–65 μm), i.e., the opposite of the above trend. Whether this is a real effort, e.g., possibly due to the anisotropic nature of BeO or to an anomaly due to their particular samples or test conditions (e.g., high load, presumably with attendant cracking) yet remains to be determined. Lawn and Wilshaws' (1975) recent analysis of cracking around indents at larger loads indicates that the extent of such cracking is inversely related to fracture energy, which might explain the results of Armstrong et al. However, the present data indicate that hardness normally increases with decreasing grain size, especially in softer materials.

4. IMPURITY AND ADDITIVE DEPENDENCE
   OF ROOM-TEMPERATURE HARDNESS

Alloying effects that increase the hardness of softer materials, such as alkali halides (e.g., Chin et al., 1973) and CaF_2 (e.g., Patel and Desai, 1970) have been clearly established. However, very little study of such effects has been made in harder ceramics. Further, limited studies of the effects of additives or impurities on the hardness of harder ceramic materials are often complicated by little or no characterization, so that effects of grain size or porosity or distribution of the additives—e.g., at grain boundaries, in solid solutions, or precipitated homogeneously or otherwise—often cannot be ascertained. Thus, as noted earlier, the variation of hardness that Kalish et al. (1966) reported as a function of stoichiometry in B_4C appears to be due primarily to grain size variation. However, again there are some fairly clear studies.

King and Yavorsky (1968) showed the Vickers hardness of MgO-stabilized ZrO_2 increased rapidly from about 1200 kg/mm^2 to approximately 1600 kg/m^2, then leveled off at about 1650 kg/m^2 as the weight percent of MgO stabilization increased from approximately 2.8 to 5.8 vol %. While these investigators did not present sufficient characterization data to unequivocally verify whether this significant rise in hardness is due predominantly or solely to an alloying type effect, the weight of evidence suggests that it in fact is due to alloying. Similarly, Wolfe and Kaufman (1967) show the Knoop hardness of UO_2 rising rapidly from about 650 kg/mm^2 as the O/U ratio increases from 2.0, it then appears to level off at about 850 kg/mm^2 as the O/U ratio approaches 2.15. Hannink and Murray (1972) have shown that the hardness of substoichiometric VC, i.e., $VC_{0.84}$, depends on the size of ordered domains, indicating an effect of substructure, and hence dependence on both composition and processing history. Such dependence is probably quite common in materials of variable stoichiometry such as carbides.

Dolomite and basalt data summarized by Rice (1971) indicate that bodies consisting of grains of two or more phases also follow a Petch-type hardness–grain size relationship. Also, Kennard (1973) has shown that the Knoop (300 gm) hardness of transverse sections of directionally solidified MgO–$MgAl_2O_4$ eutectics varies with the interlaminar spacing according to a Petch-type relation.

5. EFFECT OF TEMPERATURE ON THE MICROSTRUCTURAL
   DEPENDENCE OF HARDNESS

There have apparently been no specific studies of the effect of grain size, porosity, or impurities and additives on the temperature dependence of hardness. However, data of Atkins and Tabor (1966) and of Koester and

Moak (1967) indicates that the porosity dependence of hardness remains similar at elevated temperatures to what it is at room temperature or increases, Fig. 42. Analysis of this data indicates b values of 3–6 for NbC, similar to room temperature hardness, while TiB$_2$ data indicate $b = 11–20$ and TiC $b \sim 20$ or more. These last two b values indicate a higher porosity dependence of hardness at elevated temperatures, which would be consistent with compressive strength trends discussed later.

The indicated changing rates of porosity dependence of hardness (Fig. 42), probably depend upon the character of the porosity. At elevated temperatures, greater plastic flow increases the plastic zone size around the hardness indent, so more pores would be involved in the interaction with the plastic flow from an indent. Thus, larger isolated pores that would not frequently be close enough to lower-temperature hardness indents to significantly lower average hardness could do so at higher temperatures, increasing the porosity dependence of hardness.

While no apparent studies of the effect of temperature on the grain size dependence of porosity have been made, the correspondence of hardness and compressive strength shown in the next section indicate that increasing grain size will reduce hardness values at elevated temperature in a similar fashion as that indicated at room temperature. Further, increased plastic flow at elevated temperatures indicates hardness following a Petch relation similar to room temperature data based on increasing similarity to the behavior of ductile metals.

6. Discussion and Summary of the Microstructural Dependence of Hardness

Hardness clearly decreases with increasing porosity. Fitted to the exponential relation, the average b value is ~ 6, the same as for compressive strengths and 50% higher than for fracture energy, tensile strength, and Young's modulus. The variable porosity dependence is probably predominantly due to variation in size, shape, and location of pores. However, specifics of this porosity dependence, its variability, and the apparent increase of the porosity dependence at higher temperatures are not completely clear.

The limited data available clearly indicate that hardness in softer materials increases with decreasing grain size according to a Petch-type relationship. Still more limited data suggest that a similar though possibly lower increase in hardness with decreasing grain size exists for harder materials, with lower indent loads, i.e., with little or no cracking. A decreased dependence on grain size with increasing hardness could be quite real, since the decreasing extent of plastic flow will reduce the interaction of the flow with the grain boundaries and hence the effect of grain size. Results of hardness

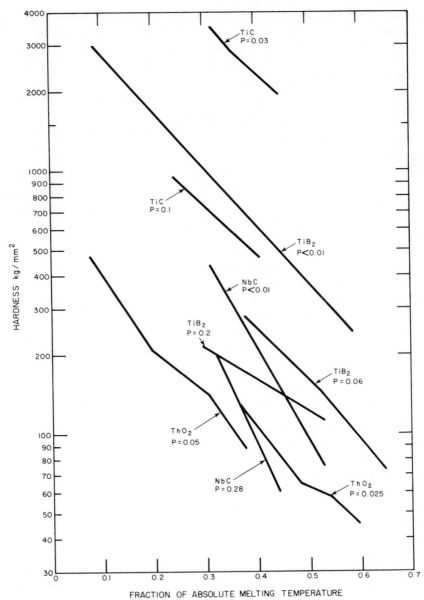

Fig. 42. Hardness of various ceramic materials differing mainly in porosity as a function of temperature.

increasing with increasing grain size with heavier loads in BeO may be due to cracking effects.

Although grain orientation can be a factor in polycrystalline hardness values, it does not appear to be a major variable, since substantially wider scatter is not seen in larger-grain bodies than in finer-grain samples. However, more data on the effect of grain orientation and on the hardness of single crystals versus orientation is needed.

Hardness is determined primarily by plastic flow at lower loads, generally explaining much of the surface, impurity and additive, and microstructural dependence. However, cracking can become significant at higher loads, especially in harder materials, thus perhaps altering microstructural dependence, e.g., adding the effect of fracture energy. A major need is to understand better the microstructural dependence and activation of both cracking and plastic flow accompanying indenting. Gilman's (1973) presentation of Rice's observations on the relative roles of primary and secondary slip systems appear to be a useful start in understanding this aspect of plastic flow on microstructural dependence. However, more work is needed, especially to understand the effects of a free surface (including pores), since bulging or other surface deformation may relax to some extent the requirement for five independent slip systems, as may twinning.

B. Microstructural Dependence of Compressive Strength

1. BACKGROUND AND THEORIES

Ceramic compressive strengths are important not only for the high level of such strengths and important resultant applications, but also for their correlation with other behavior, e.g., friction and wear. As with tensile strengths, ceramics compressively fail in a brittle fashion over much of their temperature regime. At sufficiently high temperatures, they fail plastically, with a transition between these two extremes at intermediate temperatures. The extent of each of these regimes, of course, depends upon the basic nature of the material, but it also depends very extensively on porosity, grain size, impurities, and strain rate. Qualitatively, the plastic deformation regime, may be somewhat better understood than the brittle failure regime at lower temperature. However, none of the regimes is particularly well understood in terms of quantitative predictions of microstructural dependence. Theories of brittle compressive behavior of ceramics, developed from continuum mechanics, only allow a microstructural dependence to be inferred through their relating compressive strength as a multiple of tensile strength, e.g., eight times tensile strength for the Griffith theory.

Rice (1971) has shown however, that compressive strength depends on grain size in a different fashion than tensile strength, so there cannot be a fixed ratio between compressive and tensile strength. Further, the observed ratios vary over a much broader range than is predicted by any of these theories. McClintock and Wlash (1962) predicted that variable friction along cracks in compression could increase the ratio of compressive to tensile strengths. On the other hand, Babel and Sines (1968) demonstrated analytically that spherical pores will lead to a ratio of compressive to tensile strengths of 3, as predicted by the Coulomb–Mohr theory, to higher values as the elipticity of the pore increases, reaching a value of 8 as the pore approaches an infinitely sharp crack, i.e., the same as the Griffith theory. While the above two modifications of the continuum mechanics brittle fracture theories would cover most but not necessarily all of the observed compressive to tensile strength ratios, there is no way of knowing how the two mechanisms would interact and how to predict their possible variable degree of interaction.

Rice (1971, 1973b) proposed that the brittle compressive strength behavior of dense, high-quality ceramics could be determined by crack nucleation and growth processes related to microplasticity, i.e., slip and/or twinning, especially in single-phase bodies. He showed that the yield stress for crystalline ceramics, taken as $H/3$ as in metals, is equal to or greater than the upper limit of their compressive strength. Similar use of Marsh's theory for calculating yield stresses for glass as the upper limit of their compressive strength was considered. Variations of compressive strength below these hardness-based limits were attributed to (1) nucleation and/or growth of cracks by twinning or a very predominant slip system not fully reflected in hardness and (2) other stresses or stress concentrations, e.g., at grain boundaries, due to TEA and EA, and due to pores, cracks, or inclusions anywhere in the body. This theory is consistent with the quite variable compressive-to-tensile strength ratios due to the zero intercept of brittle tensile strength Griffith-type behavior and the nonzero intercept of Petch-type compressive strength behavior as expected from microplastic mechanisms (see Section B). Further, even when tensile strengths are determined by microplastic mechanisms, they typically result from the operation of the easiest mode of microplastic deformation whereas the stability of cracks, and the high stress levels in compression activate more modes of microplastic deformation, again resulting in variable compressive-to-tensile strength ratios. Jortner (1968) considered modifying brittle theories of compressive failure by considering local plastic deformation at cracks or voids, but a comprehensive theory was not developed. A basic problem with essentially all of these compressive theories, as noted by Jortner, is that they generally do not treat the separate problems of crack nucleation and crack propagation.

As Rice (1971) has reviewed, there is considerable evidence that a number, probably a large number, of cracks generally coalesce to cause compressive failure.

There has generally been no attempt to specifically derive a theory of the porosity dependence of compressive strengths, though, as noted earlier, the exponential relation was first empirically applied to compressive data. For highly porous materials, i.e., foam plastics, Patel and Finnie (1970) present an equation $C = C'(1 - P)^m$ where C' and m are parameters dependent on foam structure, e.g., m ~ 1–2, while a value of $\frac{2}{3}$ would be obtained if structure were neglected. Whether their equation, which is based in part on plastic buckling of parts of the foam structure, is applicable to brittle ceramics remains to be determined. However, the relation of compressive strength to tensile strength and/or hardness and the fact that compressive strength must depend on elastic properties and fracture energy indicate that it should have related porosity dependence. In fact, all of these properties have similar porosity dependence, which simplifies matters and is probably a major factor in the wide use of similar porosity relations for compressive strength, as for other mechanical properties. A few investigators developing strength–porosity models (e.g., Schiller, 1958) (Section IV,A and Table IX) have specifically noted applicability to compressive strength, apparently based on expected tensile–compressive strength relations or on empirical fitting of data. Also, as noted earlier, the exponential porosity correction was first applied to compressive strength.

2. POROSITY–COMPRESSIVE STRENGTHS OF CERAMICS

While studies of the porosity effects on elastic and tensile strength behavior are far more extensive, there is nevertheless sufficient data in the literature to indicate the basic compressive strength–porosity behavior. A substantial portion of the data on the dependence of room temperature compressive strength on porosity is given in Table XVI, where again the exponential relation has been used to put data on a common basis. Some of the data are also presented in Fig. 43, showing first that it all fits the exponential relation quite well, even to very high porosities, e.g., Trostel's (1972) ZrO_2 data is linear to its upper porosity, $P = 0.9$, while his Al_2O_3 data drops off somewhat before this limit, i.e., at $P \sim 0.85$. Second, the b values average from about 6 ± 2 for essentially single-phase crystalline materials of Table XVI to about 7 ± 4 for all materials in Table XVI. These averages are in good agreement with Schiller's (1958) compressive strength data on plaster (typically $0.3 < P < 0.6$), which give b values of 5–8 (Table XVII). These averages are also consistent with approximate b values for correcting Floyd's (1965) data to that of dense Al_2O_3 (Fig. 43). These b averages are similar to those for hardness, again suggesting a

TABLE XVI

Porosity Dependence of Compressive Strength at Room Temperature

$$C = C_0 e^{-bP}$$

Material	Approximate Grain size, G (μm)	Approximate Porosity range (%)	C_0 (10^3 psi)	b	Source
Oxides					
Al_2O_3	—	4–50	500	7.4 ± 0.7	Ryshkewitch (1953)
Al_2O_3	—	29–90	100	5.1	Trostel (1972)
B_6O	—	1–23	266	5.4 ± 0.4	Ruh et al. (1975)
BeO	—	4–31	200	7.3 ± 0.5	Udy and Boulger (1949), see Lang and Schofield (1955)
BeO	—	1–6	335	7.4 ± 1.5	Elston and Labbe (1961)
SiO_2	—	0–20	190	18.2 ± 0.2	Gannon et al. (1965)
		30–50	4,000	16.3 ± 0.2	
		30–60	26	6.3 ± 0.8	
		60–80	60,000	15.2 ± 1.0	
ThO_2	60	0–17	386	1.4 ± 0.4	Curtis and Johnson (1957)
UO_2	20–50	8–23	221	6 ± 1	Burdick and Parker (1956)
ZrO_2	—	10–60	570	7.2 ± 0.4	Ryshkewitch (1953)
Other materials					
Cement	—	9–40	20	1.3 ± 0.1	Verbeck and Helmuth (1968),
Cement	—	1–20	88	6.0 ± 0.3	Roy and Gouda (1973), and Brainer (1972)
Glassy Carbon	—	30–68	200	3.3 ± 0.6	Compiled from Hucke (1972) and Yamada (1968)
POCO Graphite	—	18–32	63	6.4 ± 0.8	POCO Graphite
Graph-i-tite G (with grain)	—	15–32	20	4.7 ± 0.3	Carborundum Co.
CVD infiltrated carbon felt	—	—	30	9.5 ± 0.1	Kotlensky (1973)
TiB_2	15–20	5–15	315	4.5 ± 0.1	Mandorf et al. (1963)

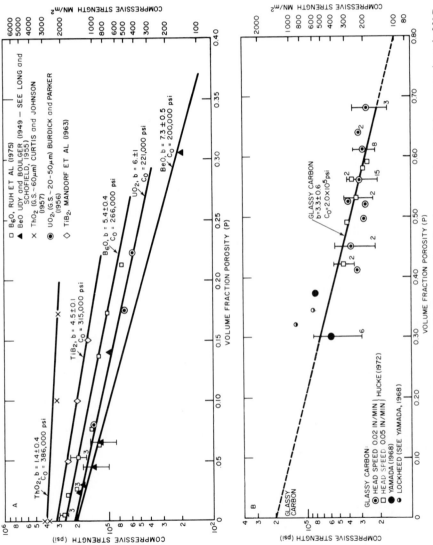

Fig. 43. Compressive strength versus volume fraction porosity of various ceramic materials at approximately 23°C.

TABLE XVII

COMPARISON OF SCHILLER'S P_{cr}
WITH EXPONENTIAL b VALUES

Material	P_{cr}	b values for strengths		
		Flexure	Tension	Compression
Coal[a]	0.385	13	—	—
	0.30	16	—	—
Plaster[b]	0.55	6	~ 5	8
	0.85	~ 4	~ 4	5
	0.65	~ 6	—	—
	0.8	~ 5	~ 5	5
	0.75	~ 5	~ 5	5

[a] Data of Millard (see Schiller, 1958).
[b] Data of Schiller (1958).

relationship with hardness, but substantially higher than (e.g., nearly double that of Young's modulus) fracture energy and tensile strength.

As with other mechanical properties the porosity dependence of compressive strength shows considerable variation, which is again attributed mainly to variation in pore size, shape, and location. This is illustrated by Ryshkewitch's (1953) ZrO_2 and Al_2O_3, where a foamlike structure, obtained by use of H_2O_2 in slip-casting specimens, having quite uniform pore size and spatial distribution, gave low scatter of compressive strengths. In his Al_2O_3 samples, a preferred orientation to the porosity correlated with an orientation dependence of the compressive strength. Similarly, the wide variation in compressive strength porosity behavior of fused silica in the studies of Gannon et al. (1965) appears to correlate with expected inhomogeneous shape and spatial distribution of porosity from their processing of bodies from different mixes of fiber and powder starting material. Finally, the inverse relation between P_{cr} and b in Schiller's data (Table XVII) also indicates higher b values due to inhomogeneous porosity. Indeed, sample trial plotting shows that for the same range of P, P_{cr} always decreases as b increases. Changes in grain size for different porosities may also be a factor in some of the variation of the compressive strength–porosity trends. Also, as discussed later, difficulty of eliminating end effects in compressive testing may be another factor.

Data on the porosity dependence of compressive strength at elevated temperatures is even more limited than at room temperature, but they nonetheless indicate some important trends. Data of Evans et al. (1972) on NiO, though scattered, suggest that the porosity dependence of compressive strength remains approximately constant to about 150°C and then begins decreasing to about 600°C (Fig. 44). Similarly data of Elston and Labbe (1961) on BeO at 20,500 and 1000°C indicate s substantial reduction in the porosity

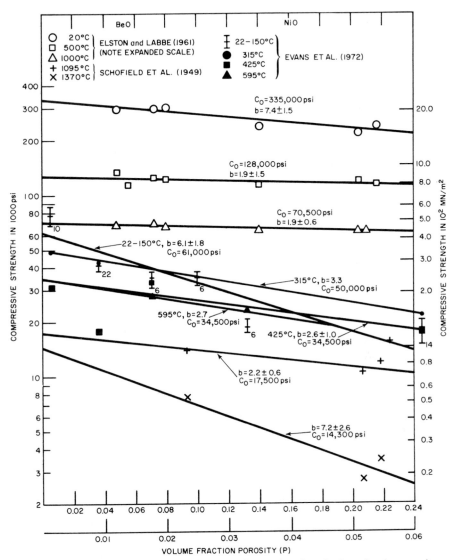

Fig. 44. Compressive strength of various ceramics as a function of volume fraction porosity at room and elevated temperatures. Note that the lower scale refers only to the BeO data of Elston and Labbe.

dependence as the temperature increases over this range. On the other hand, limited Al_2O_3 data of Shofield *et al.* (1949) indicate a substantial increase in the porosity dependence of compressive strength on going from approximately 1100°C to approximately 1370°C. Such marked increases in the porosity dependence of the compressive strength of Al_2O_3 at quite high temperatures is also indicated in Section 3, below, discussing the grain size dependence of compressive strength, and by analysis of Becher's (1971) data on high purity aluminas ($G \sim 20$ mm). Comparison of his data for a body with approximately 2% porosity with two bodies having nearly zero porosity indicates exponential b correction factors of ~ 20 at 1210°C, ~ 55 at 1420°C, and ~ 70 at 1700°C. Note also the marked dropoff seen in the compressive strength data in Ryshkewitch's compressive strength data of ThO_2 in comparison to more dense materials (Fig. 45).

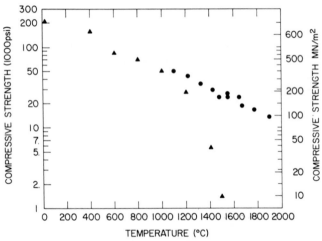

Fig. 45. Compressive strength of ThO_2 versus temperature. Note the much more rapid decrease of strength of the more porous material above about 1200°C. Data compiled by Peterson and Curtis (1970) from: (●) Yust and Poteat (grain size ~ 10 μm, $P \sim 0.025$); (▲) Ryshkewith (grain size ~ 10 μm, $P \sim 0.08$).

Thus, there are substantial indications that the porosity dependence of compressive strength increases considerably at high temperatures, and there is more limited evidence that the porosity dependence of compressive strength may decrease at intermediate temperatures. These two tendencies may not be inconsistent, since at the lower temperatures, where extensive grain boundary sliding is not expected to occur, some slight degree of plasticity may help relieve local stresses and reduce the effect of porosity.

On the other hand, at higher temperatures, e.g., $> T_m/3$, where grain boundary sliding becomes significant, pores along grain boundaries should greatly enhance the onset and extent of such sliding and hence have a more detrimental effect on compressive behavior.

As with lower-temperature compressive strength and other mechanical behavior, the character of porosity should play an important role in the high-temperature compressive strength. However, Langdon and Pask's (1971) study is one of the few to address these parameters. They showed that in MgO samples having from 0 to $\sim 1\%$ porosity, pore size and location were important, especially for the amount of plastic strain prior to failure. Thus, having some pores at grain boundaries that were comparable to the grain size precluded lower-temperature plastic deformation, while finer pores at grain boundaries, and especially within grains, enhanced lower-temperature plastic deformation. Such pore size and location effects were most significant below 1200°C, where slip intersections were difficult.

3. GRAIN-SIZE DEPENDENCE OF COMPRESSIVE STRENGTH

Many investigators have failed to provide information on the grain size of samples in their compressive strength studies, apparently assuming that grain size was not a significant factor. However, Rice (1971, 1972b) has surveyed the available compressive strengths grain-size data and has shown clearly that compressive strength generally increases with decreasing grain size according to a Petch-type relationship. The data that he compiled are summarized in Fig. 46 along with Floyd's (1965) Al_2O_3 data. Note that the lower strength of Floyd's data for 96 and 94% Al_2O_3 is consistent with the expected porosity dependence of compressive strength, with the presence of impurities, primarily SiO_2, also probably contributing some to the lower strengths of his samples compared to those for nearly theoretically dense, high-purity Al_2O_3 (top curve, Fig. 46). Compressive strength data for CdTe and PbTe (Fig. 32) clearly indicate a Petch-type behavior. Note, also that according to Evans et al. (1972), 22 μm grain size NiO always has lower strength than finer-grain bodies of similar porosity (Fig. 44) again indicating a compressive strength decreasing with increasing grain size. There is often substantial scatter in the compressive strength grain-size data, and some of the data has been obtained by correcting strengths of bodies with limited but variable porosity to zero porosity, e.g., using corrections discussed in the previous section. However, the collective data show that compressive strength clearly increases with decreasing grain size and is generally consistent with a simple Petch-type relationship.

There have been a few sufficiently comprehensive studies to indicate that compressive strength similarly follows a Petch-type relationship at elevated temperatures (Fig. 47). Note that Evans' (1963) Lucalox data

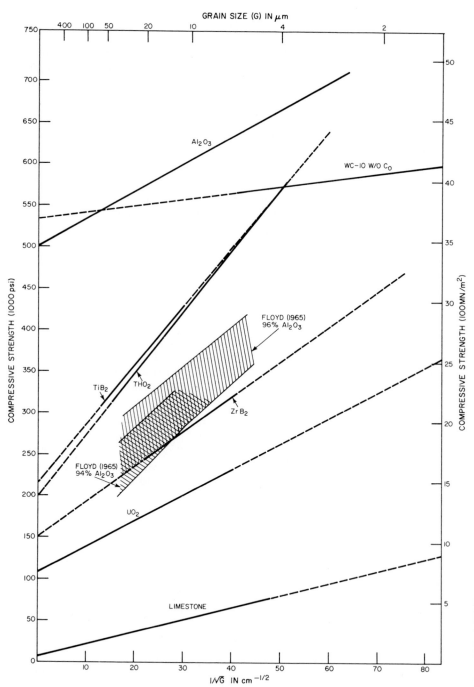

Fig. 46. Compressive strength versus inverse square root grain size of various ceramic materials at approximately 23°C. Curves are from the survey of Rice (1971). Solid lines show the range of data covered. Note that the alumina data of Floyd (1965) is consistent with the upper alumina curve for dense alumina when porosity and impurity effects are considered.

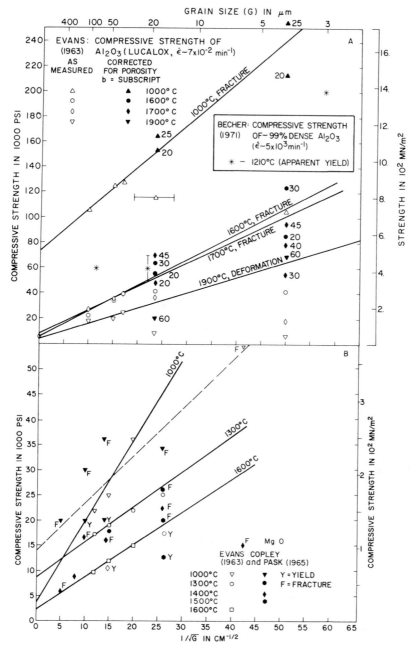

Fig. 47. High-temperature compressive strength of Al_2O_3 and MgO plotted on a common grain size scale. (A) Al_2O_3; note the *b* values adjacent to the finer-grain samples used to correct their strengths to zero porosity indicating the effect of porosity becomes increasingly severe on compressive strength as temperature increases. (B) Combined MgO data of Evans (1963) and Copely and Pask (1965); note that the combination gives a much more rational slope and intercept at 1000°C than Evans' data alone.

(see also Weil, 1964) at finer grain sizes, and limited amounts of porosity, indicate porosity corrections much higher than those found at room temperature if the compressive strength behavior is the expected linear extrapolation of larger grain size data, indicating an increasing porosity dependence of compressive strength with increasing temperature as noted in the previous section. Becher's (1971) Al_2O_3 data at 1210°C falls approximately in between Evans' 1000 and 1600°C data, as expected, while his higher-temperature data falls below that of Evans. However, this is expected, since Becher utilized a strain rate over an order of magnitude below that of Evans, so

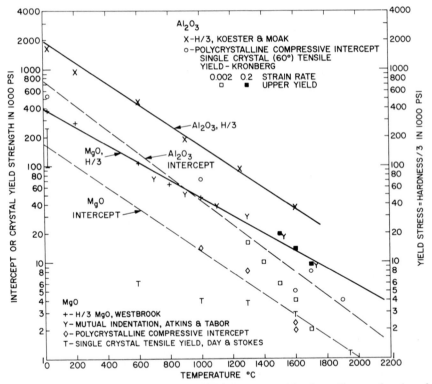

Fig. 48. Comparison of compressive strength intercept and hardness data as a function of temperature for Al_2O_3 and MgO. Note that the Petch intercepts for compressive strength data agree well with single-crystal yield data at high temperatures. The intercept data parallels the $H/3$ data, which is taken as the yield stress. The lower level of the Petch intercept data relative to $H/3$ is attributed to surface work hardening increasing the hardness and hence the apparent yield stress, the fact that hardness data was typically obtained on polycrystalline samples and was not extrapolated to infinite grain size to eliminate the grain size dependence, and the fact that compressive strength is lowered by defects and inhomogeneities scattered throughout the bulk of the samples. Data after Rice (1973b), by permission from "The Science of Hardness Testing and Its Research Applications," Copyright American Society for Metals, 1973.

greater differences between the two sets of data would be expected with increasing temperature as indicated. Note also that Al_2O_3 single-crystal data at 1600°C is in good agreement with Evans polycrystalline data. Similarly, combined data on MgO from Evans (1963) (see also Weil 1964) and Copely and Pask (1965) (Fig. 47) are in general agreement and fit a Petch relationship. Note that the combined data indicate a more sensible intercept at 1000°C than Evans' data alone would.

The observation of definite yielding in NiO, MgO, and Al_2O_3 samples, i.e., Becher's (1971) data, at elevated temperatures and in CdTe and PbTe at room temperatures clearly shows that the Petch-type behavior here can be interpreted in terms of micro- or macroplastic flow processes. This is further supported by Rice's previous demonstration that tensile and compressive strength behavior in dense pure bodies become similar at higher temperatures and also agree with single-crystal yield stress data (Fig. 48). Further, support of Rice's hypothesis that the $H/3$ limit of compressive strength in dense, high-quality ceramics even at room temperature is determined by microplastic processes has been previously demonstrated (Rice, 1973b) by the correspondence between the temperature dependence of hardness and the Petch-plot intercepts for both MgO and alumina (Fig. 48). Surface work hardening has been sighted as one of the important reasons why the hardness over three limit tends to be higher than the Petch intercept. Variable porosity and impurity results must also contribute to this.

4. DISCUSSION AND SUMMARY OF THE MICROSTRUCTURAL DEPENDENCE OF COMPRESSIVE STRENGTH

A major factor in evaluating the microstructural dependence of compressive strength data is whether the data represents true compressive strength behavior or is substantially or totally compromised by large-scale tensile stresses due to misalignment, end effects, or both (Rice, 1971). These are often manifested by longitudinal fracture, i.e., fracture parallel to the compressive axis, often on a single fracture plane as opposed to multiple fractures. Thus, for example, the low, e.g., 40,000 psi, room-temperature compressive fracture stresses Langdon and Pask (1971) report for their dense MgO in contrast to expected compressive failure stresses of about 200,000 psi (Fig. 48) combined with their longitudinal fracture along a single plane suggest such tensile failure effects. Indeed, compressive loading of transparent CVD ZnSe shows longitudinal cracking beginning at specimen ends that can lead to splitting of samples (Pohanka and Rice, to be published). However, in many tests, propagation of end cracks reduced or ceased, so that sufficiently high stresses were reached in the ZnSe and significant microcracking occurred throughout the body, which often was along grain boundary facets, especially from triple points, along with some transgranular failure, often apparently along twins. Such microcracking is believed to represent the initial stages of true compressive failure. One may frequently be able to observe intrinsic compressive behavior, indicating that even in the presence of end effects and resultant longitudinal cracking. Thus, while more work is clearly needed, the above ZnSe observations, the very high stress levels reported in many compressive strength studies, and the mode of fracture (discussed below) and the multiple fracturing often of an explosive nature reported by some authors suggest that most of the trends noted in this review probably represent basic compressive behavior.

Turning now to the mode of failure, Rice's (1971) review of compressive failure indicates predominately intergranular failure. As temperatures are raised, some of the propensity for intergranular fracture may be relaxed, at least until grain boundary sliding becomes effective, again increasing intergranular failure at higher temperatures. Thus, for example, Langdon and Pask's (1971) predominantly transgranular fracture along their longitudinal failure of MgO at 23°C is consistent with tensile effects as discussed above, in contrast to expected intergranular compressive failure. On the other hand, at higher temperature, where both their observed plastic deformation as well as their more complex fracture mode suggest true compressive behavior, failure was observed to be predominantly intergranular. The above ZnSe microcracking (e.g., along grain boundary facets) appears to be the source of larger, transgranular fractures at higher stress. More recently, Adams and Sines (1975) report that at about 40% of the compressive failure stress ($\sim$500,000 psi) of Wesgo Al995 Al_2O_3, small particles began spalling off the surfaces; as the stress increased, both the number and size of such spalled particles increased. Early spalls occurred from grain boundary inclusions and large (>50 μm) grains with associated voids. Spall fractures included both transgranular and intergranular fracture.

At high temperatures, both the type and nature of impurities can also significantly effect compressive strengths and hence the resultant microstructural dependence in much the same fashion that they effect tensile strengths. At lower temperatures, where elastic compatibility problems may be more severe in compressive than in tensile testing because of high compressive loads, there may be some differences in impurity effects. Unfortunately, little of this has been studied in a detailed and systematic fashion. Langdon and Pask (1971) did attribute the lowest fracture stresses at 800 and 1000°C and total grain boundary separation in MgO samples hot pressed with LiF to residual grain boundary contamination from the LiF similar to tensile strength effects in CaO made with LiF (Section IV,D).

There are also effects of loading rate, surrounding pressure, and possibly of environment on compressive strength. Although many loading rate effects on compressive strength are probably similar to those on tensile strength, e.g., as indicated in Fig. 47(top), definitive tests to study this are not known. Considerable compressive testing of samples under hydrostatic pressure (reviewed by Rice, 1971) shows that cracking and gross failure are inhibited, so higher stress levels can be reached that may accentuate microplastic phenomena that are probably basic to uniaxial compressive testing of dense ceramics. Charles (1959) is one of the few or only investigators to examine the effect of an environment, water vapor from 25 to 240°C, on the compressive strengths of fused SiO_2, some crystals (quartz and MgO), rocks, and alkali silicate minerals (granite, albite, spodumene, and horn-blende). All showed reduced compressive strength due to water temperature, e.g., 100–240°C. Whether this represents a true environmental effect on compressive strength is uncertain, since limited compressive strengths and longitudinal cracking, i.e., parallel with the compression axis of most of the specimens, indicate tensile failure from end friction effects rather than fundamental compressive behavior.

Finally, consider the implications of the observed microstructural dependence on the mechanism of compressive failure, Data following the Petch equation lend substantial support to the concept of microplastically induced compressive failure. However, superimposed effects of inhomo-geneous stresses from elastic and thermal expansion anisotropy must also be considered, since these would also reduce compressive strengths with increasing grain size. Indications that true compressive failure probably involves connection of many cracks before total failure occurs are consistent with both microplastic and inhomogeneous stress effects. The possible more limited dependence of hardness on grain size in harder materials is not inconsistent with the grain size dependence of compressive strength and microplastic mechanism of failure. The limited range of both very high stresses around the hardness indenters and of the range of slip in very hard materials may limit grain boundary interaction of hardness. However, the very high level of compressive stresses throughout a sample under compression would presumably eliminate any of this effect and its attendant reduction in grain size dependence.

The similar high ($b \sim 6$) porosity dependence of compressive strength and that of hardness again suggests a relationship between the two. The reasons for this apparent higher porosity dependence of compressive strength and hardness are not known. It is speculated, however, that it might result from pores acting as an integral part of flaws in tensile behavior, as discussed earlier, in contrast to their presumably acting primarily as stress concentrators in compressive testing. When pores act as an integral part of flaws the resultant crack is away from the pore wall, and the stress drops off very rapidly with distance of the pore, so there is less stress concentration from the pore; i.e., the tensile strength then is determined primarily by the combination of the pore and grain size and the porosity effect on Young's modulus and fracture energy as discussed earlier. In compression, such crack development from pores would be greatly inhibited if not prevented, at least until much higher stresses are reached, allowing pores to act as stress concentrators to activate slip or twinning. Pores could be more effective stress concentration than cracks or grain boundaries because of the large volume pores stress. Again, different porosity effects, i.e., b values, are attributed mainly to variable pore size, shape, and location.

C. Microstructural Dependence of Machining and Wear

1. BACKGROUND AND THEORIES

Machining and other abrasion and wear processes of ceramics, and hence their microstructural dependence, are quite important; e.g., machining typically is the major factor in their cost and plays a dominant role in their mechanical behavior. Abrasion and wear are important because of the many traditional and newer applications (e.g., bearings) that depend on these properties.

Recently, Koepke and Stokes (1970) developed a machining model of uncertain validity (see Rice, 1974d) that involves no material parameters, except indirectly through the machining forces. Choudry and Gielisse (1974; Grelisse, private communication) developed a model in which the grinding forces, in addition to machining parameters, depends on the ratio E/γ and inversely on a parameter associated with the flaw density. However, detailed evaluation is not possible without numerical values for several parameters. (Note also that a series of typographical errors makes it very difficult to follow their equations). Dependence only on E/γ would leave no porosity dependence. Grain-size dependence is uncertain because of the uncertainty of the dependence, but it appears that grinding forces are inversely proportional to G. Lawn and Wilshaw (1975) have recently indicated that the rate of material removal from machining or abrasion would be inversely proportional to H^{-1} or $H^{-3/2}$, which would in turn imply that the removal rate increased with porosity as e^{bP}, with b being about 6 ± 3 based on the dependence of H. Similarly, this would imply that the removal rate increased approximately as $G^{1/2}$.

Though there has been a recent increase in the study of friction and wear and machining processes in ceramics, there is still relatively limited data on the microstructural dependence of these processes. Such studies of course can be difficult due to the variety of abrasion conditions that can, for example, involve quite high local temperatures and resultant complicated thermal effects in addition to possible local chemical effects. Studies providing some definition of microstructural effects on machining abrasion and wear of ceramics are summarized below.

2. MACHINING OF CERAMICS

Recently, Rice and Speronello (1976) (see also Rice, 1974d) have conducted substantial study on the effect of grain size on diamond sawing and machining of a variety of ceramics, as well as some limited study of the effects of porosity on such machining. The major results, namely, that the inverse of the machining rate—i.e., the rate of diamond sawing a unit cross-sectional area or diamond grinding a unit area (for a fixed depth of cut)—increased with the inverse square root of grain size. Further, the rate of this increase was progressively higher as the hardness of the material increased, as illustrated in Fig. 49. More limited studies of the effect of porosities on fine (0.6–3 μm) grain MgO and Al_2O_3 indicated that the inverse of the machining rate decreased with increasing porosity, as would be expected. Fitting the limited data to the exponential relationship gives $b \sim 4$–6 for Al_2O_3 (Table XVIII), indicating a possible increase in b values with increasing hardness.

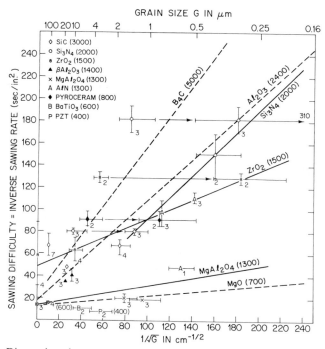

Fig. 49. Diamond sawing rate versus grain size for a number of relatively dense ceramic materials. Only dashed lines and not data points (extensive for Al_2O_3) are used for Al_2O_3 and BuC for clarity. Note that the inverse of the sawing rate referred to sawing difficulty follows a Petch type relationship, with the level of sawing difficulty and its rate of increase with decreasing grain size generally increasing with hardness. Note that four data points (two for Si_3N_4 and one each for ZrO_2 and Pyroceram) are plotted for both the large and fine grain sizes in each body, showing that the values for fine grains agree with values from other bodies. Approximate hardness values shown in parentheses. Data from Rice (1976).

TABLE XVIII

GRAIN SIZE AND POROSITY DEPENDENCE OF MACHINING AND WEAR OF Al_2O_3

Test	Range of grain size (G) (μm)	A of G^{-A}	Range of porosity (P) (volume fraction)	b of e^{-bp}	Investigator
Diamond sawing	1–∞	0.33	0–0.16	4–5	Rice and Sperinello (1976)
Diamond grinding	1–∞	0.18	—	—	Rice and Sperinello (1976)
Abrasion of grinding balls	2–60	1.7	2–15	23	Pearson et al. (1967)
Wheel abrasion tester	4–12	~0	0.05–0.15	9	Hines (1974)
Cutting tool wear	—	—	0.05–0.22	6	Whitney and Bates (1975)

Another important implication of the grain size studies was that where a substantial variation in grain sizes occurred, the rate of machining was determined by the smaller grains rather than the larger grains, the opposite of what frequently occurs in tensile strength control. This, however, is consistent with the machining process, since the rate of machining decreases with decreasing grain size, and the saw blade or grinding wheel cannot advance until all material is removed; it therefore must overcome the toughest or most difficult portion of the specimen, namely the finest-grain portions. Based on the assumption that size of the resultant flaws left in the surface of the machined sample are proportional to the forces between the abrasive tool (blade or wheel) and the sample, the resultant grain-size dependence of the machining rate provides a rational for the observation of approximately constant flaw size in a given material for a given machining process, regardless of grain size. This is based on the observation that the reduction in the rate of machining should primarily represent an increase in machining forces; i.e., because the less force opposing the tool, the faster it moves. This then implies that the machining forces increase as the inverse square root of grain size. Since both hardness and strength also increases in this fashion, the two effects should balance to leave a constant flaw size independent of grain size. Thus, this would give a changing flaw-to-grain size ratio as grain size was varied, as observed (see Section IV,C).

The quality of surface finish has long been known to depend on microstructure, since, for example, chipping around pores and pullout of grains, especially large ones, tend to produce poor surface finish. This latter effect, namely, the reduced severity of pullout problems with finer-grain bodies, has in fact been the primary reason for fine-grain hot-pressed Al_2O_3 replacing larger-grain Lucalox in the hub, rotor, and end cap components of some gyro systems. Similarly, surface relief polishing often occurs with bodies having phases of different hardness. While these general trends are widely observed, there has been little specific consideration of the details of the microstructural dependence of finishing.

It is now quite clear that abrasively machined surfaces of ceramics are work hardened. As reviewed by Rice (1974c), the depth of this work hardening is inversely proportional to hardness and decreases with the pressure or coarseness of machining. It also appears to have some microstructural dependence, e.g., decreasing some in depth with decreasing grain size. Rice (1973a) has pointed out that pore location can affect surface finish effects by affecting the type of fracture and, in cases where microplasticity is important in failure, by work hardening. Thus, for example, pores located along grain boundaries will enhance grain pullout of complete grain sections, while pores located within the grains will provide less tendency for sections to fracture out, and if they do, they should commonly be substantially less than a complete grain section. Further, in samples in which fracture depends on microplastic effects, such as in MgO, pores within grains were suggested to enhance work hardening, which enhances possible toughening of grains, whereas pores at grain boundaries would not have a similar effect.

3. ABRASION AND WEAR OF CERAMICS

Duncan *et al.* (1965) qualitatively observed that the abrasive wear of various bodies having about 45–95% Al_2O_3 content decreased with increasing alumina content as one might expect. Pearson *et al.* (1967), in their abrasion studies of 99.8% alumina balls, found nonlinear increases of wet abrasive loss W with both grain size and porosity that are much higher than in other related studies (Table XVIII), e.g., in marked contrast to Rice and Sperinello's machining rate data discussed in the last section. Similarly Hines' (1974) more recent abrasion tests of 96% Al_2O_3 (with $\sim 5\%$ porosity) while showing no effect of grain size (probably due to grain size variation in his limited range, $\sim 4-12\,\mu m$) found W proportional to e^{9P} (at $G = 4\,\mu m$). Plotting of Artamonov and Bovkuns (1974) data on the relative wear of TiC, ZrC, and WC as a function of P shows this data to be consistent with an e^{-bP} dependence with $b \sim 5 \pm 1$, 1.9 ± 1, and 4 ± 1, respectively. While part of the different results of Pearson *et al.* may be due to test difference, much appears to be due to much of their original grain boundary porosity becoming entrapped within grains as grain size increased with further sintering to reduce porosity. Thus, in view of previously discussed effects of intragranular pores increasing fracture energy and strength (and possibly their effect in increasing work hardening), the change from inter- to intragranular porosity as the total porosity is decreased should further reduce the wear along with the reduced porosity, giving an apparent increase in the porosity dependence. Similarly, such shifts of pore location should reduce the wear in large-grain bodies, giving an apparent increase in the grain-size dependence. Al_2O_3 bodies that Hines alloyed with Cr_2O_3 or $MgTiO_3$ showed improved wear resistance. Recently, Hockey (private communication) has begun a basic study of wear mechanisms of ceramic material, with preliminary results indicating as much as a twentyfold increase in wear as one goes from a dense fine grain ($\sim 3\,\mu m$ Al_2O_3) body to one having $\sim 30\,\mu m$ grain size. However, whether this is an intrinsic grain-size dependence or one resulting from differing surface finish effects due to the grain size is not yet certain. Both of the above grain-size and porosity trends are reasonable in view of the asperity concept of friction and wear, since the size of asperity would be expected to increase with increasing porosity and grain size.

Another important area related to friction and wear of ceramics is their use in machining other materials, primarily metals, e.g., cutting tool inserts. Although ceramic cutting tool microstructure appears to be an important factor in the performance of such ceramic tools, little data has been published clearly delineating the microstructural control of their machining behavior. Evans *et al.* (1967) investigated the performance of an Al_2O_3 cutting tool having grain sizes ranging from approximately 2 to over 150 μm. They found that as long as the total porosity of sample, was less than 10%, cutting tool materials having the finest grain size generally gave the best performance. Detailed grain size analysis showed that not only was fine grain size important, but it was also important to have a narrow distribution of grains, thereby indicating that large grains are detrimental to tool performance as they frequently are to tensile strength. Recently, Whitney and Bates (1975) presented data showing the cutting speed approximately doubling for a fixed 10 min. lifetime of three different Al_2O_3 tool materials as porosities decreased from about 22% to about 5%. While the limited data appears to fit a linear relationship somewhat better than the exponential relationship, *b* values were approximately 6 $\pm$ 2, i.e., similar to values of other abrasion tests (Table XVIII) and for other mechanical properties, especially hardness and compressive strengths. These investigators also note that tool performance can also depend significantly on grain size, generally being best for fine-grain Al_2O_3 bodies as well as recently developed Al_2O_3–TiC composite cutting tools, which show substantially improved machining capability.

Kim *et al.* (1973) have shown that extensive plastic deformation in the form of dislocations and twins occurs in the subsurface crater area of Al_2O_3 tools. Based on this and the inverse dependence of crater area on hardness, they suggest that plastic flow controls the rate of crater wear. Such plastic flow effects could be important in the grain-size dependence of tool performance.

Thus, the friction and wear behavior of ceramics, which plays a significant and growing role in many of their applications, depends substantially on microstructure. However, our knowledge of the microstructural dependence of this broad range of behavior is now only beginning to be definitive. A great deal more work is needed, again with much of the emphasis needed on the effects of the specific character of the microstructure, i.e., not just average grain size or average porosity, but the distribution of these.

VI. Microstructural Dependence of Creep

Only a brief summary of the microstructural dependence of creep will be presented. First, much of the work lacks adequate characterization to definitively separate microstructural effects, which are often more severe and interacting in creep than in other mechanical behavior. Second, there are a number of recently published good sources of information on creep, e.g., Landgon *et al.* (1971) and, in particular, the proceedings of a symposium on Plastic Deformation of Ceramic Material held at Pennsylvania State University, July 1974 (Bradt and Tressler, 1975), with the papers by Notis, and Cannon and Coble being of particular pertinence.

First, consider the grain-size dependence of creep. In the secondary or steady state creep region, which is the one of typical interest, the strain rate varies as G^{-A}, $A = 0, 1, 2,$ or 3.

$A = 0$ Independence of grain size occurs when deformation is controlled by the climb or glide of dislocations or by annihilation or dissolution of dislocation dipoles.

$A = 1$ Grain size dependence occurs for various grain boundary sliding mechanisms as well as for vacancy diffusion around grain boundary ledges.

$A = 2$ Dependence occurs when deformation is controlled by vacancy diffusion through the lattice.

$A = 3$ Dependence occurs when deformation is controlled by vacancy diffusion along grain boundaries.

Thus the grain size effect on creep can vary from zero to a very pronounced depending upon the mechanism of creep.

A major step forward in understanding the change from one creep mechanism to another has been made by Ashby and colleagues (e.g., Ashby, 1972) in their development of the deformation map approach. This approach has been modified for grain size effects (e.g., Notis, 1974) (Fig. 50). Such maps, whether in terms of temperature or grain size, are similar to a phase diagram, i.e., the different mechanisms of creep are represented as regions on a diagram of strain rate as a function of temperature or grain size, just as regions of a phase diagram are a function of composition and a second variable, such as pressure or temperature. If the effects of porosity and impurities are adequately understood, deformation maps based on these variables can also be made.

Since the above-noted range of grain-size dependence of creep rate is predicted by the various analytical models for each of the different mechanisms of creep, creep mechanisms are intrinsically grain-size dependent. Whether experimental results comparing bodies of different grain sizes are also influenced by different impurity or porosity contents or different grain boundary structure, i.e., changes in the ranges or types of grain boundaries as a function of grain size, is not known. However, there is clearly a pronounced increase of creep rate with decreasing grain size, much of which is intrinsic. It should be noted that these trends with grain size are opposite to those of short-term strength. Thus, if one is doing high-temperature testing at intermediate strain rates, the picture can be complicated by the

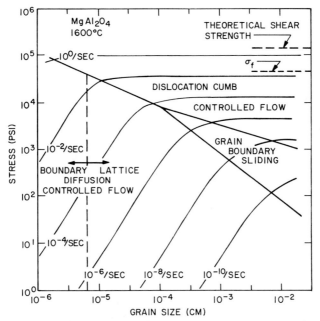

Fig. 50. Grain size deformation map of $MgAl_2O_4$ at 1600°C. Courtesy of Dr. M. Notis, Lehigh University.

opposite trends of short term and long term (i.e., creep testing) effects of grain size. This is also quite important from the practical standpoint, since if one is looking for a body with good strength at room temperature as well as with both short-term and long-term loading at elevated temperatures, important compromises often have to be made, since the latter often varies as a function of grain size opposite to that of the former two.

Turning next to the effect of porosity on creep rate, only a few investigators have analytically considered this with Langdon's (1972) note being the most recent and comprehensive. The general approach has been to modify the applied stress to obtain a local effective stress due to the porosity. This has typically been done using McClelland's (1961) equation to modify the applied stress for effects of hot pressing:

$$\text{Effective stress} = \text{applied stress}/(1 - P^{2/3})$$

Langdon then expressed the shear modulus as a function of porosity per Eq. (18) of Table II, obtaining the following general porosity dependence for the strain $\dot{\varepsilon}$ or creep rate:

$$\dot{\varepsilon} \alpha \left(\frac{\sigma_a}{G_0}\right)^n = \left[1 + \frac{bP}{1 - (b + 1)P}\right]^{1-n} \bigg/ [1 - P^{2/3}]^{-n}$$

Where this approach is applicable, there are two additional considerations that are suggested. The first is whether the suggested stress correction term is adequate, e.g., in view of the substantially higher dependence of strength on porosity. The term $(1 - P^{2/3})$ corresponds approximately to e^{-2P}, whereas strength varies on the average approximately as e^{-4P}. Second, where one assumes that the porosity effect on creep will be the combination of its effect on each of the separate parameters, i.e., shear modulus and stress, it appears much more convenient and useful to use a simple empirical relationship fitting the stress and shear modulus dependence of porosity, such as the exponential relationship. Thus, for example, if Langdon's approach is used with this modification and recognizing that the shear modulus varies approximately as e^{-3P}, one then obtains the $\dot{\varepsilon} \alpha e^{(5n-3)P}$. On the other hand, if the stress corrections is closer to that for strength so that the b values for stress correction and shear modulus are approximately the same, one obtains $\dot{\varepsilon} \alpha e^{(2n-1)bP}$.

The above approach of treating the porosity dependence of the creep rate as the combination of the porosity dependence of each of its appropriate parameters, namely, stress and shear modulus, could be quite erroneous in a number of cases, since the presence of different types and amounts of porosity could change the mechanism of creep and hence the grain size exponent with these changes being confused with porosity dependence. Thus, for example, a substantial amount of porosity at grain boundaries should enhance the possibility of grain boundaries sliding due to boundary weakening and stress concentrations. On the other hand, an equal amount of porosity present predominantly, if not exclusively, as intragranular pores could favor Nabarro–Herring creep due to resultant stresses within the grains. Also, again the effect of pore inhomogeneity might be significant, since the creep rate is quite dependent on porosity in a nonlinear fashion; a body having a mixture of more porous and less porous regions could exhibit a different creep rate than one having the same amount of porosity that was homogeneously distributed. Similarly, there could well be a pore size dependence on creep, since smaller, more homogeneously distributed pores would tend to provide a more homogeneous stress enhancement throughout the body than would large, isolated pores.

Next, consider the effects of impurities. Clearly, impurities that soften at a lower temperature than the matrix will have a significant effect on increasing creep rate, especially if they are located at grain boundaries, (e.g., see Kossowsky, 1973), while impurity particles that remain hard at high temperatures can significantly retard creep, especially if they are located as individual particles between grains. Impurities or additives in solution will have varying effects depending on how they effect dislocations and diffusion. Uncertain effects of impurities are an important factor in limiting the interpretation of a substantial amount of data, in particular, the uncertainty of the location of many of these impurities or the nature of their distribution.

Grain shape can also play a significant role in that one can have interlocking grains as in some mullite and refractory structures, which can be considerably more resistant to creep than specimens without such structures, e.g., having a more fluid phase surrounding a more equiaxial grain structure.

Understanding of the effects of more fluid grain boundary phases on the creep rate is quite important (e.g., in refractories, and more recently in glass ceramics, and SiC and particularly Si_3N_4) because of the presence of grain boundary phases in these hot-pressed materials, especially Si_3N_4. Lange's (1974) model for this, which relates void development in a fluid grain boundary phase to grain size, appears to be a useful starting point for analysis for this type of effect in the absence of the complicating porosity and multiphased nature of the refractory materials. White and Ashbee (1975) have observed void development in the deformation of glass–ceramic bodies.

Finally, it should be noted that for convenience most creep tests are done in compression or in bending, which could give different results than creep in tension. Thus for example, if creep is significantly affected by the greater extremes of porosity, grain size distribution, or impurities, tensile testing, which provides a large stress over a greater volume, would give more severe results than flexure testing. Thus, for example, White and Ashbee (1974) noted dimensional instability in fatigue tests (see Section IV,D) of a $Li_2O–ZnO–SiO_2$ glass ceramic due to a much higher yield stress in compression than tension. Also, where voids play a significant role, there is more opportunity for void closure in compression than in tension, so that porosity should increase at a faster rate in a tensile creep test than in compressive creep test.

VII. Summary, General Discussion, and Research Needs

A. General Microstructural Dependence of Mechanical Behavior

Clearly, our knowledge of the microstructural dependence of mechanical properties of ceramics has been expanding at a rapid rate, but much remains to be done. In this concluding section, general microstructural trends of mechanical behavior are first reviewed and some specific research needs noted. Then basic needs for further understanding are discussed, and finally, some of the promising or desirable microstructural developments for improved mechanical behavior are discussed.

First, a major factor is that, in contrast to most past analysis based solely on average microstructure, e.g., porosity or grain size, mechanical behavior is effected and some is dominated by the extremes of the microstructure. Thus, tensile failure is often totally determined by the largest single or group of larger grains or pores present, while the smallest grains or pores control machining. Tensile strength, not so dominated by microstructural extremes, and most other mechanical behavior, e.g., fracture energy, compressive strength, and elastic properties, though more closely related to average microstructure are still significantly affected by the distributions of the shape, size, and location of key microstructural features such as grains, pores, and second phases. The crack size relative to the microstructure can also be a factor in these variations. Thus, for example, Freiman *et al.* (1975b) indicated that the higher scatter and somewhat lower polycrystalline fracture energies in their ZnSe are probably due to large grains, which can represent a reasonable fraction of the crack front in their DCB test samples. Also, the much thinner web of the double cantilever beam type test would presumably make it more susceptible to this effect than the notch beam or work of fracture test (but not necessarily the fracture energy for the onset of failure). General porosity, grain size, and impurity effects are reviewed, next followed by a review of fracture energy.

Table XIX summarizes the overall average porosity trends of most of the porosity-dependent properties considered in this review. It is worthwhile to note that many other physical properties fit the exponential relation and show similar levels and variations of b values (e.g., Table XX), as do most other mechanical properties, e.g., machining, wear, etc. (Table XVIII). Variations in the size, location, and especially shape of pores are the major factors in the variation in b values. Data over a limited range of porosity

TABLE XIX

APPROXIMATE AVERAGE POROSITY DEPENDENCE
OF MECHANICAL PROPERTIES AT ROOM TEMPERATURE

Property	Approximate average b value
Young's modulus	4 ± 2
Bulk modulus	4 ± 2
Shear modulus	3.5 ± 1
Poisson's ratio	1.2 ± 1.2
Fracture energy	4 ± 2
Flexure (tensile) strength	4 ± 2
Hardness	6 ± 3
Compressive strength	6 ± 3

TABLE XX

ROOM TEMPERATURE POROSITY CORRECTIONS FOR NONMECHANICAL PROPERTIES

Material and reference	Porosity range (%)	Property	b (of e^{-bP})
Al_2O_3 (Morse and Hill, 1970)	0–17	Electrical strength	6.0 ± 1.5
Al_2O_3 (Grimm et al., 1971)	0.03–0.65	Optical transmission	1.3 ± 0.1
Graphite (Wagner et al., 1974)	15–34	Electrical conductivity	3.3 ± 1.4
		Thermal conductivity	4.8 ± 1.4
Graphite (Rhee, 1972)	15–32	Electrical conductivity	3.2 ± 0.1
		Thermal conductivity	4.6 ± 0.2
Ferrite (Igarashi and Okasaki, 1973)	4–42	Initial permeability	2.8 ± 0.5
PZT (Okazaki and Nagata, 1972)	1–14	Permativity	3.0 ± 0.2
		Electrical conductivity	6.1 ± 0.3
		Breakdown voltage	5 ± 2.5
Si_3N_4 (Walton, 1974)	20–32	Dielectric constant	2.6 ± 0.1

also contributes to this scatter. As discussed earlier, the apparently higher porosity dependence of hardness and compressive strength may be due to differences in stress concentration, but substantial further work is needed. In summary, Pores have any or all of the following effects on mechanical behavior:

1. Increasing stress, either just due to the removal of load bearing material or additionally due to stress concentrations effects.
2. Acting as an integral part of flaws.
3. Acting as sources of dislocation.

Pores as part of failure-causing flaws have begun to receive reasonable consideration. However, the important concept of pores acting as sources of dislocation, although suggested, generally independently by several authors with some degree of supporting evidence (e.g., Stokes and Li, 1963; King and Yavorsky, 1968; Langdon and Pask, 1971; Rice, 1972a,1973) has not been seriously investigated.

Turning next to the effect of grain size on mechanical properties, one of the dominating changes in understanding has been the recognition of the importance of grain size distribution effects noted earlier and of flaw–grain size effects. Thus, once larger grains, whether isolated in a fine-grain matrix, or as part of the main distribution of grains in the body become large enough to become comparable with the size of flaws causing failure, then they will dominate the grain-size dependence of strengths. In such cases, fracture energies controlling failure will decrease, approaching those of the low fracture energy planes of single crystals for intragranular flaws and fracture energies at or near those for bicrystals for intergranular flaws as the flaw size to grain size ratio decreases. While an estimate of grain boundary fracture energies being approximately one-half of those of single crystal energies is often used, there is little or no experimental support for this, indicating an important need to understand the energies for crack propagation along grain boundaries. A very important question is what values of the flaw size-to-grain size ratio are involved in the transition from failure being controlled by fracture energies at or close to those of single crystals or grain boundaries, i.e., bicrystal, fracture energies, to being controlled by normal polycrystalline fracture energies.

Another important grain-size development has been the recognition that fracture phenomena determined by stresses resulting from incompatible strains, e.g., from phase transformation and thermal expansion anisotropy, are controlled by the total strain energy within a grain or impurity particle rather than the stresses. While the latter are independent of grain or particle size, the former increase with increasing grain or particle size and hence explain many cracking and related phenomena that increase with grain size.

An important question related to internal stress effects is the statistical distribution of such stresses, which depends on the statistical distribution of grain orientations. In addition to the important development of recognizing the significance of flaw size–grain size effects on the fracture energy controlling failure, much of the confusion about the grain-size dependence of fracture energies appears recently to have been clarified. It now seems reasonable clear that fracture energy is generally independent of grain size in dense materials of cubic structure or limited anisotropy. However, for most tests, fracture energy of materials with considerable anisotropy first increases with increasing grain size, passes through a maximum, then decreases. While substantial work remains to be done, microcracking due to anisotropy is most likely the major factor in this changing fracture energy with increasing grain size. This probably occurs around the crack during the application of stress rather than as a general phenomenon in the body, and is dependent on TEA stresses and probably compounded by elastic anisotropy effects.

Fracture energies measured by notched beam or possibly fracture mirror techniques appear to typically show no grain-size dependence regardless of structure. This may be due to flaw sizes being of the same order of magnitude as the grain sizes, which would reduce or eliminate the indicated microcracking effects. Also, even when flaws sizes are large in comparison to the grain size, flaws that have the least possible microcracking to absorb extra energy should dominate failure because of the lower energy requirement. Pores located within grains, which is more common in larger grains, may also increase fracture energy.

Closely related to the flaw size–grain size effects are the flaw size–second-phase spacing in composites. Again, unless the maximum, second-phase spacing is comparable to, or most likely smaller than, flaws in the matrix material only; then the matrix rather than the composite fracture energy will control tensile strength. Also, the second-phase particles, fibers, etc. can act as flaws if they are sufficiently large and differ sufficiently from the properties of the matrix. Further, lower moduli of many of the second phases also limit possible strength increases. These fracture energy observations do not preclude the utility of many of the ceramic composites that have been made, even though their strength is not increased along with their toughness. Thus, there are thermal stress applications and also possibly some mechanical impact applications where cracks of substantial size in comparison to the microstructure cannot be avoided, and hence the higher toughness of these samples as measured in a fracture energy test will be pertinent. High fracture energies of porous bodies or bodies with large grains can be similarly useful.

Turning briefly to high temperatures, microstructural effects on mechanical behavior can be similar, greater, or less than at room temperature. Thus, porosity effects, especially on strength, may first decrease then increase with increasing temperature. However, grain-size dependence can be reversed, i.e., finer-grain bodies being weaker instead of stronger in creep. Clearly the effect of large grains or groups of larger grains will be greatly diminished, even with a reversal of the grain-size effect. However, effects of larger or more concentrated pores or impurity regions can be accentuated. Again, fracture energies and strengths can show opposite trends.

B. General Research Needs and Aids

Work to further expand our knowledge of mechanical properties ranges from obtaining basic data and generating concepts in areas that have limited exploration, e.g., impact and erosion phenomena, to filling in and refining many bits of experimental data and theoretical analyses indicated in earlier sections. However, in general there are two closely related major overriding needs that exist—much better characterization of test samples and sophistication in the analysis of test results.

The weighting, often greatly, of mechanical behavior toward the extreme of the microstructure giving the lowest values of such behavior places much greater emphasis on microstructural characterization. However, there has been limited microstructural characterization in most studies, and in many cases, no significant characterization has been given. Where some characterization is given, it is almost solely restricted to giving the average microstructural character, which may vary from useful to meaningless. While the need is clearly to address the admittedly difficult problem of numerically, i.e., statistically, documenting the distribution of microstructural features, a great deal of utility can be gained even from qualitative documentation of this, e.g., use of "representative" micrographs.

It is important to have more sophisticated analysis to indicate the type and degree of characterization that is needed to adequately carry out the experimental and analytical tasks of a study, since characterization simply for characterization's sake can approach an infinite task and end up obscuring the important information. Clearly, the long-range need is for the development of models to take into consideration the statistical distribution of the shape, type, and location of pores, grains, second phases, etc. Correspondingly, significant work needs to be done to be able to obtain necessary statistical information in a reasonable fashion. However, in the interim, which is likely to be a substantial period of time, one can make substantial progress by using existing models. Thus, for example, one can approximate the effect of porosity distribution on elastic properties by approximating the body as a series of mosaic blocks of differing average porosity determined approximately from representative sectioning of samples and applying the theories based on uniform size and distributed porosity to each one of these mosaic blocks.

Two simple experimental procedures can greatly facilitate the accuracy and validity of both the experimental tests and the subsequent analysis. The first is to conduct additional modified tests. All too often, tests are done entirely with a single shape, size, orientation, or surface finish of specimens. While there is a clear need for maintaining many such parameters constant throughout the bulk of the study, very often exploratory study of variations in these parameters could be quite enlightening, e.g., in the early phase of the program. Thus, for example, many fabrication processes, such as hot pressing, uniaxial cold pressing, slip casting, and sintering, can lead to preferential orientations or distributions of voids. Tests of samples of various size and particularly of various orientations cut from bodies with reference to the known processing or drying directions can be important clues as to whether or not tests are being significantly affected by inhomogeneity effects. Use of more than one type of test, e.g., measurement of elastic modulus by comparing longitudinal and transverse vibration methods, can also be revealing. Similarly, comparison of three-point and four-point flexural tests can be useful in many cases. Unfortunately, many investigators have assumed that the only really valid and acceptable way of testing samples is by four-point bending, in part, based on the assumption that this represents more dependable engineering data. This may often be true, but it can depend significantly on sample size and distribution of potential failure origins. It is submitted that the volume of most laboratory test specimens is so small in comparison to many actual engineering applications that the relative engineering utility of four-point over three-point flexural tests is often negligible. This is particularly so if one performs four-point testing on rods, since the volume under stress in a rod under four-point flexure may not be significantly higher, and in fact may be less, than three-point flexure of bars of rectangular cross section.

The most significant experimental procedure that can aid the under-standing of mechanical properties is study of fracture surfaces, especially to identify fracture origins. While this may be difficult to impossible for many samples, there are many, many samples on which this can be done, often quite easily. Further, the most important tool is a modest one, namely, an optical microscope. While the scanning electron microscope is extremely valuable for higher-magnification observation and in particular because of its much greater depth of field, by itself it is not as effective a tool in finding fracture origins as an optical microscope. Further, much of the limited depth of focus problem of the optical microscope can be eliminated for the experimenter by simply focusing up and down. Such focusing, however, does not allow obtaining good pictures to demonstrate results to others, which is in fact one of the primary and very important advantages of the scanning electron microscope. It is indeed amazing the number of mechanical properties studies conducted that were extensively concerned directly or indirectly with the size and character of flaws and microstructure from which failure originated in which no attempt was made to experimentally observe and verify the predicted or implied flaw character. Earlier attempts to do this on a broader scale would have brought much earlier recognition of the importance of the inhomogeneity of the microstructural character on mechanical properties and much of the confusion about fracture energy and its relationship in polycrystalline or composite structures to strength.

The study of fracture surface has broad utility, pervading essentially all aspects of the mechanical behavior of ceramics, because of two primary factors. First and foremost, is the fact that fracture will typically seek out the microstructural extremes that one needs to know for essentially all mechanical behavior. Thus, one fracture may reveal more of the extremes of microstructure than many, many polished sections. Further, many microstructural features, particularly microstructural extremes are not adequately detected by normal ceramographic procedures. Very frequently, clusters of large pores, large grains, impurity particles, and other micro-structural extremes are preferred regions for grain pullout and other polishing defects, which are then neglected in the microstructural characterization. Similarly, voids may be filled with debris and impurity particles removed or contaminated by polishing operations. The author has, for example, ob-served voids critical to a problem in the utilization of an alumina body by observation on the fracture surface, whereas these voids (e.g., ~5 μm diameter) were not detected by several laboratories using standard ceramographic procedures. Further, it is worthwhile to note that Treibs and Carlson (1973) conducted a round robin test among six laboratories to measure porosity in reactor fuel materials. While density measurements, i.e., average porosity determination, were in general good agreement, ceramographic analysis of the porosity varied by at least a factor of two.

Finally, it is important to note that examination of fracture surfaces is both rapid and inexpensive, since if one has mechanical test specimens that have already been fractured, there is essentially no preparation time or cost. Further, the author has very frequently simply created a fracture with a pair of pliers or by chipping with a hammer or other handy tool to effectively instantaneously prepare fracture surfaces for examination for general microstructural characterization. Clearly, microstructural characterization on fracture surfaces raises the questions of statistical sampling and measurement technique; i.e., since the surface is not flat, linear intercept techniques for measuring grain size are not perfectly valid. These are important questions that should be addressed for future refinement of this technique; however, extensive experience of the author suggests that these are generally not serious problems for the present state of quantitative analysis of microstructure. The above discussion is not meant to imply that one should not do normal ceramographic analysis, but instead, that the analysis of fracture surfaces can in many cases replace and also in many cases complement the standard analysis.

Finally, long standing standard microscopic measuring techniques can significantly aid in determining the statistics of grain size and porosity. It is readily apparent that the presence of a few larger grains or larger voids in a field of substantially finer grains or voids will not significantly alter the average value, since the numerically greater number of the small ones strongly biases the average diameter to the small size. This has long been known in quantitative microscopy, and averaging methods shown in Table XXI have been developed to obtain averages not only with respect to the grain, particle, or void diameter, but also with regard to surface area or volume. That these averages do indeed differ significantly is illustrated by two grain size examples in Table XXI. While many authors have followed Knudsen's (1959) analysis of grain size data, unfortunately few, if any, of these investigators followed his example of using average techniques biased toward the larger grain size. Such results as those of Table XXI are also consistent with those of Aboave and Langdon (1969), who measured 12,000 grains in a MgO body having a mean grain size of 25 μm and showed that there were a significant number of grains having approximately three or more times the average grain size.

TABLE XXI

DIFFERENT GRAIN SIZE AVERAGES

Average	With respect to:	Examples of each average (μ)	
		UO_2[a]	Al_2O_3[b]
$G_1 = \sum_i ng_i \Big/ n$	Grain diameter	0.9	29
$G_2 = \sum_i ng_i{}^2 \Big/ \sum_i ng_i$	—	—	38
$G_3 = \sum_i ng_i{}^3 \Big/ \sum_i ng_i{}^2$	Grain surface area	3.6	51
$G_4 = \sum_i ng_i{}^4 \Big/ \sum_i ng_i{}^3$	Grain volume	4.8	59

[a] From Burdick and Parker (1956).
[b] Authors data on Lucalox, based on diameters of 30 grains.

C. Microstructural Development for Improved Mechanical Behavior

Finally, an important question is how far can microstructural development be used to increase the mechanical properties of ceramics, in particular, the tensile strength and thermal shock resistance that dominate so much of their potential engineering utility. It is clear that tensile strength can be increased by further refinement of grain size and that, if there is any porosity in the body, it should be fine and homogeneously distributed, preferably within grains. Polycrystalline ceramic fibers indicate the potential for strengths of the order of 250,000 psi even with several percent fine porosity. Clearly, reducing porosity should compensate in part, if not substantially, for expected reductions in strength as a result of wider variations in microstructure and flaws as specimen sizes are scaled up. Important questions of how much machining or other surface contact would reduce strengths of such fine-grain samples with comparable or superior microstructures to those of high strength polycrystalline fibers remain to be determined. However, the author has measured flexural strengths of 250,000 psi on small rods of WC with very fine, e.g., $<0.1\ \mu$m, grains despite normal handling of specimens.

Two routes appear to be open for reducing or eliminating strength losses as a result of machining and handling. The first would be processing techniques that can produce samples that need no machining after final firing or related processing. In the latter case, production of samples that have a porous surface layer, presumably graded into a relatively dense if not fully dense interior, would be highly advantageous for impact and other surface damage resistance. It may in fact be possible to do some limited machining on such porous surfaces in which, again, the pores should be extremely fine, e.g., preferably <0.1 μm. The other route is to produce samples that are much tougher. Two, not necessarily independent, ways to do this are (1) develop fine homogeneous microcracks, primarily by control or utilization of internal stresses associated with phase transformations, thermal expansion anisotropy, and quite possibly by careful control of grain boundary phases in anisotropic materials, and (2) develop suitable ceramic composites.

In developing composites, it is critically important to recognize the importance of keeping the size of the second phase sufficiently small so that it does not act as a relatively low-stress flaw and the spacing between second phases sufficiently close to that flaws will be controlled by the second phase. Most of the present fiber and particle composites are inadequate for this purpose. Also, effects of stresses from incompatible strains between different phases in composites must be studied and understood. Use of precipitation phenomena, as indicated by Heuer and colleagues and especially Garvie et. al (1975) appears promising so long as detrimental grain boundary precipitation does not occur. Use of microcracking mechanisms, e.g., using immiscible phases (Al_2O_3–ZrO_2) as demonstrated by Claussen (1976) also appears quite promising. Directional solidification of in situ composites may also be promising. It is expected that there will be some maxima in strengthening effects as a function of both the sizes of second-phase regions and their spacing that must be determined. More must also be known about the dependence of crack interaction with second-phase particles and their associated stress fields or microcracking as a function of material, processing (e.g., interfacial), and microstructural parameters.

The composites approach has the important advantage that it can also give substantial improvement in creep resistance, provided appropriate choices of second phases can be made, in contrast to the other methods of strengthening, namely, reduction of grain size and introduction of microcrack, which could tend to lower creep resistance. Additionally, the composite approach can be a significant aid in controlling microstructure during processing. It should further be noted that the toughening processes can in general be used with the strengthening processes discussed above. Thus, from the broadest standpoint of ceramic applications, the composites approach, recognizing the critical parameters of composite structure combined with the possible use of porous surfaces may be most promising for the widest variety of advanced engineering applications of ceramics. It should also be noted that fine grain size and high toughness will generally favor reduced abrasion and wear and possibly erosion losses. Many traditional ceramics are of course composites, but much newer development remains to be done.

Thus, microstructural development offers substantial improvement in mechanical behavior of ceramics. This will be a challenge to processing demanding many refinements in existing technology and often development of new technology. However, the critical mechanical benefits and the large and growing importance of ceramics in many critical functions not only makes meeting the challenge important, but imperative.

ACKNOWLEDGMENTS

The author wishes to acknowledge the support of the Naval Air Systems Command, Charles Bersch, Contracting Officer, for support of porosity studies in this review, and of the Office of Naval Research for support of general microstructural studies. Contributions of many photos or graphs by various investigators was most helpful, and comments on various parts of the manuscript from Drs. S. W. Freiman, P. F. Becher, J. B. Wachtman, and D. P. H. Hasselman were very useful. It is also a pleasure to acknowledge the extensive typing of the various drafts by Ms. Riggelman, Compher, Fortin, and Pellecchia. Finally, it is a pleasure to acknowledge the considerable data handling and analysis of Dr. R. C. Pohanka and especially Mr. B. Bender.

References

Aboav, D. A., and Langdon, T. G. (1969). Metallography 1, 333.

Acquaviva, S., and Chait, R. (1972). Static and Cyclic Fatigue of Ceramic Materials. Army Mater. and Mech. Res. Center, Tech. Rep. AMMRC TR 72-9, MCIC Accession 83434.

Adams, M., and Sines, G. (1975) An Experimental Study on the Compressive Biaxial Strength of Ceramics. Final Rep. No. UCLA-ENG-7537 prepared for Naval Air Syst. Command, Contract No. N00019-73-C-0294 Mod. P00004

Adler, W. F. (1974). Analysis of Multiple Particle Impacts on Brittle Materials. Tech. Rep. AFML-TR-74-210, Air Force Mater. Lab., Air Force Syst. Com., Wright-Patterson Air Force Base, Ohio.

Agarwal, B. D., Panizza, G. A., and Broutman, L. J. (1971). Micromechanics analysis of porous and filled ceramic composites, J. Am. Ceram. Soc. 54(12), 620–624.

Ainsworth, J. H., and Moore, R. E. (1969). Fracture behavior of thermally shocked aluminum oxide, J. Am. Ceram. Soc. 52(11), 628–629.

Anderson, R. C. (1970). Thoria and yttria, In "High Temperature Oxides" (A. M. Alper, ed.), Part II, 1–40. Academic Press, New York.

Ali, M. A., Knapp, W. J., and Kurtz, P. (1967). Strength of sintered specimens containing hollow glass microspheres, Ceram. Bull. 46(3), 275–277.

Andersson, C. A., and Salkovitz, E. I. (1974). Fracture of polycrystalline graphite, In "Fracture Mechanics of Ceramics" (R. C. Bradt, D. P. H. Hasselman, and F. F. Lange, eds.), Vol. 2, pp. 509–526. Plenum Press, New York.

Armstrong, R. W., Raymond, E. L., and Vandervoort, R. R. (1970). Anomalous increase in hardness with increase in grain size in beryllia, J. Am. Ceram. Soc. 53(9), 529–530.

Artamonov, A. Ya., and Bovkun, G. A. (1974). In "Refractory Carbides," (G. V. Samsonov, ed.).

Artusio, G., Gallina, V., Mannone, G., and Sgambetterra, E. (1966). Effect of porosity and pore size on the elastic moduli of sintered iron and copper-tin, Powder Metall. 9(17), 89–100.

Ashby, M. F. (1972). *Acta Metall.* **20**, 887–896.

Atkins, A. G., and Tabor, D. (1966). *Proc. Roy. Soc.* **292**, 441–459.

Austin, A. E., Mueller, J. J., Miller, J. F., and Brog, K. C. (1974). Frabrication of BaF$_2$ Infrared Windows. Battelle, Columbus Lab. Rep. for U.S. Army Missile Command contract DAAH01-74-C-0358.

Babel, H. W., and Sines, G. (1968). *J. Basic Eng.* **June**, 285–291.

Bache, H. H. (1970). *J. Am. Ceram. Soc.* **53**(12), 654–658.

Badaliance, R., Krohn, D. A. and Hasselman, D. P. H. (1974). *J. Am. Ceram. Soc.* **57**(10), 432.

Bailey, J. E., and Barker, H. A. (1971). *In* "Ceramics in Severe Environments, Materials Science Research," Vol. 5, pp. 341–359. Plenum, New York.

Bailey, J. E., and Hill, N. A. (1970). *Proc. Brit. Ceram. Soc.* No. 15, 15–35.

Balloffet, Y, Phillips, E., and Hughes, F. (1970). Pressure Sintering of Silicon Carbide. Prepared by At. Energy of Canada Limited, Whiteshell Nucl. Res. Establishment, Pinawa, Manitoba, AECL-3673.

Bal'shin, M. I. (1948). "Poroshkouoe Metallovedenie." Moscow.

Bal'shin, M. I. (1949). *Dokl. Akad. Nauk SSSR.* **67**(5), 831–834.

Bansal, G. K., and Heuer, A. H. (1973). *In* "Fracture Mechanics of Ceramics" (R. C. Bradt, D. P. H. Hasselman, and F. F. Lange, eds.), Part 2, Microstructure, Materials, and Applications. Plenum Press, New York.

Bansal, G. K., Duckworth, W. H., and Niesz, D. E. (1974). Strength-Size Relationships in Ceramic Materials: Investigation of Pyroceram 9606. Tech. Rep. No. 3, Contract No. N00014-73-C-0408, NR 032-541, Office of Naval Res.

Barry, T. I., Lay, L. A., and Marrell, R. (1972). Refractory Glass Ceramics. Nat. Phys. Lab. Rep. NRL-IMS-17 (available from NTIS).

Becher, P. F. (1971). *Mater. Sci. Res.* **5**, 315–329.

Becher, P. F. (1974). *J. Am. Ceram. Soc.* **57**(2), 107–108.

Becher, P. F. (1976). *J. Amer. Ceram. Soc.* **59**(1–2), 59–61.

Becher, P. F., Newell, W. L., and Rice, R. W. (1975). *Am. Ceram. Soc. Bull.* **54**(4), 406.

Bentle, G. G. (1962). *J. Nucl. Mater.* **6**(3), 336–337.

Bentle, G. G., and Miller, K. T. (1967). *J. Appl. Phys.* **38**(11), 4248–4257.

Bertolotti, R. L. (1974). *J. Am. Ceram. Soc.* **57**(7), 300–302.

Bertolotti, R. L., and Fulrath, R. M. (1967). *J. Am. Ceram. Soc.* **50**(11), 558–562.

Binns, D. B., and Popper, P. (1966). *Proc. Brit. Ceram. Soc.* No. 6, 71–82.

Binns, D. B., Cooper, M. T., and Littler, E. S. (1970) *Proc. Brit. Ceram. Soc.* No. 18. Stoke-on Trent.

Birch, F. (1961). *Geophys. J. Roy. Astron. Soc.* **4**, 295; (1961). *J. Geophys. Res.* **66**, 2199.

Blakelock, H. D., Hill, N. A., Lee, S. A., and Goatcher, C. (1970). *Proc. Brit. Ceram. Soc.* No. 15, 69–83.

Boocock, J., Furzer, A. S., and Matthews, J. R. (1972). The Effect of Porosity on the Elastic Moduli of UO$_2$ as Measured by an Ultrasonic Technique. United Kingdom At. Energy Authority, Berkshire, England, p. 8.

Bowie, O. L. (1956). *J. Mater. Phys.* **35**(1), 60–71.

Brace, W. F. (1960). *J. Geophy. Res.* **65**(6), 1773–1788.

Brace, W. F. (1961). *Penn. State Univ. Mineral Exp. Sta. Bull.* **76**, 99–103.

Bradt, R. C., and Tressler, R. E. (eds.) (1975). *Proc. Symp. Plastic Deformation Ceram. Mater., July, 1974.* Plenum, New York.

Bradt, R. C., Hartline, S. D., Coppola, J. A., Weaver, G. Q., and Alliegro R. A. (1974). "Fracture of Commercial Silicon Carbides, from SiC 1973" (O. Marshall, O. Faust, and O. Ryan, eds.). Univ. South Carolina Press, Columbia, South Carolina.

Bratschun, W. R. (1965). An unusual dielectric failure of a piesoelectric ceramic, *J. Appl. Phys.* **36**(8), 2589–2590.

Brennan, J. J. (1975). Development of Fiber Reinforced Ceramic Matrix Composites. Final Rep. prepared under Contract N62269-74-C-0359, Naval Air Develop. Center, Warminster, Pennsylvania for Naval Air Syst. Command, R911848-4.

Brennan, J. J. (1976). Investigate fiber reinforced Si_3N_4, Final Report by United Technologies Research Center for Naval Air Systems Command, Contract N62269-75-C-0137, March 1976.

Brown, S. D., Biddulph, R. B., and Wilcox, P. D. (1964). *J. Am. Ceram. Soc.* **47**(7), 320–322.

Buch, F., and Ahlquist, C. N. (1974). *Mater. Sci. Eng.* **13**, 194–196.

Buckner, D. A., Hafner, H. C., and Kreidl, N. J. (1962). *J. Am. Ceram. Soc.* **45**(9), 435–438.

Budiansky, B. (1970). *J. Composite Mater.* **4**, 286–295.

Budnikov, P. P., Kerbe, F., and Charitonov, F. J. (1968). *Proc. Brit. Ceram. Soc.* **4**, 69–76.

Burdick, M. D., and Parker, H. S. (1956). *J. Am. Ceram. Soc.* **39**(5), 181–187.

Bursill, L. A., and McLaren, A. C. (1965). *J. Appl. Phys.* **36**(6), 2084–2085.

Butler, B. L. (1974). *Rev. Ceram. Tech.* No. 32.

Carnahan, R. D. (1968). *J. Am. Ceram. Soc.* **51**(4), 223–224.

Carnall, E. (1970). Eastman Kodak Co., Private communication.

Carniglia, S. C. (1965). *J. Am. Ceram. Soc.* **48**(11), 580–583.

Carniglia, S. C. (1972a). *J. Am. Ceram. Soc.* **55**(12), 610–618.

Carniglia, S. C. (1972b). *J. Am. Ceram. Soc.* **55**(5), 243–249.

Carniglia, S. C. (1973). *J. Am. Ceram. Soc.* **56**(10), 547.

Charles, R. J. (1959). *In* "Fracture" (B. L. Averbach, D. K. Felbeck, G. T. Hahn, and D. A. Thomas, eds.). MIT Press, Cambridge, Massachusetts and Wiley, New York.

Chen, C. P., and Weisz, R. S. (1972). *Ceram. Bull.* **51**(6), 532–538.

Chin, G. Y., Van Uitert, L. G., Green, M. L., Zydzik, G. J., and Kometani, T. Y. (1973). *J. Am. Ceram. Soc.* **56**(7), 369–372.

Choudry A., and Gielisse, P. J. (1973). *Mater. Sci. Res.* **7**, 149–166.

Chu, G. P. K. (1966). *In* "Ceramic Microstructures, Their Analysis, Significance & Production" Eds. R. M. Fulrath & J. A. Pask, (*Proc. Berkeley Int. Mater. Conf. 3rd*) (R. M. Fulrath and J. A. Pask, eds.), pp. 828–862. Wiley, New York.

Chung, D. H. (1963). *Phil. Mag.* **89**(8), 833–841.

Chung, D. H. (1972). *Science* **177**, 261–263.

Chung, D. H., and Buessem, W. R. (1968a). *In* "Anisotropy in Single-Crystal Refractory Compounds," (F. Vahldiek and S. Mersol, eds.), Vol. 2, pp. 217–245. Plenum Press, New York.

Chung, D. H., and Buessem, W. R. (1968b). *J. Appl. Phys.* **39**(6), 2777–2782.

Clarke, F. J. P., and Wilks, R. W. (1966). *In* "Nuclear Applications of Non-fissionable Ceramics" (A. Boltax and J. H. Handwerk, eds.), pp. 57–74. Am. Nucl. Soc., Hinsdale, Illinois.

Clarke, F. J. P., Sambell, R. A. J., and Tattersall, H. G. (1962). *Phil. Mag.* **7**(75), 393.

Clarke, T. M., Johnson, D., and Fine, M. E. (1970). *J. Am. Ceram. Soc.-Discuss. Notes* **53**(7), 419–420.

Claussen, N. (1976). *J. Am. Ceram. Soc.*, **59**(1–2) 49–51.

Claussen, N. (1969). *Mater. Sci. Eng.*, **4**, 245–246.

Claussen, N., Petzow, G., and Jahn, J. (1974). "High Melting Metal-Ceramic Eutectics" (Translated from German by C. B. Finch). Met. and Ceram. Div., Oak Ridge Nat. Lab.

Clougherty, E. V., Kalish, D., and Peters, E.T. (1968). Research and Development of Refractory Oxidation Resistant Diborides, Tech. Rep. AFML-TR-68-190. Air Force Mater. Lab., Air Force Syst. Command, Wright-Patterson Air Force Base, Ohio.

Coble, R. L., and Kingery, W. D. (1956). *J. Am. Ceram. Soc.* **30**(11), 377–385.

Cohen, L. J., and Ishai, O. (1967). *J. Compos. Mater.* **1**, 390 (1967).

Congleton, J., and Petch, N. J. (1966). *Acta Met.* **14**, 1179–1182.

Congleton, J., Petch, N. J., and Shiels, S. A. (1969). *Phil. Mag.* **19**, 795–807.

Copley, S. M., and Pask, J. A. (1965). *J. Am. Ceram. Soc.* **48**(12), 636–642.

Coppola, J. A., and Bradt, R. C. (1972). *J. Am. Ceram. Soc.* **55**(9), 455–460.

Coppola, J. A., and Bradt, R. C. (1973a). *J. Am. Ceram. Soc.* **56**(7), 392–393.

Coppola, J. A., and Bradt, R. C. (1973b). *J. Am. Ceram. Soc.* **56**(4), 214–218.

Coppola, J. A., Bradt, R. C., Richerson, D. W., and Alliegro, R. A. (1972). *Am. Ceram. Soc. Bull.* **51**(11) 847–851.

Corr, E. M., and Bartlett, R. W. (1968). Evaluation of Duplex Whisker-Crystalline Silicon Nitride Structures, Tech. Rep. AFML-TR-68-197. Air Force Mater. Lab., Wright-Patterson Air Force Base, Ohio.

Cost, J. R., Janowski, K. R., and Rossi, R. C. (1968). Elastic Properties of Isotropic Graphite, Aerospace Rep. No. TR-0158(3250-10)-13; *Phil. Mag.* **17**(148), 851–854.

Crandall, W. B., Chung, D. H., and Gray, T. J. (1961). *In* "Mechanical Properties of Engineering Ceramics" (W. W. Kriegel and H. Palmour III, eds.), pp. 349–376. Wiley (Interscience), New York.

Curtis, C. E., and Johnson, J. R. (1957). *J. Am. Ceram. Soc.* **40**(2), 63–68.

Cutler, I. B. (1957). *J. Am. Ceram. Soc.* **40**(1), 20–23.

Davidge, R. W., and Evans, A. G. (1970). *Mater. Sci. Eng.* **6**, 281–298.

Davidge, R. W., and Green, T. G. (1968). *J. Mater. Sci.* **3**, 629–634.

Davidge, R. W., and Phillips, D. C. (1972). *J. Mater. Sci.* **7**, 1308–1314.

Davidge, R. W., and Tappin, G. (1968). *J. Mater. Sci.* **3**(2), 165–173.

Davidge, R. W., and Tappin, G. (1970). *Proc. Brit. Ceram. Soc.* No. 15, 47–60.

Davidge, R. W., McLaren, J. R., and Tappin, G. (1973). *J. Mater. Sci.* **8**, 1699–1705.

Davisson, J. W., and Vaughan, W. H. (1969). *Fracture*, **4**, 425–480.

Day, R. B., and Stokes, R. J. (1966). *J. Am. Ceram. Soc.* **49**, No. 7.

Dewey, J. M. (1947). *J. Appl. Phys.* **18**, 578–581.

Dinsdale, A., and Wilkinson, W. T. (1966). *Proc. Brit. Ceram. Soc.* No. 6, 119–136.

Dinsdale, A., Moulson, A. J., and Wilkinson, W. T. (1962). *Trans. Brit. Ceram. Soc.* **61**, 259–275.

Duckworth, W. (1953). *J. Am. Ceram. Soc.* **32**(2), 68.

Duncan, J. H., Trigg, D. T., and Creyke, W. E. C. (1965). *Trans. Brit. Ceram. Soc.* **64**, 121–136.

Dutta, S. K., and Gazza, G. E. (1969). *Mater. Res. Bull.*, **4**, 791–796.

Dutta, S. K., and Gazza, G. E. (1973). *Ceram. Bull.* **52**(7), 552–553.

El-Shiekh, A. M., and Nicholson, P. S. (1974). *J. Am. Ceram. Soc.* **57**(1), 19–22.

Elston, J., and Labbe, C. (1961). *J. Nucl. Mater.* **4**(2), 143–164.

Eudier, M. (1962). *Powder Metall.* No. 9, 278–290.

Evans, A. G. (1970a). *Proc. Brit. Ceram. Soc.* No. 15, 113–142.

Evans, A. G. (1970b). *Phil. Mag.* **22**(178), 844–852.

Evans, A. G. (1973a). *In* "Fracture Mechanics of Ceramics" (R. Bradt, D. Hasselman, and F. Lange, eds.), Vol. 1, pp. 17–48. Plenum Press, New York.

Evans, A. G. (1973b). *J. Am. Ceram. Soc.* **56**(8), 405–409.

Evans, A. G. (1974). *J. Mater. Sci.* **9**, 1145–1152.

Evans, A. G., and Davidge, R. W. (1969a). *J. Nucl. Mater.* **33**, 249–260.

Evans, A. G., and Davidge, R. W. (1969b). *Phil. Mag.* **20**(164), 373–388.

Evans, A. G., and Davidge, R. W. (1970). *J. Mater. Sci.* **5**, 314–325.

Evans, A. G., and Tappin, G. (1972). *Proc. Brit. Ceram. Soc.* No. 20, 275–297.

Evans, A. G., and Wiederhorn, S. M. (1974). *J. Mater. Sci.* **9**, 270–278.

Evans, P. E., Hardiman, B. P., Marthur, B. C., and Rimmer, W. S. (1967). *Trans. Brit. Ceram. Soc.* **66**, 523–540.

Evans, A. G., Rajdev, D., and Douglass, D. L. (1972). The Mechanical Properties of Nickel Oxide and Their Relationship to the Morphology of Thick Oxide Scales Formed in Nickel. Univ. of California, Los Angeles, rep. UCLA-Eng.-7229, for ONR Contract N00014-69-A-0200-4021, NR-048-239.

Evans, P. R. V. (1963). *In* "Studies of the Brittle Behavior of Ceramic Materials" (N. A. Weil, ed.). Armour Res. Foundation of Illinois Inst. of Tech. Rep. ASD-TR-61-628, Part II, for Air Force Contract AF33(616)-7465, 164-202.

Fate, W. A. (1975). *J. Appl. Phys.* **46**(6), 2375-2377.

Ferguson, W. J., and Rice, R. W. (1971). *Mater. Sci. Res.* **5**, 261-270.

Fetterolf, R. N. (1970). Development of High Strength, High Modulus Fibers, Tech. Rep. AFML-TR-70-197. Air Force Mater. Lab., Wright-Patterson Air Force Base, Ohio.

Floyd, J. R. (1965). *Trans. Brit. Ceram. Soc.* **64**, 251-265.

Forwood, C. T. (1968). *Phil. Mag.* **17**(148), 657.

Freiman, S. W., Anoda, G. Y. Jr., and Rencin, A. G. (1974a). *J. Am. Ceram. Soc.* **57**(1), 8-12.

Freiman, S. W., McKinney, K. R., and Smith, H. L. (1974b). *In* "Fracture Mechanics of Ceramics," Vol. 2, Materials, and Applications, pp. 659-676. Plenum Press, New York.

Freiman, S. W., Becher, P. F., and Klein, P. H. (1975a). *Phil. Mag.* **31**(4), 829.

Freiman, S. W., Mecholsky, J. J., and Rice, R. W. (1975b). accepted for publication in *J. Am. Ceram. Soc.* **58**(9-10), 406-409.

Freiman, S. W., Mecholsky, J. J., Mast, P., Mulville, Da., Beaubien, L., Sutton, S., and Wolock, I. (1975c). Reentry Vehicle Materials Technology (Revmat) Program—Fracture Mechanics of Graphite and C-C Composites, Q. Progr. Rep. #5, Contract P. O. 00087.

Fryxell, R. E., and Chandler, B. A. (1964). *J. Am. Ceram. Soc.* **47**(6), 283-291.

Gallina, V., and Mannone, G. (1968). *Powder Metall.* **11**(21), 73-82.

Gannon, R. E., Harris, G. M., and Vasilos, T. (1965). *Ceram. Bull.* **44**(5), 460-462.

Garten, V. A., and Head, R. B. (1971). *Int. J. Fracture Mech.* **7**(3), 343-344.

Garvie, R. C., Hannink, R. H., and Pascoe, R. T. (1975). *Nature,* **258**, 703-704.

Gatto, F. (1950). *Alluminio (Milan, Italy)* **19**(1), 19-26; translation available AEC TR 1964.

Gazza, G. E. (1973). *J. Amer. Ceram. Soc.* **56**(12), 662.

Gazza, G. E., Barfield, J. R., and Preas, D. L. (1969). *Am. Ceram. Soc. Bull.* **48**(6), 605-610.

Gebauer, J., Krohn, D. A., and Hasselman, D. P. H. (1972). *J. Am. Ceram. Soc.* **55**(4), 198.

Gerson, R., and Marshall, T. C. (1959). *J. Appl. Phys.* **30**, No. 11.

Gilman, J. J. (1959). *In* "Fracture" (B. L. Averbach, D. K. Felbeck, G. T. Hahn, and D. A. Thomas, eds.), pp. 193-224. MIT Press, Cambridge, Massachusetts.

Gilman, J. J. (1960). *J. Appl. Phys.* **31**, 2208-2218.

Gilman, J. J. (1973). *In* "The Science of Hardness Testing and its Research Applications," pp. 51-72. Am. Soc. for Met., Metals Park, Ohio.

Glaser, F. W., and Ivanick, W. (1952). *J. Met.* **4**(4) 387-390.

Godfrey, D. J., and May, E. R. W. (1971). *Mater. Sci. Res.* **5**, 149-162.

Green, D. J., Nicholson, D. S., and Embury, J. D. (1973). *J. Am. Ceram. Soc.* **56**(12), 619-623.

Griffith, A. A. (1921). *Trans. Roy. Soc. (London)* **A221**, 163-198.

Grimm, N., Scott, G. E., and Sibold, J. D. (1971). *Ceram. Bull.* **50**(12), 962-965.

Groves, G. W. (1970). *Proc. Brit. Ceram. Soc.* No. 15, 103-112.

Groves, G. W. (1971). *In* "Strengthening Methods in Crystals" (A. Kelly and R. B. Nicholson, eds.), p. 420. Halsted Press, Wiley, New York.

Gulden, T. D. (1969). *J. Am. Ceram. Soc.* **52**(11), 585-590.

Gupta, T. K. (1972). *J. Am. Ceram. Soc.* **55**(8), 429.

Gupta, T. K. (1973). *J. Am. Ceram. Soc.* **56**(7), 396-397.

Hall, E. O. (1951). *Proc. Phys. Soc. (London)* **64B**, 747.

Hamano, K., and Lee, E. S. (1972). *Bull. Tokyo Inst. Tech.* No. 108.

Hamjian, H. J., and Lidman, W. G. (1952). "Influence of Structure on Properties of Sintered Chromium Carbide," National Advisory Committee for Aeronautics, Lewis Flight Propulsion Laboratory, Cleveland, Ohio, Technical Note 2731.

Hannink, R. H. J., and Murray, M. J. (1972). *Acta Metall.* **20**, 123-131.

Harrison, W. B. (1965). Mechanical and Electrical Properties of TiC, Second Interim Tech. Rep. prepared for Army Res. Office, Contract No. DA-11-022-ORD-3441; D/A Project No. 59901007; ORD. Project No. TB2-0002; and OOR. Project No. 2884-MET, submitted by Honeywell, Res. Center, Hopkins, Minnesota.

Hartline, S. D., Bradt, R. C., Richerson, D. W., and Torti, M. L. (1974). *J. Am. Ceram. Soc.* **57**(4), 190–191.

Hashin, Z. (1962). *J. Appl. Mech.* **March**, 143–150.

Hashin, Z. (1964). *Appl. Mechan. Rev.* **17**(1), 1–9.

Hashin, Z. (1968). "Ceramic Microstructures" (R. M. Fulrath and J. A. Pask, eds.), pp. 313–341. Wiley, New York.

Hashin, Z., and Rosen, B. W. (1964). *J. Appl. Mech.* **June**, 223–232.

Hasselman, D. P. H. (1962). *J. Am. Ceram. Soc.* **45**(9), 452–453.

Hasselman, D. P. H. (1968). "Anisotropy in Single-Crystal Refractory Compounds," (F. W. Vahldiek, and S. A. Mersol, eds.), Vol. 2, pp. 247–265. Plenum Press, New York.

Hasselman, D. P. H. (1969a). *J. Am. Ceram. Soc.* **52**(11), 600–604.

Hasselman, D. P. H. (1969b). *J. Am. Ceram. Soc.* **52**(8), 458–459.

Hasselman, D. P. H. (1969c). *J. Am. Ceram. Soc.* **52**(8), 457.

Hasselman, D. P. H. (1970a). *Ceram. Bull.* **49**(12), 1033–1037.

Hasselman, D. P. H. (1970b). *J. Am. Ceram. Soc.* **53**(3), 170.

Hasselman, D. P. H. (1970c). *J. Am. Ceram. Soc.* **53**(9), 490–495.

Hasselman, D. P. H. (1971). "Ceramics in Severe Environments" (O. Kriegel and O. Palmour, III, eds.), pp. 89–103. Plenum Press, New York.

Hasselman, D. P. H., and Fulrath, R. M. (1964). *J. Am. Ceram. Soc.* **47**(1), 52–53.

Hasselman, D. P. H., and Fulrath, R. M. (1965). *J. Am. Ceram. Soc.* **48**(11), 545.

Hasselman, D. P. H., and Fulrath, R. M. (1966). *J. Am. Ceram. Soc.* **49**(2), 68–72.

Hasselman, D. P. H., and Fulrath, R. M. (1967). *J. Am. Ceram. Soc.* **50**(8), 399–404.

Heindl, R. A., and Pendergast, W. L. (1927). *J. Am. Ceram. Soc.* **10**(7), 524–534.

Helferich, R. L., and Zanis, C. A. (1973). An Investigation of Yttrium Oxide as a Crucible Material for Melting Titanium, U.S. Naval Ship Res. and Develop. Center Rep. 3911, **January**.

Heuer, A. H. (1969). *J. Am. Ceram. Soc.* **52**(9), 510–511.

Hill, N. A., O'Neill, J. S., and Lively, D. T. (1967). *Proc. Brit. Ceram. Soc.* No. 7, Nucl. and Eng. Ceram.

Hines, J. E. Jr. (1974). "Abrasive Wear of High Density Alumina." *Ph. D. Thesis, Penn. State Univ.*

Hing, P., and Groves, G. W. (1972a). *J. Mater. Sci.* **7**, 422–426.

Hing, P., and Groves, G. W. (1972b). *J. Mater. Sci.* **7**, 427–434.

Hirth, J. P. (1972). *Met. Trans.* **3**, 3047–3067.

Hoagland, R. G., Hahn, G. T., and Rosenfield, A. R. (1973). *Rock Mech.* **5**, 77–106.

Hoagland, R. G., Marschall, C. W., Rosenfield, A. R., Hollenberg, G., and Ruh, R. (1974). *Mater. Sci. Eng.* **15**, 51–62.

Hockey, B. J. (1973). "The Science of Hardness Testing and its Research Applications," pp. 21–50. Am. Soc. Met., Metals Park, Ohio.

Hollenbeck, T. M., Nable, D. C. Walker, P. L. Jr., and Bradt, R. C. (1974). *Ceram. Bull.* **53**(8), 583.

Hrma, P., and Satava, V. (1974). *J. Am. Ceram. Soc.* **57**(2), 71–73.

Hucke, E. E. (1972). Glassy Carbon, Univ. of Michigan Rep. for Adv. Res. Projects Agency Order No. 1824.

Huffadine, J. B. (1974). The Plessey Co., Private communication.

Huffadine, J. B., Whitehead, A. J., and Latimer, M. J. (1969). *Proc. Brit. Ceram. Soc.* No. 12, 201–209.

Hulse, C., and Batt, J. (1974). The Effect of Eutectic Microstructure on the Mechanical Properties of Ceramic Oxides. United Aircraft Res. Lab. Final Rep. for Office of Naval Res. Contract N00014-69-C-0073.

Hunter, O. Jr., Korklan, H. J., and Suchomel, R. R. (1974). *J. Am. Ceram. Soc.* **57**(6), 267–268.

Igarashi, H., and Okazaki, K. (1973). *Proc. Int. Symp. Rilem/IUPAC, Prague* Preliminary Rep., Part II, 183–198.

Irwin, G. R. (1957). *J. Appl. Mech.* **24**, 361–364.

Johnson, R. G., and Tibbetts, S. J. (1974). *Rev. Sci. Instrum.* **45**(3), 446–447.

Johnson, C. A., Smyth, J. R., Bradt, R. C., and Hoke, J. H. (1974). *In* "Deformation of Ceramic Materials" (R. C. Bradt and R. E. Tressler, eds.). Plenum Press, New York.

Johnson, W. C., Stein, D. F., and Rice, R. W. (1974). *J. Am. Ceram. Soc.* **57**(8), 342–344.

Jortner, J. (1968). On the Brittle Failure of Porous Materials Under Biaxial and Triaxial Loading. Univ. of California, Los Angeles.

Jun, C. K., and Shaffer, P. T. B. (1971). Advanced Ceramic Systems for Rocket Nozzle Applications, Summary Rep. The Carborundum Co., R & D Div. Niagara Falls, New York, Contract No. 0017-71-C-4410.

Kalish, D., Clougherty, E. V., and Ryan, J. (1966). Fabrication of Dense Fine Grained Ceramic Materials, Man Labs., Inc., Final Rep. AMRA CR 67-04(F) for U.S. Army Mater. Res. Agency.

Kennard, F. L. III (1973). Directional Solidification of High Temperature Oxide Eutectics, Ph.D. Thesis in Ceramic Science, Penn. State Univ.

Kerner, E. H. (1952). *Proc. Roy. Soc. (London) Ser. B* **69**, 808–813.

Kessler, J. B., Ritter, J. E. Jr., Rice, R. W. (1974) *Mater. Sci. Res.* **7**, 529–544.

Kim, C. H., Smith, W. C., Hasselman, D. P. H., and Kane, G. E. (1973). *J. Appl. Phys.* **44**(11), 5175–5176.

King, A. G., and Yavorsky, P. J. (1968a). *J. Am. Ceram. Soc.* **51**(1), pp. 38–42.

King, A. G., and Yavorsky, P. J. (1968b). *J. Am. Ceram. Soc.* **51**(1), 38–42.

Kingery, W. D. (1961). *Science* **134**.

Kingery, W. D. (1974). *J. Am. Ceram. Soc.* **57**(2), 74–83.

Kinsman, K. R., Govila, R. K., and Beardmore, P. (1975). *In* "Deformation of Ceramic Materials," 465–482 (R. C. Bradt and R. E. Tressler, eds.). Plenum Press, New York.

Kirchner, H. P. (1969). Thermal Expansion Anisotropy of Oxides and Oxide Solid Solutions, tech. rep. No. 2, prepared under contract number N00014-66-C-0190, for the Office of Naval Res.

Kirchner, H. P., and Gruver, R. M. (1970). *J. Am. Ceram. Soc.* **53**(5), 232–236.

Kirchner, H. P., Gruver, R. M., and Sotter, W. A. (1975). *J. Am. Ceram. Soc.* **58**(5–6), 188–191.

Knudsen, F. P. (1959). *J. Am. Ceram. Soc.* **42**(8), 376–388.

Knudsen, F. P. (1961). *J. Am. Ceram. Soc.* **45**(2), 94–95.

Knudsen, F. P., Parker, H. S., and Burdick, M. D. (1960). *J. Am. Ceram. Soc.*, **43**(12), 641–647.

Koepke, B. G., and Stokes, R. J. (1970). *J. Mater. Sci.* **5**, 240–247.

Koester, R. D., and Moak, D. P. (1967). *J. Am. Ceram. Soc.* **50**(6), 290–296.

Kossowsky, R. (1973). *In* "Ceramics for High Performance Applications" 347–372 (J. J. Burke, A. E. Gorum, and R. N. Katz, eds.). Brook Hill Publ., Chestnut Hill, Massachusetts.

Kotlensky, W. V. (1973). *Sample J.* **9**(1), 7–12.

Krohn, D. A., and Hasselman, D. P. H. (1972). *J. Am. Ceram. Soc.* **55**(4), 208–211.

Kuszyk, J. A., and Bradt, R. C. (1973). *J. Am. Ceram. Soc.* **56**(8), 420–423.

Kuszyk, J. A., and Bradt, R. C. (1975). Microstructural Effects on Fracture and Thermal Shock of Magnesite-Chrome Refractories, Part II, Industrial Heating, April, pp. 24–36.

Langdon, T. G. (1972). *J. Am. Ceram. Soc.* **55**(12), 630–631.

Langdon, T. G., Cropper, D. R., and Pask, J. A. (1971). *Mater. Sci. Res.* **5**, 297–313.

Langdon, T. G., and Pask, J. A. (1971). *J. Am. Ceram. Soc.* **54**(5), 240–246.

Langdon, T. G., and Pask, J. A. (1971). *Mater. Sci. Res.* **5**, 283–296.

Lange, F. F. (1970). *Phil. Mag.* **22**, 983–992.

Lange, F. F. (1971). *J. Am. Ceram. Soc.* **54**(12), 614–620.

Lange, F. F. (1973a). *J. Am. Ceram. Soc.* **56**(10), 518–522.

Lange, F. F. (1973b). *J. Am. Ceram. Soc.* **56**(9), 445–450.

Lange, F. F. (1973c). *In* "Fracture Mechanics of Ceramics" (R. C. Bradt, D. P. H. Hasselman, and F. F. Lange, eds.), Part 2, Microstructure, Materials, and Applications. Plenum Press, New York.

Lange, F. F. (1974a). "Composite Materials" (L. J. Broutman, ed.), Vol. 5, Fracture and Fatigue. Academic Press, New York.

Lange, F. F. (1974b). *J. Am. Ceram. Soc. Ceram. Abstr.* **57**(2), 84–87.

Lange, F. F. (1974c). Strong High-Temperature Ceramics, Metall. and Met. Proc. Dept., Westinghouse Res. Lab., Tech. Rep. 11, Contract N00014-68-C-0323, Office of Naval Res., MCIC Accession 88515.

Lange, F. F. (1974d). "Deformation of Ceramic Materials," pp. 361–381. Plenum Press, New York.

Lange, F. F., and Iskoe, J. L. (1974). *In* "Ceramics for High Performance Applications" 223–238 (J. J. Burke, A. E. Gorum, and R. N. Katz, eds.). Brook Hill Publ., Chestnut Hill, Massachusetts.

Lanin, A. G., Fedotov, M. A., and Glagolev, V. V. (1968). *Sov. Powder Metall. Met. Ceram.* No. 5(65), 97–101.

Larson, D. R., Coppola, J. A., and Hasselman, D. P. H. (1974). *J. Am. Ceram. Soc.* **57**(10), 417–421.

Lawn, B., and Wilshaw, R. (1975). *J. Mater. Sci.* **10**, 1049–1081.

Liebling, R. S. (1967). *Mater. Res. Bull.* **2**, 1035–1040.

Lindvay, E., and Fock, M. V. (1969). *J. Mater. Sci.* **4**, 747–752.

Long, R. E., and Schofield, H. Z. (1955). Beryllia, Reprinted from the "Reactor Handbook" (J. F. Hogerton and R. C. Grass, eds.), Vol. 3, Materials, AECD-3647. U.S. Govt. Printing Office, Washington, D.C.

Lueth, R. C. (1974). *In* "Fracture Mechanics of Ceramics" (R. C. Bradt, D. P. H. Hasselman, and F. F. Lange, eds.), Vol. 2, pp. 791–806. Plenum Press, New York.

Lynch, J. F., and Bradt, R. C. (1973). *J. Am. Ceram. Soc.* **56**(4), 228–229.

Mackenzie, J. K. (1950). *Proc. Phys. Soc. Sect. B* **63**, 2–11.

Mandorf, V., Hartwig, J., and Seldin, E. J. (1963). *Proc. Metall. Soc. Conf., Cleveland, Ohio, April 26–27* (G. M. Ault, W. F. Barclay, and H. P. Munger, eds.). Wiley (Interscience), New York.

Manning, C. R. Jr., and Lineback, L. D. (1974). *In Mater. Sci. Res.* **7**, 473–491.

Manning, W. R. (1971). Anomalous Elastic Behavior of Polycrystalline Nb_2O_5. Ames Lab., USAEC, Ames, Iowa.

Manning, W. R., and Hunter, O. (1968). *J. Am. Ceram. Soc.* **51**(9), 537–538.

Manning, W. R., and Hunter, O. Jr. (1969). *J. Am. Ceram. Soc.* **52**(9), 492–496.

Manning, W. R., and Hunter, O. Jr. (1970). *J. Am. Ceram. Soc.* **53**(5), 279–280.

Manning, W. R., Hunter, O., and Powell, B. R. (1969). *J. Am. Ceram. Soc.* **52**(8), 436–442.

Martin, R. B. (1973). *Proc. Int. Symp. Rilem/IUPAC, Prague, Czech., Sept. 18–21* 35–52 (to be published).

Martin, R. B., and Haynes, R. R. (1971a). *ACI J.* **January**, 36–41.

Martin, R. B., and Haynes, R. R. (1971b). *J. Am. Ceram. Soc.* **54**(8), 410–411.

Matthews, R. B., Hutchings, W. G., and Havelock, F. (1973). *J. Can. Ceram. Soc.* **42**.

Mazdiyasni, K. S., and Cooke, C. M. (1974). *J. Am. Ceram. Soc.* **57**(12), 536.

McClelland, J. D. (1961). *J. Am. Ceram. Soc.* **44**(10), 526.

McClintock, F. A., and Walsh, J. B. (1962). *Proc. U.S. Nat. Congr. Appl. Mech. 4th Univ. California, Berkeley, California, June 18–21* Vol. 2.

McHugh, C. O., Whalen, T. J., and Humenik, M. Jr. (1966). *J. Am. Ceram. Soc.* **49**(9), 486–491.

McLaren, J. R., Tappin, G., and Davidge, R. W. (1972). *Proc. Brit. Ceram. Soc.* No. 20, 259–274.

McLean, A. R., Fisher, E. A., and Bratton, R. J. (1974). Brittle Materials Design, High Temperature Gas Turbine, Interim Rep. No. 6, Contract No. DAAG 46-71-C-0162. Army Mater. and Mech. Res. Center, Watertown, Massachusetts.

Mecholsky, J. J., Freiman, S. W., and Rice, R. W. (1976) *J. Matler. Sci*, **11**, 1310–1319.

Mendelson, M. I., and Fine, M. E. (1974a). *In* "Fracture Mechanics of Ceramics" (R. C. Bradt, D. P. H. Hasselman, and F. F. Lange, eds.), Vol. 2, pp. 527–539. Plenum Press, New York.

Mendelson, M. I., and Fine, M. E. (1974b). *J. Am. Ceram. Soc.* **57**(4), 154–159.

Meyer, R. A., and Zimmer, J. E. (1974). Final Rep., Failure Criteria in Graphite, Aerospace Rep. No. ATR-74(7425)-3.

Molnar, B. K., and Rice, R. W. (1973). *Am. Ceram. Soc. Bull.*, **52**(6), 505–509.

Monch, S., and Claussen, N. (1968). *Sci. Ceram.* **4**, 459–464.

Morse, C. T., and Hill, G. J. (1970). *Proc. Brit. Ceram. Soc.* No. 18. Stoke-on-Trent.

Nakayama, J. (1965). *J. Am. Ceram. Soc.* **48**, 583–587.

Nakayama, J. (1974). *In* "Fracture Mechanics of Ceramics" (R. C. Bradt, D. P. H. Hasselman, and F. F. Lange, eds.), Vol. 2, pp. 759–778. Plenum Press, New York.

Neuber, H., and Wimmer, A. (1968). Experimental Investigations of the Behavior of Brittle Materials at Various Ranges of Temperature, Tech. Rep. AFML-TR-68-23. Air Force Mater. Lab., Wright-Patterson Air Force Base, Ohio.

Neuber, H., and Wimmer, A. (1971). Mechanics of Brittle Materials Under Linear Temperature Increases, Tech. Rep. AFML-TR-71-70. Air Force Mater. Lab., Wright-Patterson Air Force Base, Ohio.

Noone, M. J., and Mehan, R. L. (1974). *In* "Fracture Mechanics of Ceramics" (R. C. Bradt, D. P. H. Hasselman, and F. F. Lange, eds.), Vol. 1, pp. 201–229. Plenum Press, New York.

Notis, M. R. (1974). *In* "Deformation Mechanisms Maps—A Review with Applications" (R. C. Bradt and R. E. Tressler, eds.). Plenum Press, New York.

Ofreimow, J. W. (1930). *Roy. Soc. London A.* **127**, 290–297.

Okazaki, K., and Nagata, K. (1972). *In* "Mechanical Behavior of Materials," Vol. 4 Concrete and Cement Paste, Glass and Ceramics, pp. 404–412. Soc. of Mater. Sci., Japan.

O'Neil, J. S. (1970). *Trans. Brit. Ceram. Soc.* **69**(2), 81–84.

O'Neil, J. S., Hill, N. A., and Livey, D. T. (1966). *Ceram. Soc.* **6**, 99–101.

Onoda, G. Y. Jr. (1976). *J. Am. Ceram. Soc.* **59**(5–6), 236–239.

Orowan, E. (1933). *Z. Physik* **82**, 235–266.

Pabst, R. F. (1974). *In* "Fracture Mechanics of Ceramics" (R. C. Bradt, D. P. H. Hasselman, and F. F. Lange, eds.), Vol. 2, pp. 555–565. Plenum Press, New York.

Padgett, G. C., and Clements, J. F. (1970). *Proc. Brit. Ceram. Soc.* No. 15, 61–67.

Parr, N. L., Martin, G. F., and May, E. R. W. (1959). *In* "Special Ceramics" (P. Popper, ed.), pp. 102–129. Academic Press, New York.

Passmore, E. M., Springs, R. M., and Vasilos, T. (1965). *J. Am. Ceram. Soc.* **48**(1), 1–7.

Pasto, A. E., Martin, M. M., and Donnelly, R. G. (1974). *Proc. 11th Rare Earth Res. Conf. Oct. 7–10, 1974*, Vol. 1, Sessions A through J.

Patel, A. R., and Desai, C. C. (1970). *J. Phys. D: Appl. Phys.* **3**.

Patel, M. R., and Finnie, I. (1970). *J. Mater. JMLSA* **5**(4), 909–932.

Paul, B. (1960). *Trans. Metall. Soc. AIME* **218**, 36–41.

Pearson, A., Marhanka, J. E., MacZura, G., Hart, L. D. (1967). Dense, Abrasion-Resistant, 99.8% Alumina Ceramic. Aluminum Co. of Am., Chem. Div.

Penty, R. A., Hasselman, D. P. H., and Springgs, R. M. (1972). *J. Am. Ceram. Soc.* **55**(3), 169–170.

Perry, J. R., and Davidge, R. W. (1973). *Ceramurgia* **3**(1), 22–28.

Petch, N. J. (1953). *J. Iron Steel Inst.* **174**, 25; (1956). *Phil. Mag.* **1**, 866.

Petersen, S., and Curtis, C. E. (1970). Thorium Ceramics Data Manual, Oxides Oak Ridge Nat. Lab. rep. ORNL-4503, Vol. 1, UC-25 Met. and Ceram.

Petrak, D. R., Ruh, R., and Atkins, G. R. (1973). Elastic Properties of Boron Suboxides and the Diametrical Compression of Brittle Spheres, Tech. Rep. AFML-TR-73-84, Air Force Mat. Lab., Wright-Patterson Air Force Base, Ohio, 102; *Ceram. Bull.* **53**(8), 569–573.

Petrak, D. R., Ruh, R., and Atkins, G. R. (1974). *Am. Ceram. Soc. Bull.* **53**(8), 569–573.

Petrak, D. R., Rankin, D. T., Ruh, R., Sisson, R. E. (1975). *J. Am. Ceram. Soc.* **58**(1–2), 78–79.

Piatasik, R. S., and Hasselman, D. P. H. (1964). *J. Am. Ceram. Soc.* **47**(1), 50–51.

Pohanka, R. C., Rice, R. W., and Walker, B. E. (1976). *J. Am. Ceram. Soc.* **59**(1–2), 71–74.

Pohanka, R. C., Rice, R. W., Walker, B. E., and Smith, P. L. (1976), *Ferroelectrics*, **10**, 231–235.

Pohl, D. (1969). *Pwd. Metall. Mater.* **1**, 26.

Porter, D. L., Bansal, G., and Heuer, A. H. (1976). *J. Am. Ceram. Soc.* **59**(3–4), 179–182.

Prochazka, S., and Charles, R. J. (1973). *Am. Ceram. Soc. Bull.* **52**(12), 885–891.

Prochazka, S., Giddings, R. A., and Johnson, C. A. (1974). Investigation of Ceramics for High Temperature Turbine Vanes, Q. Rep. # 2 from General Elec. Co., prepared under Contract N62269-74-C-0255, Naval Air Develop. Center for Naval Air Syst. Command, SRD-74-123.

Radjy, F. (1974). *J. Am. Ceram. Soc.* **57**(2), 88–89.

Rankin, D. T., Stiglich, J. J., Petrak, D. R., and Ruh, R. (1971). *J. Am. Ceram. Soc.* **54**(6), 277–281.

Rasmussen, J. J., Stringfellow, G. B., Cutler, I. B., and Brown, S. D. (1965a). *J. Am. Ceram. Soc.* **48**(3), 146–150.

Rasmussen, J. J., Stringfellow, G. B., Cutler, I. B., and Brown, S. D. (1965b). *J. Am. Ceram. Soc.* **48**, No. 3.

Rhee, S. K. (1972). *J. Am. Ceram. Soc.* **55**, No. 11.

Rhodes, W. H., and Cannon, R. M. (1974). Microstructure Studies of Polycrystalline Refractory Compounds, Summary Rep., Prepared under Contract N00019-73-C-0376, U.S. Naval Air Syst., Washington, D.C.

Rhodes, W. H., Berneburg, P. L., and Cannon, R. M. (1972). Microstructure Studies of Refractory Polycrystalline Oxides, Summary Rep., Contract N00019-71-C-0325, March 9, 1971 to January 9, 1972, AVCO Corporation.

Rhodes, W. H., Berneburg, P. L., Cannon, R. M., Steele, W. C. (1973). Microstructure Studies of Polycrystalline Refractory Oxides, Summary Rep., Contract N00019-72-C-0298, prepared for U.S. Naval Air Syst., Washington, D.C.

Rhodes, W. H., Cannon, R. M., Jr. and Vasilos, T. (1974). *In* "Fracture Mechanics of Ceramics" (R. C. Bradt, D. P. H. Hasselman, and F. F. Lange, eds.), Vol. 2, pp. 709–733. Plenum Press, New York.

Rice, R. W. (1968). *In* "Ceramic Microstructures" pp. 579–593 (R. M. Fulrath and J. A. Pask, eds.). Wiley, New York.

Rice, R. W. (1969a). *Proc. Brit. Ceram. Soc.* No. 12, 99–122.

Rice, R. W. (1969b). *J. Am. Ceram. Soc.* **52**(8), 428–436.

Rice, R. W. (1970a). *J. Am. Ceram. Soc.* **53**(12), 698–699.

Rice, R. W. (1970b). "High Temperature Oxides," Vol. 3. Academic Press, New York.

Rice, R. W. (1971). *Mater. Sci. Res.* **5**, 195–229.

Rice, R. W. (1972a). *Proc. Brit. Ceram. Soc.* No. 20, 329–363.

Rice, R. W. (1972b). *Proc. Brit. Ceram. Soc.* No. 20, 205–257.

Rice, R. W. (1972c). *J. Am. Ceram. Soc.* **55**(2), 90–97.

Rice, R. W. (1972d). The Science of Ceramic Machining and Surface Finishing, Nat. Bur. of St. Spec. Publ. 348, pp. 365–376.

Rice, R. W. (1973a). *J. Am. Ceram. Soc.* **56**(10), 536–541.

Rice, R. W. (1973b). "The Science of Hardness Testing and Its Research Applications," pp. 117–134. Am. Soc. for Met. Metals Park, Ohio.

Rice, R. W. (1974a). *Mater. Sci. Res.* **7**, 439–472.

Rice, R. W. (1974b). *In* "Fracture Mechanics of Ceramics" (R. C. Bradt, D. P. H. Hasselman, and F. F. Lange, eds.), Vol. 1, pp. 323–345. Plenum Press, New York.

Rice, R. W. (1974c). *In* "Ceramics for High Performance Applications" (J. J. Burke, A. E. Gorum, and R. N. Katz, eds.), pp. 287–343. Brook Hill Publ. Co., Chestnut Hill, Massachusetts.

Rice, R. W. (1975a). *J. Am. Ceram. Soc.* **58**(9–10), 458–459.

Rice, R. W. (1975b). Submitted to *J. Am. Ceram. Soc.*

Rice, R. W. (1975c). *J. Am. Ceram. Soc.* **58**(3–4), 154.

Rice, R. W. (1976). *J. Am. Ceram. Soc.* **59**(11–12), 536–537.

Rice, R. W., and McDonough, W. J. (1972). "Mechanical Behavior of Materials," Vol. 4, 422–431. The Soc. of Mater. Sci., Japan.

Rice, R. W., and McDonough, W. J. (1975). *J. Am. Ceram. Soc.* **58**(5–6), 264.

Rice, R. W., and Speronello, B. K. (1976). *J. Am. Ceram. Soc.* **59**(7–8), 330–333.

Rice, R. W., Hunt, J. G., Friedman, G. I., and Sliney, J. L. (1968). Identifying Optimum Parameters of Hot Extrusions, Final Rep. prepared for Nat. Aeronaut. and Space Administration, Contract NAS 7-276. The Boeing Co. and Whittaker Corp.

Richerson, D. W. (1973). *Am. Ceram. Soc. Bull.* **52**(7), 560–562, 569.

Roberts, J. T. A., and Ueda, Y. (1972). *J. Am. Ceram. Soc.* **55**(3), 117.

Rossi, R. C. (1968). *J. Am. Ceram. Soc.* **51**(8), 433–439.

Rossi, R. C. (1969a). *Am. Ceram. Soc. Bull.* **48**(7), 736–737.

Rossi, R. C. (1969b). *Ceram. Bull.* **48**(7), 736–737.

Rossi, R. C. (1971). *Mater. Sci. Res.* **5**, 123–162.

Rossi, R. C., Cost, J. R., and Janowski, K. R. (1972). *J. Am. Ceram. Soc.* **55**(5), 234–237.

Rostoker, W., and Liu, S. Y. K. (1970). *J. Mater. JMLSA* **5**(3), 605–617.

Roy, D. M., and Gouda, G. R. (1973). *J. Am. Ceram. Soc.* **56**(10), 549–550.

Ruh, R., Atkins, G. R., and Petrak, D. R. (1975). Compressive Strength of Boron Suboxide. Air Force Mater. Lab., Wright-Patterson Air Force Base, Ohio.

Ryshkewitch, E. (1953). *J. Am. Ceram. Soc.* **32**(2), 65–68.

Ryshkewitch, E. (1960). Oxide Ceramics," p. 324. Academic Press, New York.

Salak, A., Miskovic, V., Dudrova, E., and Rudnayova, E. (1974). *Powder Metall. Int.* **6**(3), 128–132.

Sambell, R. A. J., Bowen, D. H., and Phillips, D. C. (1972a). *J. Mater. Sci.* **7**, 663–675.

Sambell, R. A. J., Briggs, A., Phillips, D. C., and Bowen, D. H. (1972b). *J. Mater. Sci.* **7**, 676–681.

Sarkar, B. K., and Glinn, T. G. J. (1970). *Trans. Brit. Ceram. Soc.* **69**, 199–203.

Schiller, K. K. (1958). *In* "Mechanical Properties of Non-Metallic Brittle Materials" (W. H. Walton, ed.), pp. 35–49. Wiley (Interscience), New York (620.19 C248 64106).

Schofield, H. Z., Lynch, J. F., and Duckworth, W. H. (1949). Final Summary Rep. on Fundamental Studies of Ceramic Materials. Office of Naval Res., Contract No. N50ri-111, Battelle Memorial Inst., pp. 13, 80, 99.

Schreiber, E. (1968). *J. Am. Ceram. Soc.* **51**(9), 541–542.

Seaton, C. C., and Dutta, S. K. (1974). *J. Am. Ceram. Soc.* **57**(5), 228–229.

Seaton, C. C., and Katz, R. N. (1973). *J. Am. Ceram. Soc.* **56**, No. 5.

Shante, V. K. S., and Kirkpatrick, S. (1971). *Adv. Phys.* **20**(83–88), 325–357.

Shaw, M. C. (1973). *In* "The Science of Hardness Testing and its Research Applications" (J. H. Westbrook and H. Conrad, eds.), pp. 1–11. Am. Soc. for Met.

Shockey, D. A., and Groves, G. W. (1969). *J. Am. Ceram. Soc.* **52**(2), 82.

Simpson, F. H. (1971). Continuous Oxide Filament Synthesis (Devitrification), Tech. Rep. AFML-TR-71-135.

Simpson, L. A. (1973). *J. Am. Ceram. Soc.* **56**(1), 7–11.

Simpson, L. A. (1974). *In* "Fracture Mechanics of Ceramics" (R. C. Bradt, D. P. H. Hasselman, and F. F. Lange, eds.), Vol. 2, pp. 567–577. Plenum Press, New York.

Simpson, L. A., and Merrett, G. J. (1974). *J. Mater. Sci.* **9**, 685–688.

Simpson, L. A., and Wasylyshyn, A. (1971). *J. Am. Ceram. Soc.* **54**(1), 56–57.

Sinha, M. N., Lloyd, D. J., and Tangri, K. (1973). *J. Mater. Sci.* **8**, 116–122.

Smith, C. F., and Crandall, W. B. (1964). *J. Am. Ceram. Soc.* **47**(12), 624–627.

Soga, N., and Schreiber, E. (1968). *J. Am. Ceram. Soc.* **51**(8), 465–466.

Soroka, I., and Sereda, P. J. (1968). *J. Am. Ceram. Soc.* **51**(6), 337–340.

Speck, D. A., and Miccioli, B. R. (1968). Advanced Ceramic Systems for Rocket Nozzle Applications, Carborundum Co. Rep.; "Transition Metal Carbides and Nitrides," Vol. 7, Refractory Materials. Academic Press, New York.

Spinner, S., Stone, L., and Knudsen, F. P. (1963). *J. Res. Nat. Bur. St.* **67**C(2), 93–100.

Spriggs, R. M. (1962). *J. Am. Ceram. Soc.* **45**(9), 454.

Spriggs, R. N., and Brissette, L. A. (1962). *J. Am. Ceram. Soc.* **45**(4), 198–199.

Spriggs, R. W., and Vasilos, T. (1963). *J. Am. Ceram. Soc.* **46**(5), 224–228.

Spriggs, R. M., and Vasilos, T. (1966). *In* "Nuclear Applications of Non-fissionable Ceramics" (A. Boltax and J. Handwerk, eds.), p. 381. Am. Nucl. Soc., Interstate Printer, Danville, Illinois.

Spriggs, R. M., Brissette, L. A., and Vasilos, T. (1962). *J. Am. Ceram. Soc.* **45**(8), 400.

Spriggs, R. M., Mitchell, J. B., and Vasilos, T. (1964). *J. Am. Ceram. Soc.* **47**(7), 323–327.

Spriggs, R. M., Vasilos, T., and Brissette, L. A. (1966). *Mater. Sci. Res.* **3**, 313.

Steele, B. R., Rigby, F., and Hesketh, M. C. (1966). *Proc. Brit. Ceram. Soc.* No. 6, 83–94.

Stokes, R. J., and Li, C. H. (1963). *J. Am. Ceram. Soc.* **46**(9), 423–434.

Swanson, A. W., and Pappis, J. (1975). "Application of polycrystalline ZnSe prepared by chemical vapor deposition to high power ir laser windows," AFML Technical Report TR-75-170.

Swanson, G. D. (1972). *J. Am. Ceram. Soc.* **55**(1), 48–49.

Tabata, T. (1967). *Proc. Int. Conf. Low Temp. Sci., Sapporo, 1966* **10**, 490–496.

Tashiro, M. (1966). *Glass Ind.* **47**, 428–435.

Tattersall, H. G., and Tappin, G. (1966). *J. Mater. Sci.* **1**, 296–301.

Thompson, D. S., and Pratt, P. L. (1966). *Proc. Brit. Ceram. Soc.* **6**, 37–43.

Toth, L. E. (1971). "Refractory Materials," Vol. 7. Academic Press, New York.

Treibs, H. A., and Carlson, M. C. J. (1973). Mixed Oxide Ceramography Round Robin. Hanford Eng. Develop. Lab., HEDL-TME 73-33 UC 79b.

Tressler, R. E., Langensiepen, R. A., and Bradt, R. C. (1974). *J. Am. Ceram. Soc.* **57**(5), 226.

Trostel, L. J. Jr. (1962). *J. Am. Ceram. Soc.* **45**(11), 563–564.

Tsuge, A., Kudo, H., and Komeda, K. (1974). *J. Am. Ceram. Soc.* **57**(6), 269–270.

Udy, M. C., and Boulger, F. W. (1949). Variation of Modulus of Elasticity of Beryllia with Density. The Properties of Beryllium Oxide, BMI-T-18, Battelle Memorial Inst., Columbus, Ohio.

Verbeck, G. J., and Helmuth, R. A. (1969). *Proc. 5th Int. Symp. Chem. Cement, Tokyo, 1968*, **3**, 1–32.

Virkar, A. V. (1973). Fracture Behavior of Zirconium—Zirconia Composites. Ph. D. Thesis Northwestern Univ. (See Nucl. Sci. Abstr. 21531, 30(8)2176, 1974).

Wachtman, J. B. Jr. (1969). "Mechanical and Thermal Properties of Ceramics" (J. B. Wachtman Jr., ed.), NBS Spec. Publ. 303.

Wachtman, J. B. Jr. (1972). *In* "Mechanical Behavior of Materials," Vol. IV, Concrete and Cement Paste Glass and Ceramics, pp. 432–442. The Soc. of Mater. Sci., Japan.

Wagner, P., O'Roure, J. A., and Armstrong, P. E. (1972). *J. Amer. Ceram. Soc.* **55**(4), 214–219.

Walker, B. E., Rice, R. W., Pohanka, R. C., and Spann, J. R. (1975). *Bull. Am. Ceram. Soc.* **55**(3), 274–278.

Walton, J. D. Jr. (1974). *Am. Ceram. Soc. Bull.* **53**(3), pp. 255–258.

Walton, J. D. Jr., and Bowen, M. D. (1961). "Mechanical Properties of Engineering Ceramics" (W. W. Kriegel and H. Palmour III, eds.). Wiley (Interscience), New York.

Weber, J., Greer, R., Voight, B., White, E., and Roy, R. (1969). *J. Ultrastruct. Res.* **26**, 355–366.

Weil, N. A. (1964). *Proc. Int. Symp. High Temp. Technol., Stanford Res. Inst., California, 1963* pp. 189–233. Butterworths, Washington, D.C.

Weinstein, M. (1964). *Trans. Metall. Soc. AIME* **230**, 321–328.

White, D., and Ashbee, K. H. G. (1974). *J. Mater. Sci.* **9**(6), 895–898.

Whitney, E. D., and Bates, R. R. (1975). *Am. Ceram. Soc. Bull.* **54**(4), 400.

Wiederhorn, S. M. (1966). *Mater. Sci. Res.* **3**, 503–528.

Wiederhorn, S. M. (1970). *In* "Ultrafine-Grain Ceramics" (J. J. Burke, N. L. Reed, and V. Weiss, eds.). Syracuse Univ. Press, Syracuse, New York.

Wiederhorn, S. M. (1974). *In* "Fracture Mechanics of Ceramics" (R. Bradt, D. Hasselman, and F. Lange, eds.), Vol. 2, pp. 613–646. Plenum Press, New York.

Wiederhorn, S. M., Hockey, B. J., and Roberts, D. E. (1973). *Phil. Mag.* **28**(4), 783–796.

Williams, E. C., Reid-Jones, R. C., and Dorril, D. T. (1963). *Trans. Brit. Ceram. Soc.* **62**, 405–413.

Williams, L. S. (1961). *In* "Mechanical Properties of English Ceramics (W. Kriegel and H. Palmour III, eds.) pp. 245–302. Wiley (Interscience), New York.

Wittmann, F. H., and Zaitsev, Ju. (1972). "Mechanical Behavior of Materials," *Proc. Int. Conf. Mechan. Behavior Mater.* **4**, 84–95.

Wolfe, R. A., and Kaufman, S. F. (1967). Mechanical Properties of Oxide Fuels, Better Atomic Power Lab. Rep. WAPD-TM-587 under contract, AT-11-1-GEN-14.

Wurst, J. C. (1973). Thermal, Electrical, and Physical Property Measurements of Laser Window Materials. Univ. of Dayton Res. Inst. Progr. Rep. No. 5, UDRI-QPR-73-12 for Air Force Contract F33615-72-C-1257.

Yamada, S. (1968). A Review of Glasslike Carbons. Tokai Electrode Manu. Co. Rep. for Defense Ceram. Informat. Center Rep. 68-2.

Microstructure and Ferrites

G. P. RODRIGUE

School of Electrical Engineering
Georgia Institute of Technology
Atlanta, Georgia

I. Introduction

Ferrimagnetic oxides, or ferrites, have advanced to a position of techno-logical prominence in the time period from the late 1940s to the present. Their development has filled demonstrated needs and led to entirely new applications. This chapter deals with the properties and applications of ferrites and the role that microstructure plays in determining their charac-teristics. Attention is here focused on magnetic materials having moderately low coercive fields and therefore classified as "soft" magnetic materials.

The content of this chapter is keyed to those generally knowledgeable in science and/or engineering, but without specific expertise in magnetism.

The material begins with a background discussion of magnetism and the chemical and crystallographic properties of ferrites. Since microstructure is largely determined and controlled by preparation procedures, a brief review of methods of preparation of polycrystalline ferrites is given. Single-crystal ferrites are not considered in this treatment, and a discussion of their growth from a melt is also excluded. Some discussion of possible control of magnetic properties in spinel and garnet ferrites is given before the description of the applications of ferrites.

Microwave applications, based on the tensor permeability of a saturated body, are described first. The basic formulation of ferromagnetic resonance is outlined as well as the principal types of devices. The role of microstructure in determining both low-power loss and high-power performance is treated in the light of spinwave theory, and some basics of this theory are also included.

Lower frequency applications utilizing unsaturated media are described next. This discussion involves magnetic domains and hysteresis. The importance of microstructure in determining square loop and initial permeability values is discussed, as well as the basic operation of such devices.

The question of systems of units is a bothersome one in the field of magnetism. Engineers by and large use the MKS system, yet magnetic fields are measured with a "gaussmeter," and data are usually tabulated in "oersteds" and "gauss." In this work the MKS system is used so that

$$B = \mu_0(H + M),$$

where B is magnetic flux density in webers per square meter, μ_0 the permeability of free space in henrys per meter, H the magnetic field intensity in amperes per meter, and M the magnetization in amperes per meter. The comparable CGS expression is

$$B = H + 4\pi M,$$

with B and $4\pi M$ in gauss, and H in oersteds. Tabulated values commonly found in the literature list magnetization as $4\pi M$ in gauss. Thus, for example, a material having a $4\pi M$ value of 3000 G has an M value of 2.387×10^5 A/m, the conversion factor being 79.577 or $10^3/4\pi$. Similarly a magnetic field of 2000 Oe is equal to one of 1.59×10^5 A/m. The value of B in gauss is multiplied by 10^{-4} to convert to B in webers per square meter.

This treatment is intended as an introduction to ferrite device technology and a review of current technology. Details of the many facets of material preparation and application cannot be given here, but references to original works are given for a more ambitious and thorough treatment of specific subjects.

A. Background

The existence of ferrites has been known for centuries (the lodestone, or ferrous ferrite, was used by Thales of Miletus), but little engineering application had been found until the post-World War II era. Until that time virtually all magnetic devices used metals. However, with the use of magnetic materials in devices operating at ever increasing frequencies, eddy current effects limited their performance. For example, in applications wherein the magnetic material is used as a transformer core, it is well known that the alternating magnetic field induces, in the core of the transformer, an alternating electric current that is proportional to the time rate of change of magnetic flux. These eddy currents increase with operating frequency and seriously hamper the performance of the core. To get around this difficulty laminated cores were used to interrupt the current paths. The ferrite materials, being ionic crystals with no free electrons, are natural insulators, and eddy currents are eliminated. Ferrites then are ideally suited to high-frequency transformer applications. Their extremely high resistivities (of the order of 10^6 Ω-m) make them suitable to applications at the upper radio frequency and microwave ranges. With the advent of television, high-speed computer cores, and communication at UHF, VHF, and microwave frequencies, the need for magnetic insulators provided impetus for this thrust of materials technology.

There are two principal groups of magnetically "soft" ferrites, those materials having the spinel crystal structure and those having the garnet crystal structure. Both classes have cubic unit cells, and their structure and methods of preparation are described below. Before discussing these particular materials, however, the basics of magnetism will be reviewed in a brief, qualitative, and general treatment that emphasizes those aspects most pertinent to the ensuing discussion.

B. Magnetism

The origin of magnetism can be rigorously treated only along the lines of quantum mechanics, but such rigor and detail are neither appropriate nor necessary to the present purpose. Instead, a discussion from a more physical or intuitive approach will be used.

Magnetism can be considered as originating in the motion of electrons about the nucleus of the atom. This motion can be divided into two parts, an orbital motion of the electron about the nucleus and a spinning motion of the electron about its own axis. Each type of motion contributes to the total angular momentum, and the magnetic moment is proportional to the total angular momentum. These two contributions are referred to as the spin and orbital contributions. The relative importance of spin and orbital motion varies considerably from one type atom to the next, depending on the electronic configuration of the atom and its environment. In solid crystalline materials the local crystalline electric fields can be thought of as causing a rapid precession of the plane electronic orbits about the nucleus, thus averaging out the orbital motion of most ions so that only the spin contributions are operative. In some cases, however, notably the rare earth ions, the orbital motion of the magnetic electrons is shielded from crystalline fields by outer, closed, shells (5s and 5p) of electrons and is not averaged out, or quenched. Then orbital motion can still contribute to the magnetic properties of crystalline substances and produces tight coupling between the magnetization and the crystal lattice.

While each *electron* has a magnetic moment associated with its spin (and possibly orbital) motion, an atom or ion may or may not be magnetic depending on how the contributions from different electrons combine. If the contributions from the different electrons are exactly paired off so that the moment of each electron is balanced by the magnetic moment of another, the *ion* or *atom* as a whole will be nonmagnetic, or more properly *diamagnetic*. It may be noted that in every *filled* electronic shell there is an exact pairing off of electronic magnetic moments; only incomplete shells thus have a net unbalance of spins. Those atoms or ions having such incomplete shells, and therefore having a net unbalance in electronic spin, will be magnetic or again more properly called *paramagnetic*. As the elements build up through the periodic table, the most highly paramagnetic ions, those with the largest net unbalance of angular momentum, occur in the longest periods when the largest incomplete electronic shells are being filled. The first important group occurs in the build up of the 3d shell, the so-called transition elements from the potassium through zinc. Further into the periodic table the very highly paramagnetic rare earths occur when the 4f shell is being filled.

Table I lists these magnetically significant atoms and ions and their electronic configurations. Those with 22–28 electrons and approximately half-filled 3d shells and those with 60–68 electrons and partially filled 4f shells are the most magnetically significant. Note from this table that for ions of the 3d group the magnetic electrons are outermost and fully exposed to crystalline electric fields, while for ions of the rare earth group a shielding of the magnetic 4f electrons by filled 5s and 5p shells is found. Thus, the rare earth ions do not experience orbital "quenching," while the 3d group is generally quenched (with the exception of Co^{2+}) and has "spin only" contribution to its magnetic moment.

When atoms or ions combine to form a solid material the magnetic properties of the material, as a whole, are determined not only by the type of constituent ions but also by the manner in which their magnetic moments interact. A material composed exclusively of diamagnetic ions will obviously

TABLE I

ELECTRONIC CONFIGURATION OF 3d AND 4f MAGNETIC ATOMS OR IONS

Atomic number	Atom or ion	Electronic shells														Unpaired electrons
		1s	2s	2p	3s	3p	3d	4s	4p	4d	4f	5s	5p	5d	6s	
19	K	2	2	6	2	6		1								1
20	Ca	2	2	6	2	6		2								0
21	Se						1	2								1
22	Ti						2	2								2
23	V						3	2								3
24	Cr						5	1								6
	Cr^{3+}						3	0								3
25	Mn						5	2								5
	Mn^{2+}						5	0								5
26	Fe						6	2								4
	Fe^{2+}		Argon core				6	0								4
	Fe^{3+}		18 electrons				5	0								5
27	Co						7	2								3
	Co^{2+}						7	0								3
28	Ni						8	2								2
	Ni^{2+}						8	0								2
29	Cu						10	1								1
30	Zn	2	2	6	2	6	10	2								0
55	Cs	2	2	6	2	6	10	2	6	10		2	6		1	1
56	Ba	2	2	6	2	6	10	2	6	10		2	6		2	0
57	La											2	6	1	2	1
58	Ce										1	2	6	1	2	2
59	Pr										2	2	6	1	2	3
60	Nd										3	2	6	1	2	4
61	Pm										4	2	6	1	2	5
62	Sm										5	2	6	1	2	6
63	Eu		Xenon core								6	2	6	1	2	7
64	Gd		54 electrons								7	2	6	1	2	8
	Gd^{+3}										7	2	6	0	0	7
65	Tb										8	2	6	1	2	7
66	Dy										9	2	6	1	2	6
67	Ho										10	2	6	1	2	5
	Ho^{+3}										10	2	6	0	0	4
68	Er										11	2	6	1	2	4
69	Tm										12	2	6	1	2	3
70	Yb										13	2	6	1	2	2
71	Lu	2	2	6	2	6	10	2	6	10	14	2	6	1	2	1

be diamagnetic, since the ions have no magnetic moments to interact. However, a material containing paramagnetic ions may be paramagnetic, ferromagnetic, antiferromagnetic, or ferrimagnetic, depending on the type of coupling between paramagnetic ions.

If the interaction between paramagnetic ions is only very weak so that thermal agitation is sufficient to keep the spins randomly oriented, then the material as a whole is called *paramagnetic*. In the absence of an applied magnetic field there is no net macroscopic magnetization for a paramagnetic material. Upon the application of an external magnetic field however, the unpaired spins of a paramagnetic substance partially align themselves along the direction of the applied field, and these materials exhibit a large and positive susceptibility, $\chi = M/H$, where M is the induced magnetization and H is the applied field.

In the ferromagnetic, antiferromagnetic, and ferrimagnetic materials so-called exchange interaction exists which couples nearest neighbor spins together with an extremely large energy. This energy is so great that it would generally take about 10^8 A/m of an externally applied field to duplicate the torque exerted by typical exchange energies. The presence of these exchange forces makes orientation of the unpaired spins of ions no longer random, and the substance as a whole possesses some net magnetic order. The origin of these exchange forces is quantum mechanical, but the essential elements are that their existence depends on ionic angular arrangement and interionic spacing.

Exchange forces can act either directly on nearest magnetic neighbors or alternatively through an intermediate nonmagnetic ion such as oxygen. This latter type of interaction, called superexchange, requires a particular angular configuration of magnetic–oxygen–magnetic ions. All exchange forces diminish very rapidly with increasing interionic spacing, and superexchange depends strongly on the angle formed by the oxygen and magnetic ions. While a variety of exchange mechanisms can exist, as described by Anderson (1963a), the most significant for the ferrites of technical interest is an interaction that falls off rapidly as the metal–oxygen–metal angle decreases from 180°.

The distinction between ferromagnetic, antiferromagnetic, and ferrimagnetic substances lies in the different ways in which exchange forces can couple magnetic ions. The principal different possible configurations are shown in two dimensions in Fig. 1. Figure 1a indicates the spin orientations for a *paramagnetic* material. *Ferromagnetic* substances are those in which the unpaired spins of all ions, whether of identical or different elements, are parallel to each other as shown in Fig. 1b. Thus, the net magnetization of the material at the absolute zero temperature, when thermal agitation ceases, is simply the sum of all unpaired spins of ions at the various atomic sites of the material. Examples of such materials are the familiar metals, iron, nickel, and cobalt.

In an *antiferromagnetic* material the spins of ions at the various lattice points are coupled through exchange forces in such a way that the net spin

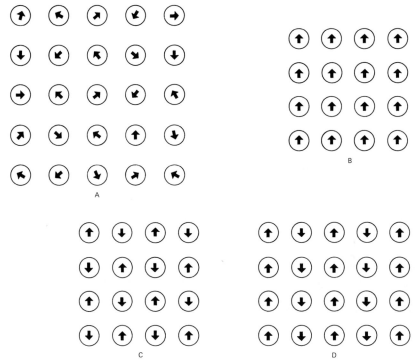

Fig. 1. Possible spin configurations for principal classifications: (A) paramagnetic, (B) ferromagnetic, (C) antiferromagnetic, and (D) ferrimagnetic.

at each lattice point is antiparallel to its neighbors. Thus while the individual ions possess a net moment, the spins of the solid substance as a whole are exactly balanced and the material has no net magnetization. This state, as shown in Fig. 1c, is distinctly different from both the diamagnetic state where the net moment at each site is zero and paramagnetic state where there is no ordering of spins. The antiferromagnetic substance shows a decidedly different and characteristic susceptibility when an external field is applied.

The *ferrimagnetic* class (Fig. 1d) of materials is macroscopically similar to the ferromagnets. In a ferrimagnetic material there exists at least two different types of paramagnetic ions or two different types of lattice sites and, in either event, the material may be considered as composed of two or more sublattices of magnetic ions. Exchange forces then cause the spins in each sublattice to be parallel but opposite to those of the other sublattice. This condition is also similar to antiferromagnetism, and the antiferromagnetic state can be considered as the special case of ferrimagnetism wherein the two sublattices have equal moments. The ferrimagnetic materials, because of the incomplete cancellation that occurs between sublattices, exhibit a net magnetization that is much smaller than would be obtained by simply summing the unbalanced spins of all ions within the material. The net moment of the two or more sublattices, however, does behave macroscopically in the same way as the moment of a ferromagnetic material, and outwardly there are only subtle differences between these two states of magnetism. It is this last class of magnetism, the ferrimagnetic state, to which the spinel and garnet ferrites belong.

In all cases where an exchange field exists the parallel or antiparallel alignment discussed is rigorously true only at absolute zero. At any finite temperature, thermal agitation will tend to disturb the rigid alignment of spins. In fact, as the temperature is raised the orientation of spins becomes more and more random, until finally at the Curie (or Néel) temperature the thermal energy is sufficient to overcome the exchange forces, and all these materials revert to a paramagnetic state. By the same token many materials which are paramagnetic at room temperature become spontaneously magnetized at lower temperatures.

It should be mentioned that, in addition to the parallel and antiparallel orientations discussed, triangular and screw spin configurations, as described by Anderson (1963b), also exist in nature but not in ferrites.

C. Magnetic Domains

Before proceeding further the concept of *magnetic domains* must be introduced. Exchange forces cause a spontaneous alignment of the spins of ions at the various lattice sites, and as a result, ferromagnetic and ferrimagnetic materials are spontaneously magnetized. It is true, however, that large ferromagnetic bodies are not usually found to be magnetized as a whole; that is, the entire magnetic moment of a lump of "magnetic" material does not point in the same direction in the absence of an external magnetic field. Complete alignment of all spins represents an energetically unfavorable configuration, because it has a very large magnetostatic energy associated with it.

In order to reduce magnetostatic energy the material breaks up into many small domains. Within each domain the material is spontaneously magnetized, and all spins point in the same direction. Adjacent domains, however, have their spins pointed in different directions such that the lines of magnetic flux tend to close on themselves within the material, as indicated in Fig. 2, and the magnetostatic energy is reduced.

The relative size of these individual domains is determined by a number of factors including the intrinsic properties of the material and its microstructure.

Between magnetic domains is found a transition region where the magnetization vector gradually rotates from one orientation to the other, as

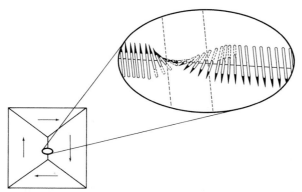

Fig. 2. A configuration of four domains with a closed magnetic flux path. Insert shows magnetic domain wall.

indicated in Fig. 2. Domains and domain walls largely determine the character of "hard" and "soft" magnetic materials, and their influence on the performance of devices is described in Section III. Domains form in order to lower the total of four constituent energies: magnetostatic, exchange, magnetocrystalline anisotropy, and magnetostrictive energies. These four contributions to the total energy and their influence on domain formation are described below.

1. MAGNETOSTATIC ENERGY

Magnetostatic energy is that energy resulting from the interaction of the magnetization of the medium with the magnetic field in which it is immersed. Formally magnetostatic energy is expressed as

$$W_m = \int_V \left(\int_0^B \mathbf{H} \cdot \delta\mathbf{B} \right) dV, \tag{1}$$

where B is the final value of magnetic flux density, $\mathbf{H}$ the magnetic field intensity, and the volume integral extends over all regions of space where the integrand is nonzero.

Magnetic fields $\mathbf{H}$, arise from the surface divergence of $\mathbf{M}$ at the interface between air and the magnetic medium. From the configurations depicted in Fig. 3 it is intuitively seen that the magnetostatic energy will be reduced by the formation of domains as the volume integral of $\mathbf{H}$ is reduced. The configuration of Fig. 3c has no external magnetic field (no surface divergence of $\mathbf{M}$), while the configuration of Fig. 3b has some intermediate magnetostatic energy. It is obvious from these sketches that the price paid for minimizing magnetostatic energy is the creation of domain walls. It will be shown below that such domain walls themselves have an associated energy per unit area, and thus a compromise must be reached.

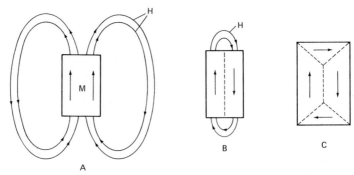

Fig. 3. Reduction of magnetostatic energy through domain wall formation; (A) single domain, (B) two domains with 180° domain wall, (C) four domains with 180° and 90° walls.

Before discussing the other forms of energy it is worth noting that when the field **H**, arises from the magnetic moment of the sample, as in Fig. 3, the magnetostatic energy per unit volume of saturated sample (Fig. 3a) can be expressed as

$$W_{\mathrm{m}} = \tfrac{1}{2}\mu_0 \mathbf{H} \cdot \mathbf{M} \qquad (\mathrm{J/m}^3) \qquad (2)$$

If the sample is ellipsoidal in shape, then the magnetic field can be approximated by the demagnetizing field expressed as a product of the magnetization and a demagnetizing factor N_{d}, as described by Osborn (1945). Thus,

$$W_{\mathrm{m}} = \tfrac{1}{2}\mu_0 \mathbf{H} \cdot \mathbf{M} = \tfrac{1}{2}\mu_0 \mathbf{H}_{\mathrm{d}} \cdot \mathbf{M} = (\mu_0/2)N_{\mathrm{d}}M^2 \qquad (\mathrm{J/m}^3) \qquad (3)$$

This expression will be used subsequently in estimating the critical particle size below which domains do not form.

2. Exchange Energy

Exchange energy is a measure of the coupling between adjacent magnetic moments (or spins) that holds them aligned. Exchange is of quantum mechanical origin, and its strength can be related to the experimentally observable Curie temperature, T_{c}. The usual representation for exchange energy is in terms of the exchange integral J as:

$$W_{\mathrm{ex}} = -2J\mathbf{S}_{\mathrm{i}} \cdot \mathbf{S}_{\mathrm{j}}, \qquad (4)$$

when $\mathbf{S}_{\mathrm{i}}$ and $\mathbf{S}_{\mathrm{j}}$ are neighboring magnetic spins and **J** is the exchange integral, which, in an order of magnitude estimate, is given by $J \cong kT_{\mathrm{c}}/2$. This expression indicates that the exchange energy is lowest for a ferromagnet ($J > 0$) if $\mathbf{S}_{\mathrm{i}}$ and $\mathbf{S}_{\mathrm{j}}$ are parallel, for then W_{ex} is a maximum negative number. In the ferrimagnetic ferrites J is negative ($J < 0$), and thus exchange energy is lowest if $\mathbf{S}_{\mathrm{i}}$ and $\mathbf{S}_{\mathrm{j}}$ are antiparallel. In the ensuing discussion the ferrimagnetic character will be suppressed, and only the *net* magnetic moment of the ferrimagnetic system will be treated. Instead of describing the domain picture for two or three sublattices of magnetization, the discussion will deal with a simple, net magnetic vector (recognizing that it is the result of two or more constituents).

The exchange energy will be raised above its lowest level if the exact alignment of nearest neighbors is disturbed. Thus any departure from exact alignment by an angle θ produces an *increase* in exchange energy *per pair* of spins of

$$\Delta W_{ex} \doteq 2JS^2 - 2JS^2 \cos \theta, \tag{5}$$

where the spins S_i and S_j are now assumed equal. The $\cos \theta$ may be expanded in a power series, with only the first two terms included for small angles. The increase in exchange energy per pair of spins for small misalignments of neighboring spins is then

$$\Delta W_{ex} \doteq JS^2\theta^2. \tag{6}$$

Exchange energy produces very strong torques, comparable to those which would be produced on the magnetic moment by interaction with fields of the order of 10^8 A/m. As a result, the angle of inclination between spins θ is always quite small, and the approximations made above are valid.

3. MAGNETOCRYSTALLINE ANISOTROPY ENERGY

Magnetocrystalline anisotropy energy arises from the interaction of the magnetic moments with electric fields in the crystal and with one another (Kanamori, 1963). The origin of these fields is quite complex, but it is sufficient here to say that the result of such interactions is a preferential direction of magnetization in the material. The so-called "easy" direction of magnetization (for cubic materials) will be along [100] or [111] directions of the unit cell for materials with positive or negative anisotropy, respectively. For cubic crystals the anisotropy energy can be expressed phenomenologically as

$$W_{AN} = K_1(\alpha_1^2\alpha_2^2 + \alpha_2^2\alpha_3^2 + \alpha_3^2\alpha_1^2) + K_2\alpha_1^2\alpha_2^2\alpha_3^2 \quad (\text{J/m}^3) \tag{7}$$

where K_1 and K_2 are the first-and second-order anisotropy constants determined experimentally, and $\alpha_1, \alpha_2, \alpha_3$ are the direction cosines of the magnetization with respect to the three crystal axes.

This directionally dependent energy produces a torque:

$$\frac{\partial W_{AN}}{\partial \theta} = \mathbf{T} = \mu_0(\mathbf{M} \times \mathbf{H}^{AN}) \tag{8}$$

that tends to align the magnetization along "easy" directions of the crystal. The torque is frequently described in terms of an effective anisotropy field, $\mathbf{H}^{AN}$. This fictitious field is used to represent the anisotropy energy and is

TABLE II

ROOM TEMPERATURE VALUES OF FIRST-ORDER ANISOTROPY CONSTANTS FOR SOME FERRITES

Composition	K_1 (joules/m^3)	$K_1/\mu_0 M$ (A/m)	Reference
Fe_3O_4	-1.1×10^4	-1.83×10^4	Bickford (1950)
$NiFe_2O_4$	-5.1×10^3	1.60×10^4	Healy (1952)
$CoFe_2O_4$	$+3.82 \times 10^5$	$+7.64 \times 10^5$	Shenker (1957)
$MnFe_2O_4$	-2.2×10^3	-4.23×10^3	Tannenwald (1955)
$MgFe_2O_4$	-2.5×10^3	-1.8×10^4	Folen and Rado (1956)
$Ni_{0.7}Co_{0.004}Fe_{2.2}O_4$	-1.0×10^3	-2.5×10^3	Bozorth *et al.* (1955)
$Co_{0.012}Fe_{2.988}O_4$	~ 0	~ 0	Bickford *et al.* (1955)
$Co_{0.05}Mn_{0.95}Fe_2O_4$	~ 0	~ 0	Seavey and Tannenwald (1956)
$Y_3Fe_5O_{12}$	-6×10^2	-3.4×10^3	Rodrigue *et al.* (1960)
$Gd_3Fe_5O_{12}$	-5×10^2	-1×10	Rodrigue *et al.* (1960)
$Er_3Fe_5O_{12}$	-8×10^2	-8×10^3	Rodrigue *et al.* (1960)
$Yb_3Fe_5O_{12}$	-5×10^2	-3.3×10^3	Rodrigue *et al.* (1960)

assigned a value such as to produce a torque equal to that arising from the anisotropy energy.

Anisotropy constants for ferrites can be either positive or negative, although the vast majority of cubic ferrites have negative first order anisotropy constants. Values for some representative ferrites are given in Table II. The total anisotropy of a material may be compensated. In solid solutions containing proper ratios of ferrites with positive and negative first-order anisotropy constants a zero value of anisotropy can be attained. The mixed cobalt ferrites of Table II illustrate this at room temperature. The first-order anisotropy "constant" K_1 is, however, a strong function of temperature so that such compensation occurs over only narrow ranges of temperature.

4. MAGNETOSTRICTIVE ENERGY

Magnetostrictive energy also arises from a coupling between the magnetic spins and the lattice. A magnetostrictive material undergoes a change in length when the material is magnetized from zero to saturation. The phenomenon is in many respects similar to piezoelectricity and is the basis of operation of magnetic transducers.

Magnetostriction is usually defined in terms of the coefficient

$$\lambda = \Delta l/l \tag{9}$$

which represents the fractional change in length that occurs as the material is magnetized to saturation from the demagnetized state. In general, λ is a function of the direction of the magnetization with respect to the crystal

TABLE III

Room Temperature Magnetostrictive Constants for Some Ferrites

Composition	$\lambda_{111}(\times 10^6)$	$\lambda_{100}(\times 10^6)$	Source
$MnFe_2O_4$	+4.5	−24	Miyata and Funatogawa (1962)
Fe_3O_4	+78.0	−20	Bozorth et al. (1955)
$CoFe_2O_4$	+130.0	−730	Folen and Rado (1956)
$NiFe_2O_4$	−24.5	−40	Folen and Rado (1956)
$Co_{0.8}Fe_{2.2}O_4$	+120	−590	Bozorth et al. (1955)
$Co_{0.3}Zn_{0.2}Fe_{2.2}O_4$	+110	−210	Bozorth et al. (1955)
$Co_{0.3}Mn_{0.4}Fe_{2.0}O_4$	+65	−200	Bozorth et al. (1955)
$Mn_{0.98}Fe_{1.86}O_4$	−1	−35	Bozorth et al. (1955)
$Mn_{0.6}Zn_{0.1}Fe_{2.1}O_4$	+14	−14	Bozorth et al. (1955)
$Ni_{0.8}Fe_{2.2}O_4$	−4	−36	Bozorth et al. (1955)
$Ni_{0.3}Zn_{0.45}Fe_{2.25}O_4$	+11	−15	Ohta (1960)
$Mn_{0.1}Fe_{2.9}O_4$	+75	−16	Miyata and Funatogawa (1962)
$Mn_{0.6}Fe_{2.4}O_4$	+45	−5	Miyata and Funatogawa (1962)
$Mn_{0.95}Fe_{2.05}O_4$	+5	−22	Miyata and Funatogawa (1962)
$Y_3Fe_5O_{12}$	−2.4	−1.4	Iida (1966)
$Gd_3Fe_5O_{12}$	−3.1	0	Iida (1966)
$Tb_3Fe_5O_{12}$	+12.0	−3.3	Iida (1966)
$Eu_3Fe_5O_{12}$	+1.8	+21	Iida (1966)

axes, λ_{100} and λ_{111} represent fractional elongations for the magnetization along [100] and [111] unit cell directions, respectively. Table III lists some representative values of magnetostrictive coefficients for cubic ferrites. Again, both positive and negative coefficients are observed, and by preparing solid solutions of ferrites with positive and negative coefficients it is possible to compensate the magnetostrictive coefficient.

In most materials operated with no externally applied field the magnetization within each domain lies along "easy" directions of magnetization, thus the magnetostrictive coefficient of that particular direction is the significant one. For materials with negative K_1 values, λ_{111} is the parameter of interest in describing magnetostriction.

In describing domain formation and wall energy the magnetostrictive energy can be regarded as a correction to the magnetocrystalline anisotropy energy, and thus it will be largely disregarded in the following simplified description of domain walls. It has, however, important consequences in determining hysteresis loop properties of materials.

5. Formation of Magnetic Domains

The four energies cited above place conflicting demands on the magnetic moments in a solid. Magnetostatic energy favors no surface divergence of **M** (no component of **M** normal to a surface). Exchange energy favors having all spins parallel, and anisotropy energy favors spins pointing only in "easy" directions. Clearly, nature must strike a compromise. The compromise solution is the creation of domains and domain walls. The existence of domains allows a sample to have an internally closed magnetic path (Fig. 3c). Within each domain the magnetization is everywhere parallel and along easy directions, thus minimizing exchange and anisotropy energy. The region between domains is one of gradual rotation of the magnetization vector, the domain wall, as sketched in the insert of Fig. 2.

In this wall region the magnetization vectors are neither aligned with each other nor along easy directions, thus both anisotropy and exchange energies are raised. Anisotropy energy favors a rapid rotation within the wall, so that few spins point in "hard" directions. Exchange favors gradual rotation to minimize the angle θ between spins. The wall thickness then is also determined by a balance of anisotropy and exchange energies. It is found that the number of spins that make up the wall thickness can be expressed as

$$n \doteq (kT_c/K_1 a^3)^{1/2} \tag{10}$$

where k is Boltzmann's constant; T_c the Curie temperature; K_1 the first-order anisotropy constant; and a is the spacing between magnetic moments. Typically, values for these parameters yield wall thickness equivalent to 100–200 spacings, a. As seen in Eq. (10) high Curie temperatures, or large exchange energy, lead to thick walls, while large anisotropy leads to thin walls.

The increase in energy due to the presence of the wall can be expressed for a unit wall area as (for a 180° wall):

$$W_{\text{wall}} = 2(kT_c K_1/a)^{1/2} \quad (\text{J/m}^2) \tag{11}$$

This expression demonstrates that the wall energy increases as both exchange and anisotropy increase.

For sufficiently small samples the formation of a domain wall will represent a higher energy state than does the saturated body with its attendant external fields. Figure 4 indicates two possible states. The total magnetostatic energy will be given by

$$E_m = \frac{1}{2} \mu_0 N_d M^2 \left(\frac{4\pi}{3} r^3 \right) = \frac{2\mu_0 M^2 \pi r^3}{9} \quad (\text{J}) \tag{12}$$

for a spherically shaped sample (where $N_d = \frac{1}{3}$) of radius r. Assuming four closure domains the total wall energy becomes:

$$E_{\text{wall}} = (kT_c K_1/a)^{1/2} 2\pi r^2 \quad (\text{J}) \tag{13}$$

Since wall energy varies as r^2 while magnetostatic energy varies as r^3, it is clear that at large particle radius $E_m > E_{\text{wall}}$, and wall formation corre-

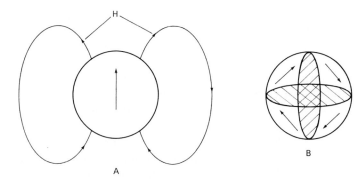

Fig. 4. Possible domain patterns for small spherical particles; (A) single domain, (B) multidomain.

sponds to a lower energy state. As the radius becomes small, however, the magnetostatic energy will fall below that of the wall. At the "critical" radius the two energies are equal, and for smaller particle sizes, domains will not form. The expression for the critical radius of single domain particles is obtained from this somewhat simplified analysis as

$$r_{\text{crit}} = \frac{9(kT_cK_1)^{1/2}}{\mu_0 M^2(a)^{1/2}} \tag{14}$$

Single domains sizes vary over wide ranges with material properties K_1, T_c, and M. For cubic ferrites, as a general rule, particles are multidomain whenever their diameters are of the order of 1 μm or greater.

D. Ferrimagnetic Oxides

All oxidic ferrimagnets have a crystal structure that consists of a basic crystal framework of oxygen ions with relatively large ionic radii. At the interstices of these ions are located the smaller metallic ions. The different types of oxidic ferrimagnetic materials are distinguished by the type of lattice formed by the oxygen ions.

The magnetic spinel and garnet materials are ferrimagnetic substances usually composed of oxides of iron, other 3d transition ions, yttrium, or the 4f rare earth ions. Some aspects of these two crystal structures are described below.

1. GARNET CRYSTAL STRUCTURE

Crystallographically, the garnet materials are isostructural with the naturally occurring minerals spessartite ($3\text{MnO}\cdot\text{Al}_2\text{O}_3\cdot3\text{SiO}_2$) and grossularite ($\text{Ca}_3\text{Al}_2\cdot3\text{SiO}_4$) and belong to the O_h^{10} (Ia3d) space group. Several other naturally occurring minerals exist that have the garnet structure, and they are all silicates. Yoder and Keith (1951) succeeded in synthetically preparing the first silicon-free garnet. Starting with spessartite, they substituted yttrium for manganese and aluminum for silicon to achieve a chemical composition $Y_3Al_5O_{12}$; today this is used in single crystal form as a laser host material.

The first ferrimagnetic garnets were prepared and identified in 1956 by Bertaut and Forrat (1956) and by Geller and Gilleo (1957a). Since their discovery they have received intensive study. Particular practical interest has been focused on the yttrium iron garnet. The rare earth iron garnets have achieved considerable notoriety from a fundamental standpoint, while for the engineer the doping of the yttrium iron garnet with rare earth ions has been of interest. The most outstanding attributes of single-crystal yttrium iron garnet are its extremely narrow ferrimagnetic resonance line-width (Dillon, 1957), its transparency to visible light (Dillon, 1958), and its extremely high mechanical Q (Spencer and LeCraw, 1959).

The crystal structure of the garnets (Geller and Gilleo, 1957b) is quite complex and can be discussed in terms of yttrium iron garnet. The crystal structure is principally determined by the large oxygen ions, and the metal ions present have only a small influence on the arrangement of ions and their interatomic distances. There are eight chemical formula units of $Y_3Fe_5O_{12}$ per unit cell of the material. Thus there are 24 yttrium or rare earth ions

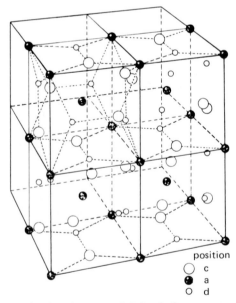

position
○ c
⬤ a
○ d

Fig. 5. Arrangement of cations in c, a, and d sites in four octants of the garnet unit cell (after Winkler, 1971).

present in each unit cell, 40 iron ions, and 96 oxygen ions. The entire unit cell then contains 160 ions, is understandably complicated, and not readily amenable to a brief description. One half of a garnet unit cell is illustrated in Fig. 5. For present purposes it is sufficient to consider only the most elementary points regarding this crystal structure.

In Fig. 5 the balls representing the various ions are not drawn to scale. Within the unit cell there are three types of lattice sites available to the metallic ions. These sites can be differentiated by the oxygen coordination of each site. Sixteen of the iron ions are located on so-called a sites. The coordination of these a sites is shown in Fig. 6a. The distance between all oxygen ions and the iron ion located at the center is 2.0 Å. The remaining 24 iron ions of the unit cell are located on the so-called 24 d sites. The coordination of d sites is shown in Fig. 6b, and in this case each oxygen ion is located 1.88 Å from the iron ion located at the center. The yttrium or rare earth ions are centered in a cluster of 8 oxygen nearest neighbors located at the corners of an irregular polyhedron which can be thought of as a somewhat twisted cube. This coordination is shown in Fig. 6c. The distance between the yttrium ion and nearest oxygen ions in this arrangement is either 2.37 or 2.43 Å. There are 24 such c sites.

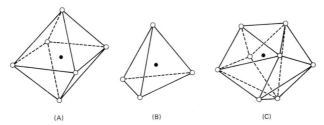

(A)   (B)   (C)

Fig. 6. Coordination of metallic ions in the garnet structure; (A) 16 a sites—octahedrally coordinated by oxygen neighbors, (B) 24 d sites—tetrahedrally coordinated by oxygen neighbors, and (C) 24 d sites—dodecahedrally coordinated by oxygen neighbors (after Geller and Gilleo, 1957b).

It can be seen from the interionic distances quoted in Table IV that the larger ions such as yttrium and the various rare earths tend to go into the c sites, while iron and ions such as aluminum and gallium with relatively small ionic radii are found in a and d sites where the distances to coordinating oxygen ions are smaller. Any of the trivalent rare earths with ionic radii less than approximately 1.00 Å can be substituted for yttrium into the c sites. These possible substitutional ions are shown in Table V together with their ionic radii and net unbalanced spin. It is seen in column 5 that the type of ion present on the c site has little influence on the lattice constants of the

TABLE IV

INTERIONIC DISTANCES IN ANGSTROM UNITS FOR NEAREST NEIGHBORS
IN YTTRIUM–IRON GARNET

	Nearest neighbors			
	Y^{+3} at c site	Fe^{+3} at a site	Fe^{+3} at d site	Coordinating O^{-2} ions
From each ion				
Y^{+3} at c site	—	4 at 3.46 Å	2 at 3.09 Å 4 at 3.79 Å	4 at 2.37 Å 4 at 2.43 Å
Fe^{+3} at a site	2 at 3.46 Å	—	6 at 3.46 Å	6 at 2.00 Å
Fe^{+3} at d site	2 at 3.09 Å 4 at 3.79 Å	4 at 3.46 Å	—	4 at 1.88 Å

iron garnets. Because of the large unbalanced moment of the rare earth ions
and the contribution of their orbital moment to the total magnetization,
their magnetic systems are strongly coupled to the crystal lattice, and thus
exhibit large magnetocrystalline anisotropy and large magnetic damping.

TABLE V

METALLIC IONS FOUND IN THE GARNET STRUCTURE

Positive ions (Me)	Ionic radius (Å)	Number of unpaired spins	Preferred site in garnet	Lattice constant of Me–iron garnet
Fe^{+3}	0.66	5	$16a$ and $24d$	—
Ga^{+3}	0.65	0	$16a$ and $24d$	—
Cr^{+3}	0.64	3	$16a$ and $24d$	—
Al^{+3}	0.55	0	$16a$ and $24d$	—
Si^{+4}	0.42	0	$24d$	—
Mg^{+2}	0.75	0	$24c$ and $16a$	—
Ca^{+2}	1.05	0	$24c$	—
Mn^{+2}	0.83	5	$24c$	—
Na^{+}	1.00	0	$24c$	—
Fe^{+2}	0.80	4	$24c$	—
Sm^{+3}	1.00	5	$24c$	12.529
Eu^{+3}	0.98	6	$24c$	12.498
Gd^{+3}	0.97	7	$24c$	12.471
Tb^{+3}	0.93	6	$24c$	12.436
Dy^{+3}	0.92	5	$24c$	12.405
Ho^{+3}	0.91	4	$24c$	12.375
Er^{+3}	0.97	3	$24c$	12.347
Yb^{+3}	0.930	1	$24c$	12.302
Y^{+3}	0.95	0	$24c$	12.376
V^{+4}	0.59	1	$24d$	—

2. Magnetic Properties of Garnets

While the oxygen coordination and interionic spacing determines which ions can fit in this crystal structure, it is the relative positioning of these ions that determines their magnetic properties. The direct exchange interaction discussed earlier occurs between nearest-neighbor magnetic ions or atoms. In the oxidic ferrimagnets the magnetic ions are separated by oxygen ions, and under these conditions the direct exchange coupling cannot exist. But superexchange, first conceived by Kramers (1934), can quite adequately interpret the existence of ferrimagnetism in such structures. The exchange force here arises from the interaction of the 2p electrons of the oxygen ions with the electronic distribution of the neighboring magnetic ions. For the dominant superexchange mechanism in the ferrites, coupling of the spins of magnetic ions through an intermediate oxygen ion increases sharply with decreasing interionic spacing and is greatest for Me_1–O–Me_2 angles near 180°. The various interionic angles in the yttrium iron garnet are given in Table VI. Since the overall crystal arrangement is negligibly disturbed when rare earth ions are substituted for the yttrium ion, these same angles will hold for yttrium or rare earth garnets.

TABLE VI

INTERIONIC ANGLES IN YTTRIUM–IRON GARNET

Ions (Me–O–Me)	Angles (θ)	Distances a and b (Å)
$Fe^{+3}(a)$–O^{-2}–$Fe^{+3}(d)$	126.6	2.00 and 1.88
$Fe^{+3}(a)$–O^{-2}–Y^{+3}	102.8	2.00 and 2.43
$Fe^{+3}(a)$–O^{-2}–Y^{+3}	104.7	2.00 and 2.37
$Fe^{+3}(d)$–O^{-2}–Y^{+3}	122.2	1.88 and 2.43
$Fe^{+3}(d)$–O^{-2}–Y^{+3}	92.2	1.88 and 2.37
Y^{+3}–O^{-2}–Y^{+3}	104.7	2.37 and 2.43
$Fe^{+3}(a)$–O^{-2}–$Fe^{+3}(a)$	147.2	4.41 and 2.00
$Fe^{+3}(d)$–O^{-2}–$Fe^{+3}(d)$	86.6	3.41 and 1.88
$Fe^{+3}(d)$–O^{-2}–$Fe^{+3}(d)$	78.8	3.68 and 1.88
$Fe^{+3}(d)$–O^{-2}–$Fe^{+3}(d)$	74.7	3.83 and 1.88
$Fe^{+3}(d)$–O^{-2}–$Fe^{+3}(d)$	74.6	3.83 and 1.88

The angles shown in Table VI indicate that strongest magnetic coupling should occur between iron ions on the a sites and iron ions on the d sites where the appropriate angle is 126.6°, and the iron to oxygen distances are 2.0 and 1.88 Å. All interactions between iron ions on the *same* type site are very weak because of the excessively large iron to oxygen distance (as in the case of the iron ions located on the a sites) or because of angles close to 90° (as in the d–d interactions).

It is also seen from this table that yttrium (or rare earth) ions located at c sites experience a moderate interaction with iron ions located on the d sites. The most important interaction occurs for an angle of 122.2°, and a yttrium (or rare earth) to oxygen distance of approximately 2.4 Å with an oxygen–iron distance of 1.88 Å. The somewhat large interionic distance and the relatively unfavorable angle of this c–d interaction make it considerably weaker than the a–d interaction. The a–d interaction is little influenced by the type of ion occupying the c sites.

This superexchange interaction is known to be negative for ions with electronic shells that are at least half full, so that iron ions located on the 24 d sites have spins oppositely directed from those of the 16 iron ions on the a sites in each unit cell. Therefore, the net magnetic moment of YIG (at O°K) corresponds to that of one iron ion per molecule (8/unit cell) or 5 Bohr magnetons per molecule (40/unit cell), the trivalent iron ions having five unbalanced spins and a moment of 5 Bohr magnetons.

The distance and angle data of Table VI indicate that when rare earth ions are located at the c sites, a strong interaction should exist only with the iron on the d sites. At absolute zero then, the saturation magnetization per molecule of rare earth garnets corresponds to the difference between the magnetization of rare earth ions on the c sites and the resultant magnetization of iron ions on the a and d sites. Thus,

$$M_T = (M_d - M_a) - M_c$$

It is possible to fully characterize the two iron sub-lattices in terms of the yttrium iron garnet properties, and, to a first approximation, to attribute differences between the rare earth iron garnets and the yttrium iron garnet solely to the rare earth ions on the c sites. Figure 7 shows the variation of saturation magnetization with temperature for the yttrium iron garnet, and also shows the deduced saturation magnetization curves of the individual a and d sublattices.

When magnetic rare earth ions occupy c sites it is still possible to treat the a and d sites as a single sublattice with a magnetization equal to the resultant of the magnetization of the a and d sublattices. Because the interaction between the d and c sites is relatively weak, the c sublattice is only weakly magnetized at the higher temperatures (room temperature and above), and

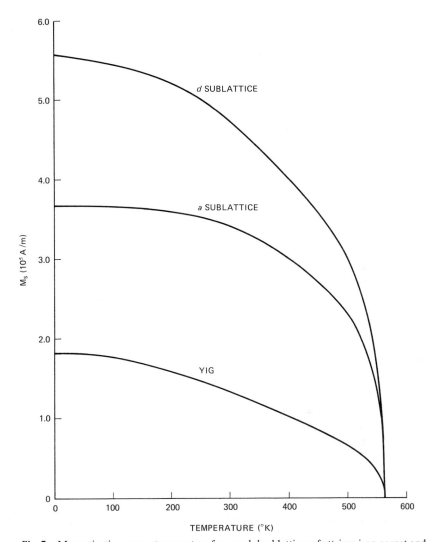

Fig. 7. Magnetization versus temperature for *a* and *d* sublattices of yttrium iron garnet and for material as a whole.

it only slightly decreases the total magnetization of the material. As the temperature is lowered, however, the *c–d* coupling overcomes thermal agitation, and the *c* sublattice approaches saturation. The *c* sublattice magnetization then begins to overtake the net magnetization of the *a* and *d* sublattices. In the case of many of the rare earth garnets (for example gadolinium with its relatively large number of unbalanced spins) there is a compensation point where the magnetization of the c sublattice is exactly equal and opposite to the resultant magnetization of the a and d sublattices. At this point the macroscopic magnetization of the material vanishes, and it is in a sense a completely compensated ferrimagnet with properties closely resembling an antiferromagnet.

Figure 8 shows curves of the saturation magnetization of a variety of rare earth iron garnets as a function of temperature. In every case the saturation

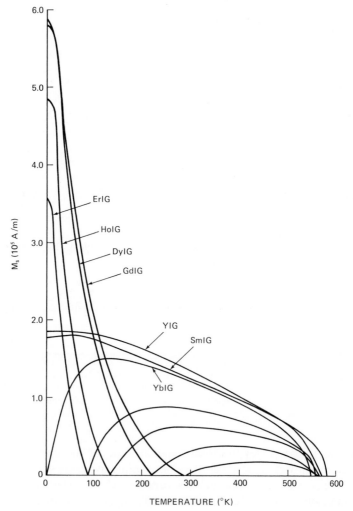

Fig. 8. Magnetization versus temperature for yttrium and rare earth iron garnets (after Pauthenet, 1956).

magnetization of the c sublattice can be deduced by subtracting from these curves the saturation magnetization of yttrium iron garnet which, in effect, represents the resultant magnetization of the a and d sublattices. In those cases where compensation points are observed the magnetization is actually reversed below that point, and the c sublattice is more magnetic than the resultant of the a and d sublattices.

It can be deduced from curves such as those of Fig. 8 that by carefully selecting the ions appearing on the c sites the saturation magnetization versus temperature curve can be controlled. Notice, for example, in Fig. 8, that the saturation magnetization of holmium iron garnet is relatively constant over the temperature range of practical interest, 250–400°K. This flatness of the magnetization curve is highly desirable in many practical devices and is achieved through the compensating action of the magnetization of the holmium c sublattice and the iron a and d sublattices. Holmium ions, however, have large unquenched orbital contributions to the total magnetic moment and as a result impart a strong damping force on the magnetic system. Such strong damping is undesirable in most practical applications, and for this reason pure holmium iron garnet is of little practical interest. A small slope in the saturation magnetization curves can also be achieved by a *partial* substitution of other magnetic ions, notably gadolinium for yttrium, on the c sublattice. Since free holmium ions have four unpaired spins and free gadolinium ions have seven, it is not surprising that by substituting approximately half the nonmagnetic yttrium ions with gadolinium a curve of saturation magnetization versus temperature very like that of holmium can be achieved.

By the same token if one substitutes nonmagnetic aluminum for iron on the dominant d sublattice the net magnetization (d minus a) of the yttrium iron garnet will be decreased without affecting its general shape. Since this last substitution affects the total number of a–d interactions, it also lowers the Curie temperature. Ionic substitutions on the c sublattice, however, leave the Curie temperature unaffected. By extending such reasoning ionic substitutions can be determined that will control not only the room-temperature saturation magnetization of the material, but also the variation of magnetization with temperature. Figure 9 shows a family of curves of saturation magnetization as a function of temperature for various gadolinium- and aluminum-substituted yttrium iron garnets, exemplifying such control.

The ability to substitute a wide variety of ions in the garnet structure facilitates other controls as well. By partial substitution with rare earth ions having large unquenched orbital motion it is possible to control the damping of the spin system and the breadth of the ferromagnetic resonance absorption line. Manganese doping can be used to reduce magnetostriction. It is thus possible to synthesize a garnet material with predictable and widely variable

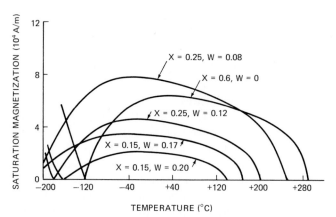

Fig. 9. Variation of saturation magnetization with temperature in a family of garnets of the general formula $[(1 - x)Y \cdot (x)Gd]_3 \cdot [(1 - w) Fe \cdot (w) Al]_5 \cdot O_{12}$ (after Rodrigue, 1969).

characteristics by appropriately doping the basic yttrium iron garnet structure.

3. SPINEL CRYSTAL STRUCTURE

Ferrites of the spinel crystal structure are isostructural with the naturally occurring mineral spinel $MgAl_2O_4$. This cubic crystal belongs to the space group O_h^7 (Fd3m) and has in each unit cell eight chemical formula units of the form $M^{2+}Fe_2^{3+}O_4^{2-}$, where M is usually a divalent metal ion of the 3d transition group. In this structure, shown in Fig. 10, the large oxygen anions form a face centered cubic lattice. The smaller metallic cations are located at the interstices of the oxygen ions. Two types of interstitial sites are available, the octahedrally coordinated B sites and the tetrahedrally coordinated A sites.

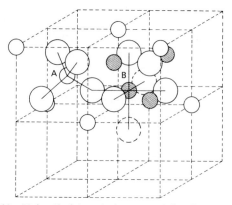

Fig. 10. Unit cell of the spinel structure (after Gorter, 1954).

Within a unit cell there are 64 tetrahedral sites and 32 octahedral sites available. Of this number, however, only 8 tetrahedral and 16 octahedral sites are occupied by magnetic ions. The interatomic distances and angles are listed in Table VII. The first super-exchange coupling listed between magnetic ions on the A sites and those on the B sites is strongest, and the coupling is negative (as is generally the case for electron shells that are at least half filled). The total exchange energy, and hence the Curie temperature, is determined by the total number of exchange couplings between A and B site magnetic ions.

TABLE VII

INTERIONIC ANGLES AND DISTANCES IN SPINEL FERRITES

Ions Me( )–O–Me( )	Angles (θ)	Distances d and b (in units of lattice constant, a)
Me(B)–O–Me(A)	125°	0.25a and 0.216a
Me(B)–O–Me(A)	154.5°	0.25a and 0.415a
Me(B)–O–Me(B)	90°	0.25a and 0.25a
Me(B)–O–Me(B)	125°	0.25a and 0.433a
Me(A)–O–Me(A)	79.5°	0.216a and 0.415a

It might be supposed from the above chemical formula that in a unit cell the 8 divalent atoms are located on the 8 occupied tetrahedral sites, and the 16 trivalent ferric ions located on the 16 occupied octahedral sites. This is termed a "normal" spinel. Most technically interesting spinels are "inverted." That is, the 16 ferric ions are equally divided among the A and B sites, and 8 divalent ions occupy the remaining 8 filled octahedral sites.

4. MAGNETIC PROPERTIES OF SPINELS

In inverted spinels the magnetic moment of the ferric ions cancel one another, and the net magnetic moment is that of the divalent ion. For example, in nickel ferrite which is inverted the net magnetic moment per unit cell at absolute saturation is that of the 8 nickel ions. On the other hand, zinc ferrite ($ZnFe_2O_4$) normally has the Zn^{2+} ions on the 8 A sites and the 16 ferric ions all located on B sites. Since Zn^{2+} is diamagnetic, no A–B interactions exist between magnetic ions. The B–B interaction is extremely weak, and thus pure zinc ferrite is paramagnetic at room temperature.

Many ferrite compositions are neither purely "normal" nor purely "inverted," Solid solutions of virtually all spinel compositions can be fabricated, and a wide range of magnetic properties thus realized. As in the case of the garnets, the saturation magnetization can be controlled (raised by Zn^{2+} substitution, and lowered by Al^{3+} substitution). Figure 11 shows a set of curves of magnetization versus temperature for a family of Zn-substituted nickel ferrites, and Fig. 12 has a similar set for Al-substituted nickel ferrites.

The magnetostriction of spinels can be controlled by Mn^{2+} additions. The spinel structure cannot contain the unquenched rare earth ions, but Co^{2+} substitution can be used to vary the anisotropy and damping.

It is worth noting that other than divalent ions can be substituted in the spinel chemical formula. One of the most technically important exceptions is Li^+ that occurs in lithium ferrite as $Li_{0.5}^+ Fe_{2.5}^{3+} O_4^{2-}$. This material is of interest because its large A–B exchange coupling produces an extraordinarily high Curie temperature of $943°K$.

TABLE VIII

IONIC RADII OF SOME IONS ENCOUNTERED
IN SPINEL FERRITES

Valence 1+	Valence 2+	Valence 3+	Valence 4+
Li 0.70	Ca 1.05	Al 0.55	Mn 0.52
	Cd 0.99	Co 0.65	Pb 0.70
	Co 0.78	Cr 0.70	Sn 0.65
	Cu 0.70	Fe 0.67	Ti 0.68
	Fe 0.80	Ga 0.65	V 0.57
	Mg 0.75	In 0.95	Zr 0.80
	Mn 0.83	Mn 0.67	
	Ni 0.74	Sc 0.83	
	Pb 1.18	Ti 0.70	
	Pt 0.52	V 0.75	
	Sn 1.02	Y 0.95	
	Ti 0.76		
	Zn 0.83		

Table VIII lists some of the ions that go into a spinel crystal structure, and Table IX lists the chemical formula, unit cell dimensions, and magnetic properties of some typical spinel materials. Solid solutions of all these compositions are readily made, and typical device materials have additions of one or more other elements.

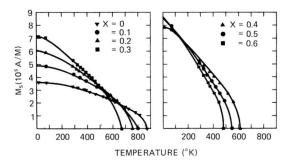

Fig. 11. Saturation magnetization of polycrystalline $Ni_{1-x}Zn_xFe_2O_4$ ferrites as functions of temperature (after Von Aulock, 1965).

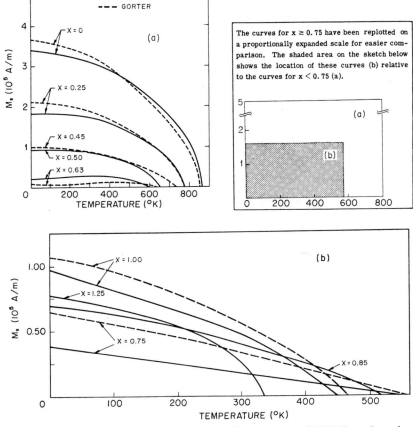

Fig. 12. Saturation magnetization of polycrystalline specimens of $NiAl_xFe_{2-x}O_4$ as functions of temperature (after Von Aulock, 1965).

TABLE IX

CRYSTALLOGRAPHIC AND MAGNETIC PROPERTIES OF SOME TERNARY SPINELS

Formula	Unit cell edge (Å)	Cation distribution		Room temperature magnetization (A/m)	T_c (°C)	Source
		(A)	[B]			
$NiFe_2O_4$	8.325	(Fe)	[NiFe]	2.7×10^5	585°	Hastings and Corliss (1953)
$MgFe_2O_4$	8.36	$(Mg_{.1}Fe_{.9})$	$[Mg_{.9}Fe_{1.1}]$	1.1×10^5	440°	Gorter (1954)
$CoFe_2O_4$	8.38	(Fe)	[CoFe]	4.22×10^5	520°	Gorter (1954)
Fe_3O_4	8.394	(Fe^{3+})	$[Fe^{2+}Fe^{3+}]$	4.8×10^5	585°	Abrahams and Calhoun (1953)
$ZnFe_2O_4$	8.44	(Zn)	$[Fe_2]$	—	—	Hastings and Corliss (1953)
$MnFe_2O_4$	8.507	$(Mn_{.8}Fe_{.2})$	$[Mn_{.2}Fe_{1.8}]$	3.98×10^5	300°	Gorter (1954)
$Li_{.5}Fe_{2.5}O_4$	8.33	(Fe)	$[Li_{0.5}Fe_{1.5}]$	3.1×10^5	670°	Braun (1952)

E. *Methods of Preparation and Control of Microstructure*

A number of preparation processes can be used for ferrites, and the relative merits of various methods are influenced by the volume, raw materials, and end use of the final product. The many alternatives cannot be detailed here, but a brief review of three preparation methods will be given.

1. WET MILLING PROCESS

Polycrystalline ferrites are normally prepared by the conventional ceramic process utilizing wet mixing of solid constituents. Figure 13 is a flow chart for a conventional, wet milling process as used in large-scale production.

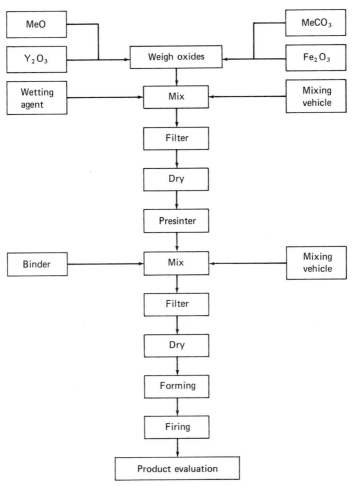

Fig. 13. Flow chart for typical production process.

Raw oxides (or carbonates) are weighed out in proper ratios with allowance made for water absorption, loss on ignition, etc. Selection of raw materials is based on chemical purity, particle size and size range, and refractivity. While chemical purity is generally a reliable measurable property, the other properties are not dependably constant from one raw materials batch to another. Changes in particle size and surface conditions can significantly affect sintering properties. Adjustments in formulation and processing are generally needed to compensate for batch-to-batch variations. Such adjustments are usually made on the basis of test firings of sample batches and/or physical and chemical measurements at intermediate steps in the process.

The constituent oxides (or carbonates) are next wet mixed in a ball milling stage using rubber lined milling jars and stainless steel balls. The mixing agent is usually water, but other carriers such as Varsol have been used in cases of water soluble constituents. Ball milling times of several hours are required and are generally accomplished in automatically timed, overnight cycles. Mixing times can be shortened through the use of attritor mills, but the long milling time is generally not a problem in production schedules. Some allowance in initial composition must also be made for iron pick-up in the milling stage, so that an initial nonstoichiometric ratio of oxides with a small deficiency of iron ($1-2\%$) is commonly used.

After the mixing operation the slurry is vacuum filtered to prevent settling and segregation of constituents, and subsequently oven dried.

The dried powder ideally consists of a molecular mixing of oxides and/or carbonates of the metallic ions in proper proportions. This powder is next presintered (in an air or oxygen atmosphere) to start the solid-state reaction that produces the desired garnet or spinel crystal structure. Substantial shrinkage occurs in the presintering stage and this, in fact, is the chief reason for this operation, i.e., to reduce shrinkage that occurs on final firing. Presintering temperatures in the range of $900-1300°C$ are used, generally some $200°C$ below the final sintering temperature.

The presintered material is a brittle agglomerate of particles, which is broken up in a granulator before wet milling again. This second milling stage serves to separate particles, to homogenize the powder (that generally is not uniformly reacted in the presintering operation), and, in some cases, to reduce particle size. At this stage an organic binder (e.g., polyvinyl alcohol) is often used to add strength to the green pressed piece. This second milling time is considerably shorter than the earlier mixing stage. The milled slurry is next spray dried, and the powder is stored in air-tight containers until the forming step.

Several types of dies are used for forming. Simple shapes are generally dry pressed in double action dies under pressures of $5-25 \times 10^6$ kg/m^2. Hydrostatic or isostatic pressing of powders in a flexible mold is also widely used in the industry, and extrusions are sometimes employed for rods and bars.

The final formed pieces are next fired to temperatures of 1100–1500°C to form the final sintered part. The temperature employed depends on the relative volatility of the constituent ions. Ferrites containing zinc or lithium, for example, are fired at relatively low temperatures, while those containing more refractory ions, like aluminum, are fired at temperatures near the upper end of the range. In general, the garnets are fired at higher temperatures then the spinels.

In this sintering operation substantial grain growth occurs. Lowest free energy of the substance corresponds to minimum surface area of particles. Thus, in the sintering process solid-state diffusion occurs, large grains grow at the expense of small ones, and pores are filled with material or migrate to the exterior. In sintering, densification and grain growth occur at the same time. Particle sizes of the green (unfired) piece are of the order of a few microns or less. The final ferrite may have grain diameters of 20 μm or more. Both the duration and the temperature of the final firing influence grain growth and density achieved. Generally the temperature used is one at which solid-state diffusion occurs, but without loss through volatization. The time is then adjusted to allow for the desired grain growth and densification. For most technically useful materials densities of greater than 96% of the theoretical value (or porosities of only a few percent) are necessary.

Two basically different types of furnace are used in production. Most commonly, an electric globar heated furnace (box kiln) having a controlled heating/cooling cycle is used. The pressed piece is first raised to about 200°C and held there an hour or so to drive off the organic binder. The temperature is next raised to the final temperature, held there for 4 to 8 hr, and subsequently cooled, relatively slowly to prevent thermal shock of the ceramic part. Such firing cycles may take three days, but very large quantities of ferrite parts can be fired simultaneously.

The second furnance type, a tunnel kiln used for extremely large-scale production, is maintained with a constant temperature profile along its length. The ferrite part is placed on a continuous belt (in appropriate trays) that slowly moves the parts through the furnace. The firing cycle is accomplished by variations in temperature along the axis of the furnace. A similar temperature vs time cycle is employed in the continous duty furnace, and parts can be continuously on and off loaded. It is generally found that the continuous duty furnace produces a more uniform product, free of the batch-to-batch idiosyncracies of the batch-type furnace.

After firing, the final shaping is accomplished by grinding (with diamond wheels or compounds). Annealing at 700–800°C is frequently used to relieve stresses induced by the grinding or shaping operations.

The process described above is the most widely used in production. The variations in this process are legion, often proprietary, and frequently keyed to a particular composition. One notable variation that has direct impact on the final microstructure is that of hot pressing.

2. HOT PRESSING

As the name implies the hot-pressing technique involves the simultaneous application of high pressure and temperature. In the conventional process the pressing is done at room temperature and the firing at atmospheric pressure; in the hot-pressing technique the application of pressure makes possible the use of lower sintering temperatures. For example, DeLau (1968) reports nickel zinc ferrite hot pressed at 1045°C and lithium ferrite hot pressed at 870°C, both at pressures of about 10^7 kg/m². This method yields ferrites of extremely small grain size and high density. Grain diameters of 0.5 μm are reported, with densities of more than 99.5% of theoretical values. The hot-pressing scheme permits densification without the normal accompanying grain growth.

The apparatus used is shown in schematic form in Fig. 14. The powder to be pressed is contained in an aluminum oxide die D which is heated to the firing temperature by the wire windings W. The upper and lower punches of the die exert a constant pressure of about 10^7 kg/m². The powder P in a zone of maximum temperature about 1 cm in height shrinks as it sinters. By gradually lowering both upper and lower plungers, at constant pressure, powder is continuously fed into the sintering zone. The upper plunger is occasionally removed to replenish the powder supply. Sintering rates of about 3 cm per hour have been reported.

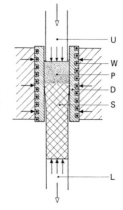

Fig. 14. Principle of the continuous hot-pressing method as developed by Gruintjes and Oudemans (1965).

This method has been applied to those materials for which high density and very small particle size are of distinct advantage. Grain size influences both low-frequency permeability and power handling ability in microwave devices, as will be described later.

Because the rate of production of final product is relatively small when compared to capital equipment required, this continuous hot-pressing technique has not been applied to large-scale production.

More recently, a two-stage sintering operation has been described that utilizes gas isostatic hot pressing as the second stage. Härdtl (1975) reports the fabrication of both garnets and spinel ferrites with densities in excess of 99.6% of theoretical values. The success of the technique depends on conventional firing to a density equal to or greater than 93% of theoretical, so that subsequent firing under gas pressures of about 2×10^6 kg/m^2 can be done on a sample without connected pores.

3. LIQUID MIXING

Another basic approach to ferrite preparation involves dissolving the constituents and subsequent mixing in solution. This process has not yet been applied to production, but has been used for smaller scale, laboratory preparations of materials.

A flow chart for this process is shown in Fig. 15. It is built upon the co-precipitation technique (Wolf and Rodrigue, 1958) used earlier for the preparation of both garnets and ferrites, but uses the advantage of spray drying and a rotary calcining (presintering) stage to yield a continuous process. The thrust of this process (Bomar, 1971) is to improve material reproducibility by removing batch-to-batch variations in raw materials.

The constituent oxides, carbonates, or metals are dissolved in nitric or sulfuric acid, and the liquid solutions then mixed in proper amounts to provide the correct metallic ion ratios. The mixture is diluted with water and then fed into a spray drier operated at an inlet temperature of about 370°C. By combining in solution, a molecular mixing is achieved such that each drop of solution fed into the spray drier contains the proper metallic ion ratio. The spray-dried product consists of hollow spheres with diameters of 2–10 μm.

The spray-dried powder is next fed into an inclined rotary calcining tube maintained at approximately 900–1100°C. As the material swirls through this tube it is converted to oxide form, and the garnet or spinel phase crystallizes. Residence time in the tube is 20–30 min. The greatly reduced presintering time is achieved because of the intimate mixing of ions. The slow solid-state diffusion process required to presinter ball-milled powders is no longer necessary. Presintered powders are typically spherical in shape and with diameters of 0.4–5 μm.

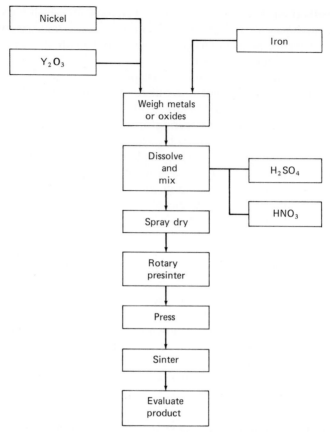

Fig. 15.   Flow chart for liquid mixing process.

This presintered powder is then ready for forming and final firing as in the more conventional process described earlier. Final particle sizes are determined by firing temperature and soak time, and typically are in the range of 5–20 μm or more. Finer grained ferrites could presumably be prepared from this powder by the hot-pressing techniques described earlier.

II.   Applications of Saturated Materials (Microwave Devices)

A principal use of ferrite materials is in UHF and microwave devices, where the medium is saturated as a single domain and no domain walls exist. The influence of microstructure on domain wall motion, taken up in the next section, is then inconsequential, but microstructure still has an important influence on device performance. In brief, microstructure determines the degree of coupling to nonuniform modes, or spin waves, and this coupling, in turn, partially determines the damping of the system at both low and high power levels. To facilitate a fuller discussion of these effects it may be helpful to first review briefly the basic operation of microwave ferrite devices and the concept of spin waves. The following remarks are greatly simplified to introduce the most pertinent concepts. More complete discussions are contained in the references.

A. Fundamentals of Microwave Ferrite Device Operation

In a saturated ferrite medium the large number of coupled spins (or magnetic moments) makes the system amenable to a classical analysis, and from a magnetic standpoint the material can be modeled as a magnetization vector $\mathbf{M}$. The influence of the medium on electromagnetic fields is determined by its permeability,

$$\mu = \mathbf{B}/\mathbf{H} \tag{15}$$

or susceptibility

$$\chi = \mathbf{M}/\mathbf{H} \tag{16}$$

These two quantities are directly related by

$$\mu = \mu_0(1 + \chi)$$

The influence of magnetic media on the performance of microwave devices is described in terms of the magnetic susceptibility. The following development outlines the approach to an analytical expression for that susceptibility.

For the medium as a whole the motion of the magnetization is governed by

$$d\mathbf{M}(\mathbf{r}, t)/dt = \gamma[\mathbf{M}(\mathbf{r}, t) \times \mathbf{B}_{\text{eff}}(\mathbf{r}, t)] + \text{damping term} \tag{17}$$

where γ is the effective magnetomechanical ratio. $\mathbf{M}$ and $\mathbf{B}_{\text{eff}}$ are the sum of the microscopic magnetic moments and fields and will, in general, vary from point to point in the medium and hence be functions of both time and position. In conventional solutions of Eq. (17) the magnetization is assumed to be uniform throughout the medium, completely independent of position. As shown in Figs. 16a and 16b for an infinite medium and sphere, respectively, this corresponds to assuming that all spins are parallel. This assumption is valid at low excitation levels, if an ellipsoidal sample (or infinite medium) is placed in uniform dc and rf magnetic fields, for then the demagnetizing fields (see Fig. 16) are also uniform. In this case, where all spins are parallel, exchange effects and dipolar fields arising from a volume divergence of $\mathbf{M}$ may be neglected in the solution of the problem. The entire system of spins then precesses as a unit, and Fig. 16 may be replaced by Fig. 17. Then Eq. (17) becomes

$$\frac{d\mathbf{M}(t)}{dt} = \gamma[\mathbf{M}(t) \times \mathbf{B}_{\text{eff}}(t)] - \alpha\left\{\frac{[\mathbf{B}_{\text{eff}}(t) \cdot \mathbf{M}(t)]\mathbf{M}(t)}{M^2(t)} - \mathbf{B}_{\text{eff}}(t)\right\} \tag{18}$$

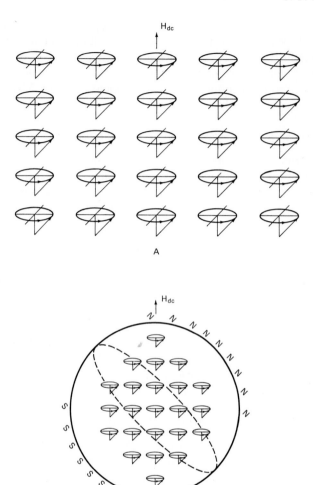

Fig. 16. (A) Uniform precession in an infinite medium. All spins are parallel, and $B_{eff}/\mu_o = H_{eff} = H_{dc} + h_{rf}$. (B) Uniform precession in an ellipsoid. All spins are parallel, and demagnetizing field is set up by surface divergence of M. $B_{eff}/\mu_o = H_{eff} = H_{dc} + h_{rf}$.

where the Landau–Lifshitz form of damping term is assumed (Clarricoats, 1961), and α is the phenomenological damping constant related to the ferromagnetic resonance linewidth through

$$\alpha = \frac{\gamma M \, \Delta H}{2 H_{res}}. \tag{19}$$

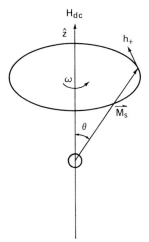

Fig. 17. Uniform precession of $\mathbf{M}_s$ about H_{dc} at frequency ω_0.

The value of $\mathbf{B}_{\mathrm{eff}}$ in Eq. (18) is determined not only by the applied dc and rf magnetic fields but by the demagnetizing fields as well (into which the crystalline anisotropy field might be incorporated in a more rigorous treatment).

In the infinite medium (Fig. 16a), the effective field contains only the applied magnetic fields, and the resonant frequency is given by the Lamor frequency

$$\omega_0 = \gamma B_0 \qquad (20)$$

where $B_0 = \mu_0 H_{\mathrm{dc}}$, and H_{dc} is the applied dc magnetic field.

In an ellipsoid (Fig. 16b) the demagnetizing fields set up by the surface divergence of $\mathbf{M}$ alter this somewhat, and the resonant frequency is then the Kittel frequency

$$\omega_0 = \gamma \mu_0 \{[H_0 + (N_x - N_z)M][H_0 + (N_y - N_z)M]\}^{1/2}, \qquad (21)$$

where N_x, N_y, and N_z are the static demagnetizing factors whose sum is 1.0. Thus, for the uniform precessional mode of conventional linear theory, there is a single resonant frequency of the system of spins. For a given applied field this resonant frequency is a function of sample shape.

A physical picture of the motion of the magnetization vector is given in Fig. 17. The length of the magnetization vector $\mathbf{M}$ is constant and represents the saturation magnetization of the material. Under the influence of the torque arising from the applied dc magnetic field, $\mathbf{M}$ precesses about the z axis with a frequency given by Eq. (21). The direction of the applied dc magnetic field is taken as the z direction, and rf fields are assumed to have a time dependence of the form $e^{j\omega t}$. In the absence of damping this precession would continue indefinitely, and the angle θ between $\mathbf{M}$ and H_{dc} would conserve its initial value. Because of the ever present damping forces, however, **M** spirals in, and aligns itself along the z axis. If a positive circularly polarized rf field h_+, is applied normal to H_{dc}, it will exert a torque such as to fan the magnetization vector **M** out from the z direction counteracting the effects of damping. As the rf frequency is varied, a resonance phenomenon occurs when the applied frequency equals the natural frequency of the system given by Eq. (21). The largest excursion of **M** from the z direction (and largest absorption of rf power) obviously occurs at resonance, and the maximum angle θ_{max} observed at resonance for any given applied field will be dependent upon the damping of the system; the smaller the damping, the larger will be this angle.

The solution of Eq. (18) yields expressions for the components of the magnetization vector in terms of the fundamental parameters. In the linear, or small-signal, solution of the equation of motion the applied rf fields h_x and h_y are assumed to be very small compared to the dc applied field, and second and higher order terms in time-varying components are neglected. In terms of internal fields then, the solution can be written as

$$m_x = \chi_{xx}h_x + j\chi_{xy}h_y \tag{22}$$

$$m_y = -j\chi_{xy}h_y + \chi_{xx}h_y \tag{23}$$

$$m_z \doteq 0, \tag{24}$$

where

$$\chi_{xx} = \frac{(\gamma^2\mu_0^2H_0^2 - \omega^2)(M_z\gamma^2\mu_0^2H_0\chi_0^2) + 2\omega^2\alpha^2\mu_0^2\chi_0}{\chi_0^2(\gamma^2\mu_0^2H_0^2 - \omega^2)^2 + 4\omega^2\alpha^2\mu_0^2}$$

$$- j\frac{\alpha\mu_0^2\omega\chi_0^2(\gamma^2\mu_0^2H_0^2 + \omega^2)}{\chi_0^2(\gamma^2\mu_0^2H_0^2 - \omega^2)^2 + 4\omega^2\alpha^2\mu_0^2} \tag{25}$$

$$\chi_{xy} = \frac{M_z\gamma\mu_0\omega\chi_0^2(\gamma^2\mu_0^2H_0^2 - \omega^2)}{\chi_0^2(\gamma^2\mu_0^2H_0^2 - \omega^2)^2 + 4\omega^2\alpha^2\mu_0^2}$$

$$- j\frac{2\omega^2\mu_0^2\gamma\alpha H_0\chi_0^2}{\chi_0^2(\gamma^2\mu_0^2H_0^2 - \omega^2)^2 + 4\omega^2\alpha^2\mu_0^2} \tag{26}$$

in which χ_0 is the static susceptibility ($\chi_0 = M_z/H_{dc}$; $H_0 = H_{dc}[1 + (\alpha^2/\gamma^2M^2)]^{1/2} \approx H_{dc}$; H_{dc} is the applied dc magnetic field; and M_z is the z component of magnetization ($M_z \approx |\mathbf{M}|$).

This description assumes internal rf applied fields (N_x, N_y, N_z neglected) for simplicity. The basics of operation of ferrite devices can be interpreted on this model. The complications introduced by considering finite shapes and more general polarizations are of no importance for the present discussion.

The components of the susceptibility tensor expressed in Eqs. (25) and (26) can be combined, for circularly polarized rf fields, into an effective scalar susceptibility for positive or negative circularly polarized waves as

$$\chi_+ = \chi_{xx} + \chi_{xy} = \chi_+{}' - j\chi_+'', \qquad \chi_- = \chi_{xx} - \chi_{xy} = \chi_-{}' - j\chi_-'' \quad (27)$$

Since $\mu_\pm = \mu_0(1 + \chi_\pm)$, the effective scalar permeability which the material presents to positive circularly polarized waves will differ from that seen by negative circularly polarized waves. These are complex permeabilities whose real and imaginary parts are plotted in Fig. 18. In the cross-hatched

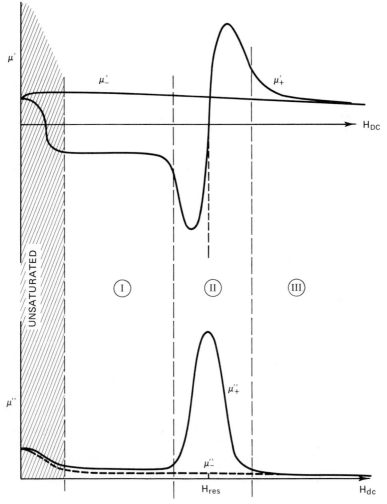

Fig. 18. Real and imaginary parts of the scalar effective permeabilities for positive and negative circularly polarized fields.

region the applied dc field is less than that needed to saturate the material, $H_{dc} < N_z M$, and this "infinite medium" theory does not apply.

Three distinctly different regions are obvious: region I, "below" resonance, region II, "near" resonance, and region III, "above" resonance. In interpreting device performance it should be recalled that the imaginary part of the permeability is associated with loss, while the real part determines the velocity of propagation of the wave. The propagation constant β is determined by the real part of the permeability, $\beta \sim (\mu'\varepsilon)^{1/2}$, and thus the wavelength, or phase change per unit length, in the medium is a function of μ'.

In a ferrite material the sense of polarization of a wave is conventionally defined with respect to the direction of the magnetic moment, or H_{dc}, *not* by the direction of propagation of the electromagnetic wave. Thus, a wave that has positive circular polarization when propagating in one direction will have negative circular polarization when propagating in the opposite direction. Referring now to Fig. 18 it can be seen that if the material is biased with a dc field of H_{res}, it will be *at* resonance. For this condition μ''_+ is quite large and μ''_- is small. Therefore, a positive circularly polarized wave will be strongly attenuated, and a negative circularly polarized wave should experience negligible attenuation on passing through the material. If these two senses of polarization are associated with wave propagation in different directions, then the device will be nonreciprocal. A wave propagating in the forward direction can be transmitted with essentially no loss, while one propagating in the reverse direction will be strongly absorbed. Such a device is a resonance isolator and is used to isolate microwave sources from the effects of reflections on a transmission line.

The strength of the absorption is determined by the value of μ'' at resonance. Since, for a fixed value of saturation magnetization, a decrease in width of the resonant absorption line ΔH results in a higher peak value of μ''_+, it is not surprising that the figure-of-merit (i.e., the reverse-to-forward attenuation ratio) of such a device varies as ΔH. Thus, any influence of microstructure on resonance linewidth will have an impact on resonance isolator figure-of-merit.

Devices operating off-resonance, in region I or III, make use of the fact that two waves with differing senses of polarization will experience differing amounts of phase shift. Such devices are called differential phase shifters, or nonreciprocal phase shifters. Circulators, phase shifting devices that also operate in these regions, have the property of directing an input signal in a particular nonreciprocal fashion around a transmission line junction, as indicated in Fig. 19. A signal in port 1 emerges at port 2, that into 2 emerges at 3, in 3 and out 1.

All devices operating in regions I and III have as a measure of merit the ratio of the differential phase shift experienced to the attenuation in the

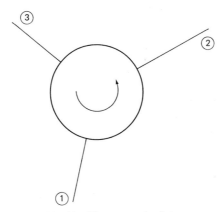

Fig. 19. Three-port circulator.

device. Thus, for highest figure of merit, the loss in region I and III should be as low as possible. Microstructure plays a role in determining this loss by providing a coupling mechanism from the uniform mode (normally driven by the rf fields) to spin waves, and thus influencing the damping constant α.

B. Spin Waves

Spin wave existence is not included in the linear theory of ferromagnetic resonance outlined above, but emerges in a more general treatment. The uniform precessional mode shown in Fig. 16, in which all moments are parallel, is only one of the possible configurations. This mode is generally excited in microwave waveguides because the applied rf magnetic fields are essentially uniform over the magnetic material dimensions. However, it is entirely possible for the spins to precess about the dc magnetic field when they are not all parallel, and in this case the more general equation of motion [Eq. (17)] must be applied.

The simplest nonuniform case is that shown in Figs. 20a and 20b in which the magnetic moment vector of the spins in each successive plane of the lattice advances by an angle ϕ around the precession circle from the position of those in the preceding plane. ϕ is given by $\phi = ka$, where a is interspin spacing in the material, and $k = 2\pi/\lambda_k$, λ_k being the wavelength of the periodic variations. This is the physical picture of a plane wave (two-dimensional "spin wave") propagating through the material in the y direction. A spin in one of these planes will be acted upon by an effective field that has contributions from the dc, rf, and demagnetizing fields, as in uniform precession; but, in addition, the effective field will now have contributions from an exchange field since spins in adjacent planes are no longer parallel.

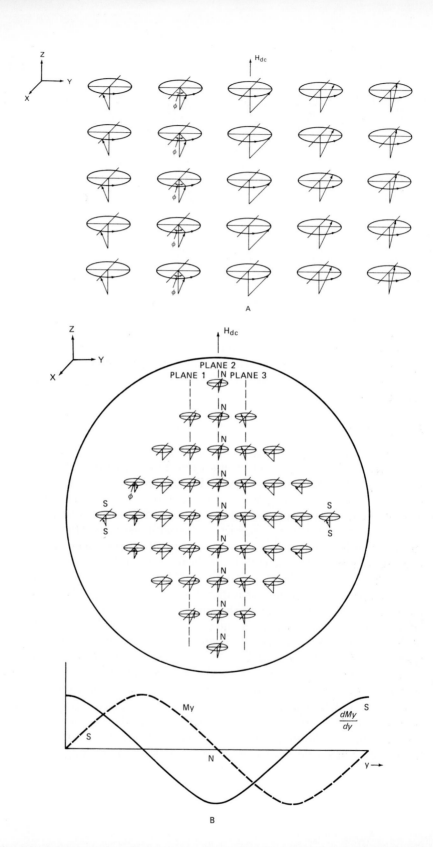

A dipolar field (arising from the volume divergence of **M**) must also be considered.

Since the effective field in Eq. (17) is now altered from the uniform case, the resonant frequencies of these nonuniform modes are displaced from the Kittel resonance [Eq. (21)]. The magnitude of the exchange and dipolar fields, and hence the resonant frequency of the modes, will obviously depend on the angle ϕ between spins in successive planes, and it can be shown that the frequency spectrum is that shown in Fig. 21. Since ϕ may vary continuously, there is a continuous spectrum of resonant frequencies of such spin waves.

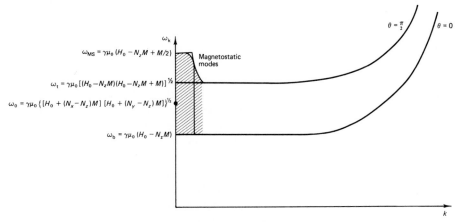

Fig. 21. The spinwave spectrum. In the cross-hatched region, spin waves are not strictly correct modes of the system and magnetostatic modes should be used.

As the wavelength of spin waves becomes long (or k values become small), boundaries of the sample become significant. For samples of finite dimensions, boundary conditions must be imposed on the normal flux and tangential field at the surface of the sample. As in all boundary value problems, solutions become quantized and only certain specific values of ϕ are allowed. The continuous spectrum is thus broken up into a discrete spectrum, and the normal modes of the system are a discrete set of nonuniform modes each occurring at a particular resonant frequency in a given dc magnetic field. These discrete higher order modes have been observed by White and Solt (1956) as anomalous absorption lines appearing when the sample is placed in nonuniform rf fields.

Fig. 20. (a) Nonuniform precession in an infinite medium where precessional angle of spins in successive lattice planes differs by $\emptyset = ka$ (45° for case illustrated.) Now $B_{\text{eff}}/\mu_0 = H_{\text{eff}} = H_{\text{dc}} + h_{\text{rf}} + H_{\text{dipolar}} + H_{\text{ex}}$. (b) Nonuniform precession in ellipsoid showing interior poles arising from volume divergence of **M**. Density of interior poles is proportional to $d(M_y)/dy$.

A rigorously general solution of Eq. (17) for these nonuniform modes would require that boundary conditions be satisfied at the sample–air interface and that the effective field include terms from exchange fields, demagnetizing or dipolar fields, and dc and rf applied magnetic fields. No one has yet been able to solve this completely general case. However, Walker (1957) and Suhl (1957) have obtained analytic solutions in two simplified cases.

For simplicity Suhl has assumed that the nonuniform modes are of extremely short wavelengths (short compared to the sample dimensions) so that the boundary conditions at the surface may be neglected, and the continuous spectrum of plane wave disturbances (spin waves) may be taken as normal modes of the system.

Walker has found an analytic solution for the nonuniform modes when their wavelength is long (of the order of the sample dimensions). For these long wavelength modes, where λ is greater than $1-10$ μm, the magnetization will be essentially constant over a distance of several lattice constants, and exchange fields can therefore be ignored. Propagation effects are also ignored as the sample is assumed to be small compared to the electromagnetic wavelength within it. The problem then reduces to a magnetostatic one with each spin moving under the influence of the applied dc magnetic field and the resultant dipolar (or demagnetizing) field of the other spins. Since the nonuniform spatial distribution of $M(r, t)$ is unknown, these demagnetizing fields must be determined from Maxwell's equations. The solution of Eq. (17) with damping neglected can then be considered a boundary value problem whose solution will yield the allowed spatial distributions of M and the frequency spectrum of these nonuniform modes. The uniform precessional mode of conventional theory must be contained in this long wavelength mode spectrum, and it is identified as the $(1, 1, 0)$ mode in Walker's notation.

At low signal levels the excitation of nonuniform modes requires that the rf exciting field be nonuniform over the sample, as was the case in the experiments of White et al. In order to excite any particular mode at low signal levels, the exciting field must have the symmetry of the desired mode, a situation analogous to the problem of the excitation of higher order modes in waveguides (or in a drum head).

Figure 21 indicates the range of frequencies for which spin waves and longer wavelength nonuniform modes exist. Note that this manifold of resonant frequencies for nonuniform modes is a function of the dc applied magnetic field. As the field is increased the manifold moves upward on the frequency axis.

The existence of these modes is a chief source of damping or line broadening in ferrites. Energy coupled into the magnetic system from electromagnetic fields is either reradiated or dissipated as heat in the crystal lattice.

The degree of coupling of the magnetic system to the lattice determines the damping of the magnetic system. Such coupling arises from spin–lattice coupling effects. When the magnetic moment of an atom or ion contains appreciable contributions from orbital angular momentum—as in the case of the 4f rare earths (except for Gd) and the incompletely quenched Co of the 3d transition group—then large spin–lattice coupling will exist. It is the *orbital* motion that interacts with the lattice, and hence *orbital* contributions to the magnetic moment are accompanied by large magnetocrystalline anisotropy and large damping.

Spin waves offer a parallel path from the magnetic system to the lattice. As depicted in Fig. 22, energy coupled into the uniform mode is fed to the lattice directly via spin–lattice coupling. It can also be coupled through spin waves and then to the lattice. Such spinwave coupling increases the damping.

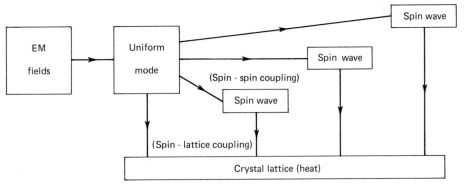

Fig. 22. Schematic diagram of possible loss mechanisms.

How does microstructure determine damping? Microstructure controls the coupling from the uniform precession to spin waves and determines the type of spin waves excited. In order to excite spin waves to an appreciable amplitude at low rf field strength, two conditions must be met:

1. The spinwave resonant frequency must be degenerate with the applied frequency.

2. Some disturbance of the uniform precession must be present whose periodicity approximates the spinwave wavelength λ_k.

The first condition is determined by the applied dc magnetic field. Figure 23 indicates the position of the spinwave manifold with respect to the operating frequency for regions I, II, and III of Fig. 18. No degeneracy of spin wave and operating frequencies exists for region III operation, while for operation in region I only short wavelength spin waves are degenerate with the operating frequency.

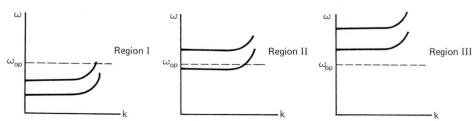

Fig. 23. Position of spinwave manifold for different values of dc magnetic field.

The second condition is determined by imperfections or discontinuities in the sample. Obviously, grain boundaries provide such a discontinuity in the uniform magnetization of the sample. As a result, the uniform mode will be coupled to those spin waves whose wavelengths correspond to the grain dimensions. Thus, fine-grained ferrites will have heavy coupling between the uniform mode and high $k(=2\pi/\lambda_k)$ spin waves. Those ferrites with larger grain sizes will couple energy from the uniform mode to medium to low k values.

When these two conditions are taken together it is seen that near resonance (region II) some degeneracy with spin waves always exists, and microstructure or grain size determines which k values are excited. For operation below resonance (region I) only high-k spin waves are degenerate with the applied frequency and therefore eligible for excitation. These will be most effectively

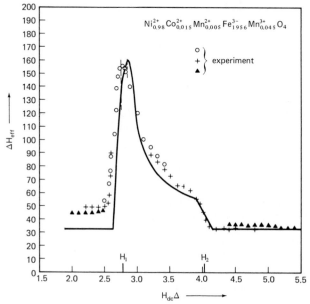

Fig. 24. The effective linewidth as a function of external bias field H_o for a spinel ferrite (after Vrehen *et al.*, 1969).

excited by fine-grained particles. As a result, lower loss in this below resonance region will be observed for larger particle size samples, all else being equal. In the above resonance region (III) no spinwave degeneracy exists for sufficiently large applied fields, and microstructure is less important.

Finally, it should be noted that experimental evidence suggests that the coupling of spin waves to the lattice, measured by the spinwave linewidth ΔH_k, is a function of k value (Schlömann *et al.*, 1960). In fact, ΔH_k appears to vary directly as k. Thus the high-k spin waves are more closely coupled to the lattice (i.e., more heavily damped) than are the low-k values. The fine-grained particles then provide coupling to more heavily damped spin waves and therefore enhance loss. Evidence to support this concept of the role of spin waves in determining loss is found in the experimental measurements of damping constant as a function of applied fields shown in Fig. 24 for a sample with large grain sizes. Clearly, the effective damping constant increases in the region between H_1 and H_2 which delineates the region of spinwave degeneracy for low to medium k spin waves.

C. High-Power Effects

The preceding discussion of spin waves dealt with their existence, their coupling to the uniform mode, and their role in determining damping of the magnetic system at low rf field strengths. It is found experimentally that the permeability μ is also a function of the strength of the rf magnetic fields. Such power dependence is not contained in the linear theory outlined above. Original experimental results by Bloembergen and Wang (1954) and by Damon (1953) show that at high rf power levels the imaginary part of the permeability increases at low dc field values and the main resonance curve saturates, as indicated in Fig. 25. These high-power effects degrade device

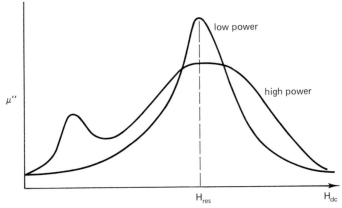

Fig. 25. Variation of loss component (μ'') of permeability with applied dc bias field at low and high rf power levels.

performance in region I by increasing undesirable loss. In region II the saturation at high powers decreases the figure of merit of resonance isolators. Such high-power effects have been successfully interpreted (Damon, 1963) as arising from the parametric coupling of spin waves to the uniform precession by nonlinear terms in the equation of motion.

Spin waves are always excited, to some extent at least, by thermal excitation. When their existence is taken into account it is found that at high rf power levels certain pairs of spin waves become heavily coupled to the uniform precession and produce the observed increase in loss. Suhl's analysis determines which spin waves are most tightly coupled and the critical signal level for exponential growth of these waves. To simplify the general problem it is assumed that the nonuniform modes are of short wavelength so that requirements at the boundaries may be neglected. Under this stipulation the proper modes of the system are the infinite plane wave disturbances or spin waves discussed in connection with Fig. 21. The magnetization can be expressed as a Fourier series of spin waves with time varying coefficients of the form

$$\mathbf{M}(\mathbf{r}, t) = \sum_k \mathbf{M}_k(t) e^{j\mathbf{k} \cdot \mathbf{r}} \tag{28}$$

where $\mathbf{k}$ denotes the direction of propagation of the kth spin wave and $\mathbf{r}$ denotes position in space. $\mathbf{M}_k$ is a measure of the amplitude of excitation of the kth spin wave. As these spin waves can have short wavelengths, exchange fields H_{ex} must be included in the effective field as well as the applied and dipolar (or demagnetizing) fields $H^{(d)}$. This latter field can be found from Maxwell's equation div $\mathbf{B} = 0$ for the higher order waves, and is taken as the usual static demagnetizing field for waves of $k \doteq 0$. These fields are then expressed in terms of the applied field and the magnetization as

$$\mathbf{H}_{ex} = H_e(a^2/M) \sum k^2 \mathbf{M}_k(t) e^{j\mathbf{k} \cdot \mathbf{r}} \tag{29}$$

$$\mathbf{H}^{(d)}_{k>0} = - \sum_{k>0} \frac{\mathbf{k}(\mathbf{k} \cdot \mathbf{M_k})}{k^2} e^{j\mathbf{k} \cdot \mathbf{r}} \tag{30}$$

$$h^{(d)}_0 = -f(N_x M_{x0}, N_y M_{y0}, N_z M_{z0}) \tag{31}$$

where H_e is the effective exchange field depending on the exchange constant, ionic spin, and the lattice constant a; and the f of Eq. (31) denotes the usual demagnetizing function. The expansion (30) in terms of spin waves is not rigorously valid for small k, as boundary effects then come into play, and the basic assumptions of this formulation are violated. For these long wavelength spin waves expansion should be made in terms of long wavelength, magnetostatic modes. However, the problem would then become extremely difficult, if solvable. The fact that the theoretical predictions obtained using Eq. (30) agree well with experimental results indicates that no great error is incurred in making the plane wave assumption.

It can be seen that when Eqs. (29)–(31) are inserted into Eq. (17), spin waves will be coupled by demagnetizing and exchange fields in second and higher order terms to the uniform precessional mode. Moreover, it is found that the coupling coefficient is dependent on the amplitude of the uniform mode precession and the amplitude of the spinwave precession. Suhl has shown that certain pairs of spin waves grow exponentially at the expense of the uniform precession when the rf field exceeds certain "threshold" values.

It is found that spin waves with $\omega_k = \omega/2$ are inherently more tightly coupled (through second-order terms) and will have a threshold field for growth of spin waves that varies directly as ΔH_k, the spinwave linewidth, of the material. This ΔH_k is a direct measure of spin–lattice coupling. The excitation phenomenon is a discontinuous one—a parametric oscillation of spin waves. That is, the observed uniform precession loss maintains its low power value until the threshold rf field strength (or power level) is reached where coupling to particular spinwave pairs exceeds their loss through spin–lattice coupling. The rf power level at which nonlinearities set in is determined by the spinwave linewidth ΔH_k. As noted above the spinwave linewidth varies with the k value of the particular spin waves involved. As a result, it is found that by using fine-grained ferrites with enhanced coupling to high k (short wavelength) spin waves, the pertinent spinwave linewidth is increased. Thus, fine-grained materials are able to handle higher rf power levels before entering the nonlinear region with its increased loss.

Figure 26 shows some typical results by Blankenship and Hunt (1966) indicating the increase in rf power handling achieved through the use of

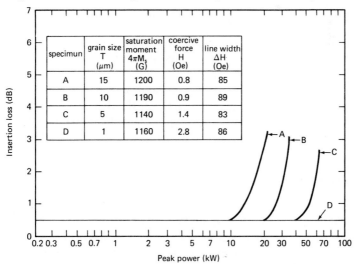

specimun	grain size T (μm)	saturation moment $4\pi M_s$ (G)	coercive force H (Oe)	line width ΔH (Oe)
A	15	1200	0.8	85
B	10	1190	0.9	89
C	5	1140	1.4	83
D	1	1160	2.8	86

Fig. 26. The effect of grain size on microwave properties of a gadolinium substituted yttrium–iron garnet. The insertion loss is measured on a phaseshifter with 360° of differential phase shift (after Blankenship and Hunt, 1966).

fine-grained ferrites in microwave digital phase shifters. These devices operate below resonance (region I) and show increased insertion loss at high rf powers. The four samples tested demonstrate the correlation between grain size and high-power handling ability.

D. Summary

Microstructure plays a role in determining high-power handling ability as well as low-power loss in saturated microwave devices. Fine-grained ferrites have somewhat higher low-power loss, and are able to handle higher rf power levels before the onset of nonlinearities. As is often the case some compromise must be reached between the conflicting demands of small loss at low rf power levels and good high-power handling ability. It is found, however, that in many cases the use of fine-grained ferrites can significantly increase tolerance to high power without appreciably increasing the loss at low powers. The data of Fig. 26 are a case in point. Another means of controlling ΔH_k is that of doping the ferrite with cobalt or a rare earth element having large spin–orbit coupling, and hence rapid relaxation to the lattice. Much work has been done along these lines, and is discussed extensively in the literature (Schlomann *et al.*, 1960, von Aulock, 1965).

III. Applications of Unsaturated Materials

At the lower frequencies ferrites are almost invariably employed as unsaturated materials, and device performance is based on the existence of magnetic domains in the material. The operation of devices depends on a change in the relative size of the domains, or in the orientation of the magnetic moment within the domains, under the influence of an externally applied field. This is reflected by a change in operating point on the hysteresis loop (B–H loop) shown in Fig. 27. Ferrite devices such as radio and TV transformers, antennas, computer cores, and magnetic tapes, disks, drums, and recording heads all operate by virtue of the dynamics of the hysteresis loop. In most of these cases the motion of magnetic domain walls is important, and a discussion of magnetic domains and domain walls is reviewed first.

A. Domains and Hysteresis

The hysteresis loop (Fig. 27) observed for magnetic materials is largely determined by the motion of domain walls, whose existence was described in Section I. When a magnetic field H is applied to a demagnetized sample (here assumed to be a single crystal), those domains with moments oriented along the applied field will grow at the expense of other domains. Figure 28 diagrams such a sequence of events. As the favorably oriented domain

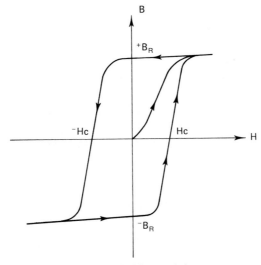

Fig. 27. Typical hysteresis loop.

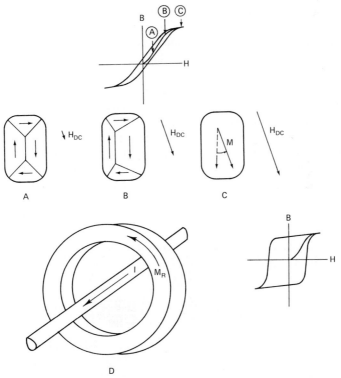

Fig. 28. Motion of the domain wall under the influence of an externally applied field increases the net magnetization, (A) and (B). Under strong external fields the saturated sample has its magnetization rotated from "easy" direction (C). In toroidal configuration single domain closes on itself with no magnetostatic energy, and large remanence is possible (D).

volume is increased, the net magnetization of the sample increases and is reflected by an increase in $\mathbf{B} = \mu_0(\mathbf{H} + \mathbf{M})$. Ultimately, all domain walls will be swept out, and only a single domain will remain. Finally, the magnetization of the domain will rotate into alignment with the applied field and away from the easy direction. On reduction of the applied field the magnetization will rotate back to an easy direction, and domain walls will nucleate and move through the material to again demagnetize the sample.

For a toroidally shaped sample a single domain around the circumference of the toroid (Fig. 28D) is a domain of closure, and magnetostatic fields are minimized. Thus, no demagnetizing fields exist to force the formation of domains (though magnetostrictive effects may). Toroids, then, may maintain substantial net magnetization after the applied field has been reduced to zero.

Reverse domains will be formed when a reverse field is applied. The coercive field H_c is defined as the field required to reduce the magnetization of a (once saturated) sample to zero. The shape of the hysteresis loop is obviously a function of sample shape, material properties, and, as described below, microstructure.

The description of the magnetization process outlined above is valid for single crystal samples and for polycrystalline samples when grain boundaries are taken into account. Since exchange, which plays a major role in domain wall formation, is an extremely short-range force, it is effective only over distances of a few angstroms. It is not effective across grain boundaries. Thus, domain walls must start and stop within each crystallite of a polycrystalline sample. The distance a domain wall may move is limited by the dimensions of the particles which make up a polycrystalline body.

On the other hand, magnetostatic forces are long range. When adjacent particles have their magnetic moments parallel, no demagnetizing fields are present at the interface and, on a macroscopic scale, $\mathbf{M}$ and $\mathbf{B}$ are continuous.

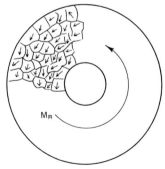

Fig. 29. In a polycrystalline toroid the magnetic moment of each crystallite points in "easy" direction of that crystal. Net magnetic moment is then the remanent value M_R.

Thus, a magnetized polycrystalline toroid might be envisaged as in Fig. 29. Within each grain the easy axes are uniquely defined, but from grain to grain they have random orientations. As a result, with no applied field and after initial saturation, the magnetization in each crystallite relaxes to point in the nearest easy direction. The residual net magnetization, or remanent magnetization, can be shown to be theoretically 0.87 times the saturated value. This theoretical value is arrived at by assuming a negative anisotropy, and hence eight equivalent "easy" directions.

The theoretical value of 0.87 is never achieved in practice. One effect that reduces remanence is magnetostriction. Magnetostrictive effects tend to produce only two preferred directions, and hence the average over nearest preferred directions is reduced. Porosity in the sample also acts to reduce remanence. Demagnetizing fields will exist in voids and nonmagnetic inclusions in the toroid and hence act to reduce the remanence.

Through the use of special compositions the magnetostrictive effects can be compensated for (at least over narrow temperature ranges), and, by control of microstructure, voids and inclusions can be minimized. Thus, in carefully prepared samples remanence ratios of 0.6–0.7 can be achieved.

B. Initial Permeability

The magnetization process described above is valid at dc or for low-frequency applied fields, and must be modified when higher frequencies are used. In determining the response to rapidly varying applied magnetic fields the dynamics of domain wall motion must be taken into account. A domain wall not only has an associated energy, as described in Section I, but also has an inertia associated with its motion. As a result its ability to respond to high-frequency signals is limited.

The initial permeability

$$\mu_i = \left[\frac{dB}{dH}\right]_{H \to 0} \tag{32}$$

is the initial slope of the $B-H$ curve in the limit of low fields. It is determined by two processes: domain wall motion and domain rotation. At low frequencies domain wall motion is the dominant mechanism. An applied field causes domain walls to shift, and this motion results in a change in net magnetization and hence B.

Two possible motions of domain walls can be distinguished: reversible and irreversible. These can be interpreted with the aid of Fig. 30 which indicates the domain wall energy as a function of position in the grain or crystal. This wall energy profile is determined by the microstructure of the

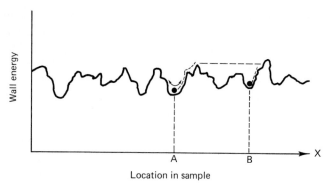

Wall energy

Location in sample

Fig. 30. Representation of wall energy in an imperfect sample as a function of position. At a given wall position (eg., A) restoring force on wall is determined by dE_{wall}/dx. With field applied, wall moves to next higher maximum (A to B).

sample. Local imperfections, voids, etc., lead to such variations in wall energy. A perfect single crystal would have no preferred location for the wall; nothing in the wall energy expression is sensitive to location. But local defects in an imperfect crystal tend to "snag" the domain wall. Walls will preferentially nucleate at voids in the sample and will be captured by the demagnetization fields that arise at such voids. Reversible wall motion corresponds to small displacement of the wall, within a valley, such as A in Fig. 30. Applied fields rock the wall within the local valley, and upon removal of the applied field the wall returns to its original position. The restoring force for reversible motion is determined by the slope of the curve of wall energy versus position, evaluated at the point of interest.

Wall inertia is determined by the intrinsic properties of the material. Irreversible motion corresponds to the wall leaving a local valley, as in moving from A to B in Fig. 30. On the *B–H* loop irreversible motion corresponds to a portion of a minor hysteresis loop.

Since that part of initial permeability determined by wall motion is directly proportional to the distance the wall moves, highly perfect samples will have larger low-frequency permeabilities. Highly perfect samples have, however, very small restoring forces and therefore are unable to follow rapidly varying fields. Thus, it is found that with increasing structional perfection the low-frequency permeability *increases*, and the relaxation frequency for this permeability *decreases*. The curves of Fig. 31 illustrate this for a family of nickel ferrites in which porosity is varied (Pippin, 1957). It is clearly seen that the highest density samples have largest values of low-frequency initial permeability and lowest relaxation frequencies (as determined by the peak in the μ'' curves).

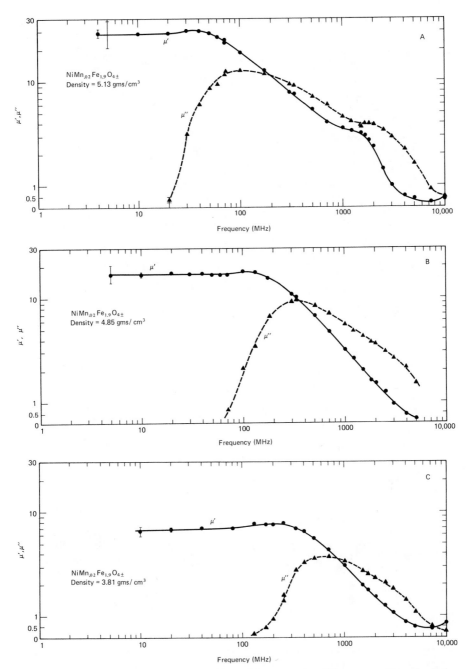

Fig. 31. Initial permeability as a function of frequency for different porosities of spinel nickel ferrite (after Pippin, 1957).

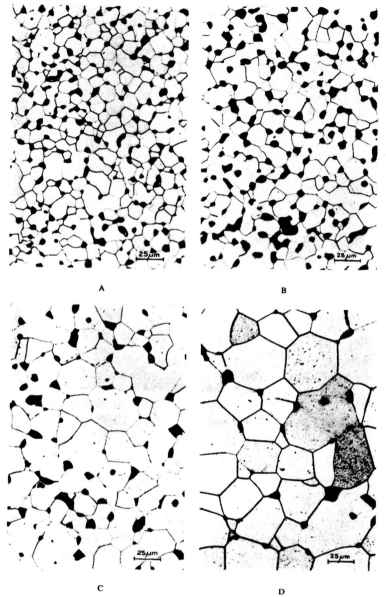

Fig. 32. Photomicrographs of Mn–Zn ferrites having increasing initial magnetic permeability (after Perduijn and Peloschek, 1968). (A) $\mu_i = 6{,}500$, (B) $\mu_i = 10{,}000$, (C) $\mu_i = 16{,}000$, (D) $\mu_i = 21{,}500$.

Porosity control is achieved by variations of sintering conditions and is accompanied by some change in grain size. Figures 32 and 33 summarize results of Perduijn and Peloschek (1968) on manganese–zinc ferrite, where increased grain size is closely correlated with increased initial permeability.

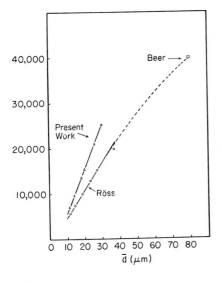

Fig. 33. Initial permeability versus average grain diameter for Mn–Zn ferrites (after Perduijn and Peloschek, 1968).

The second source of initial permeability, domain rotation, is largely responsible for the high-frequency permeability. At sufficiently high applied frequencies the domain wall inertia precludes any appreciable wall motion, but the magnetization can rotate within each domain.

This rotational mechanism is the same as described in Section II as ferromagnetic resonance. Within each domain of the unmagnetized material the magnetization is aligned in an easy direction and experiences a torque arising from the anisotropy energy. This torque can be modeled as arising from an effective anisotropy field as indicated in Eq. (8). When an rf field is applied, rotation occurs because the magnetization precesses about this effective anisotropy field. The rotation is appreciable if the applied frequency is near the Larmor frequency of this anisotropy field, $\omega = \gamma \mu_0 H^{AN}$. Anisotropy field values are determined by the ratio $K_1/\mu_0 M$, and are tabulated in Table II for typical ferrites. For a material with an anisotropy field of 4×10^3 A/m, resonance will occur at 140 MHz, and maximum response from domain rotation would be expected at that frequency. Figure 34 shows permeability spectra (Pippin, 1957) for two ferrite materials where anisotropy is varied by cobalt addition. The cobalt control of anisotropy is seen to control the high-frequency end of the initial permeability curves.

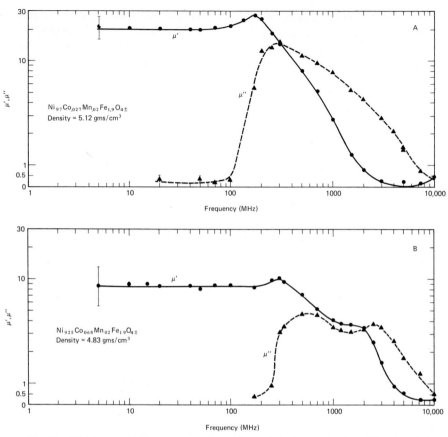

Fig. 34. Initial permeability spectrum for nickel cobalt ferrites having different cobalt contents and thus different anisotropy. Larger anisotropy field of (B) composition causes increases permeability at higher frequencies.

Microstructure is relatively ineffective in controlling this high-frequency peak—though internal demagnetizing fields arising at voids might be thought of as influencing the resonant frequency. Note that in Fig. 31 the high-frequency permeability is quite insensitive to large changes in porosity.

C. Applications of Unsaturated Materials

1. Radio Frequency Applications

At radio and video frequencies ferrites are used as core materials in transformers and in filter coils in resonant circuits. They are also used as cores in magnetic loop antennas—ferrite rod antennas—to make antenna size compatible with modern miniaturized transistor designs. In all of these applications small rf fields are applied to a demagnetized sample, and the initial permeability is of primary concern. The goal then is to achieve a high initial permeability and a low magnetic loss tangent. For optimum performance the slope of the $B–H$ curve as $H \rightarrow 0$ should be large, and the hysteresis loop narrow.

As described in Section III,B, initial permeability arises from wall motion and domain rotation. Large values of initial permeability are realized at low frequencies (below $\sim$ 20 MHz) primarily through wall motion. In order that the loss tangent (μ''/μ') be small, the operating frequency must be below the relaxation frequency for wall motion. The following criteria are then established for large values of initial permeability at low frequency.

1. Large saturation magnetization. Initial permeability varies directly with the magnetization within the domains.

2. Small anisotropy—since anisotropy contributes to wall inertia, small anisotropy leads to more easily moved domain walls and larger initial permeability.

3. Large grain sizes—domain walls can only move within a domain. Large grains then promote large wall displacement and high initial permeability. It has been observed (Perduijn and Peloschek, 1968) that permeability varies directly with grain size.

4. Pores should occur only between grains—pores and inclusions snag domain walls, and any pores occurring within grains will reduce the permeability.

5. Stability of M and H^{AN} over operating temperature ranges—usually achieved by proper choice of material composition.

6. High resistivity—realized through ferrite chemical composition and control of stoichiometry.

The most widely used low-frequency ferrites are from the spinel families of nickel–zinc and lithium ferrites. Manganese and copper are frequently used to reduce magnetostriction and promote grain growth, respectively. The addition of the Zn increases saturation magnetization in the spinels, but also increases sensitivity to temperature. The lithium series is characterized by very high Curie temperature and good temperature stability. Some garnet materials also have favorable low-frequency permeability properties, but the added cost of their raw materials has prevented their widespread use in these applications.

For application at frequencies of 50 MHz and above operation should be above the relaxation frequency for wall motion. The initial permeability is in this range largely determined by rotational effects. If walls are not somehow immobilized, their relaxation frequencies will spread over wide ranges of frequency. In every case where wall pinning is employed some sacrifice in permeability is made, but the sacrifice is necessary to preclude large values of μ'' and unacceptably large values of magnetic loss tangent, $\tan \delta_m = \mu''/\mu'$. Wall pinning can be achieved by the addition to Co^{2+} (Iida, 1960). Walls will also be immobile if they do not exist. The use of fine-grained ferrites—particle sizes below 1 μm—should preclude magnetic wall losses.

Figure 35 shows results obtained on hot-pressed nickel–zinc ferrites. The normally sintered material has a high loss tangent at 1 MHz from domain wall motion. This loss can be eliminated by excluding domain walls, that is by using single domain particles as are obtained in fine-grained hot-pressed material. In such a case the removal of domain walls reduces the value of μ' at low frequencies, but more importantly it greatly reduces the value of μ'' from 0.1 to about 5 MHz. The high-frequency peak (~ 20 MHz) in μ'' results from the rotational resonance in the anisotropy field and is independent of particle size.

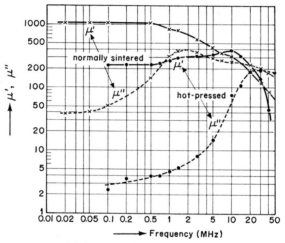

Fig. 35. Real part, μ', and imaginary part, μ'', of the magnetic permeability as a function of the frequency for a hot-pressed and a normally sintered Ni–Zn ferrite (after DeLau, 1968).

To achieve an acceptably high permeability at frequencies of 10–100 MHz the effective internal field in which wall resonance occurs must be sufficiently large. Anisotropy fields can be increased by Co^{2+} addition. As shown in Fig. 36, the introduction of an air gap in the core introduces a demagnetizing field that will also support rotational resonance and contribute to high-frequency permeability.

The ultimate in high-frequency permeability to date has been achieved in the hexagonal ferrites of the Z and Y crystal structures. These specialized materials are outside the scope of this chapter, but are mentioned briefly for completeness. Figure 37 shows the permeability spectra of CoZnZ and Zn_2Y

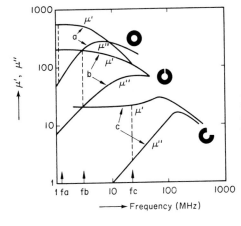

Fig. 36. Permeability spectra of $Ni_{0.36}$·$Al_{0.64}Fe_2O_4$; (*top*) toroidal sample, (*middle*) the same sample with a small air gap, (*bottom*) the same sample with a large air gap (after Verweel, 1957).

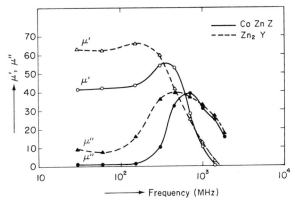

Fig. 37. Permeability spectrum of oriented samples of CoZnZ (solid lines) and Zn_2Y (broken lines) (After Verweel, 1971).

materials indicating acceptably high permeabilities well over 100 MHz, without serious increase in tan δ_m. A large, planar anisotropy is responsible for their unusual properties.

2. SQUARE LOOP FERRITES

In contrast to the high-permeability devices described above, square loop ferrites operate with sizeable time varying fields applied and require an appreciable width to their hysteresis loop. Their operation is based on the fact that after a driving field with maximum amplitude H_m has been reduced to zero, a properly fabricated material will retain an appreciable remanent flux density, $\pm B_R$, as indicated in the drawing of Fig. 27. Thus, information can be stored in the magnetic system in either digital or analog form.

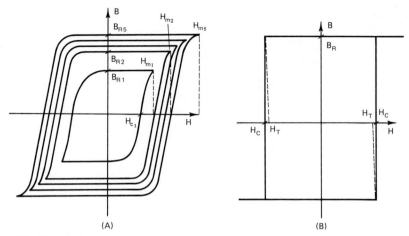

Fig. 38. Ideal hysteresis loops for analog (A) and digital (B) information storage.

Analog storage is required in voice or sound recordings on tape. In this case it is desirable that the $B–H$ loop have minor hysteresis loops such that the remanent magnetization, characterized by B_R, is roughly proportional to the amplitude of the driving field H_m, as indicated in Fig. 38a. For digital data handling only "1" and "0" are stored, and the ideal $B–H$ loop will have a B_R value that is approximately independent of the magnitude of H_m, providing that H_m is larger than the coercive field H_c, (Fig. 38b). The pertinent characteristics of square loop ferrite materials are the values of drive and coercive fields, H_m and H_c, respectively, the core's threshold for switching, H_T in Figure 38b, and the drive, remanent, and saturation values of flux, B_m, B_R, and B_s, respectively. In addition, the switching coefficient S_W determines the time t_s required to switch a square loop core. These parameters will be used below in discussing some applications of square loop ferrites.

a. Recording tapes. In recording on tapes the recording head converts sound variations to current and hence to magnetic fields H_m. Highest fidelity of reproduction requires that the remanent magnetization have a proportional relation to the initial sound. Such tapes are normally made of thin coatings (up to 100 μm) of fine particles (<1 μm) of Fe_3O_4 or γ-Fe_2O_3 immersed in a suspension and fixed onto plastic tape. The proportionality is attained by the depth of penetration of the fields in the film of single domain particles. Large coercive fields are desirable to provide immunity to demagnetization on the reel of tape. Coercive fields up to 4000 A/m are employed.

Ferrites are often used as recording heads. Such heads are transformers with a thin air gap (a few microns) whose fringing fields magnetize the tape. The tape moves across the gap at high speed, and, to prevent damage to the tape, the head must be highly polished. The gap width and head finish must not deteriorate with use. High-density nickel–zinc ferrites have been widely used in such applications. The mechanical hardness of the ferrites makes them highly resistant to wear, and low porosity makes the ferrite amenable to polishing and resistant to breakout of crystallites.

b. Computer tapes. For digital memory storage, as on computer tapes or cores, a very square hysteresis loop is desirable. In the case of coincident current memories the ratio of H_T/H_m becomes of central importance. Figure 39 shows a matrix for coincident switching of cores. A given core, e.g., "A" in Fig. 39, will be switched only when both X and Y leads threading it are pulsed simultaneously. Each pulse carries a drive current producing a field $\frac{1}{2}H_m$. If both are driven simultaneously the total field is H_m, and the core is switched to the $+B_R$ state. If a reverse pulse is present only on the X leads, a field of $-\frac{1}{2}H_m$ is impressed on the core. Ideally, the core should not be disturbed. It should remain in the $+B_R$ state. Such will be the case if $H_T > \frac{1}{2}H_m$. The threshold field H_T is a measure of the field necessary to disturb the magnetization. In practice H_T/H_m must be at least 0.6 to ensure adequate protection against single pulse switching. Reversal of magnetization must occur only when X and Y current pulses are applied simultaneously.

It is desirable for clean switching and adequate readout pulses that the ratio of B_R/B_m be as high as possible. Values of 0.9 or greater are not uncommon. Note that this is B_R/B_m, *not* B_R/B_s, which is substantially lower—0.6 to 0.7 in typical cases. In other words, the applied field H_m does not drive the core into saturation, but only well past its coercive field H_c. The squareness of a core's hysteresis loop, its value of B_R/B_m, and its H_c are determined by the intrinsic properties of the material and by its microstructure.

Once the drive field H_m is removed, a closed toroidal core ideally remains magnetized because no demagnetizing fields are present. In practice, however, demagnetizing fields do exist at small voids and nonmagnetic second

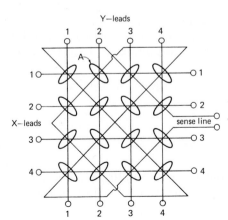

Fig. 39. A 4 × 4 core matrix for switching by coincident currents from row (X) and column (Y) inputs. Flux reversals are detected by the induced voltage in the sense line.

phases. Any magnetostriction present will also tend to reduce the remanent magnetization. Demagnetizing fields are counteracted by the magnetocrystalline anisotropy of the material which acts to hold the magnetization along easy directions of magnetization. With four easy [111] axes the magnetocrystalline anisotropy would, as mentioned earlier, provide a remanence ratio, M_R/M_s, of 0.87. In fact, remanence ratios expressed as a fraction of saturation magnetization are much lower, 0.6 to 0.7. When expressed as a fraction of drive magnetization M_m, however, much higher ratios are realizable, $M_R/M_m \doteq 0.90-0.95$. The maximum drive level of magnetization usually reflects a complete displacement of domain walls, but very little rotation of the magnetization within domains. Thus the driven state M_m can be visualized as having the magnetization within each domain pointing along the easy direction most nearly coincident with the drive-field direction. On removal of the field the magnetization remains undisturbed, and remanence ratios expressed as a fraction of maximum drive level approach 1.0.

The anisotropy field supports remanent magnetization. It also is the chief determining factor for coercive field H_c. A compromise is often required between a low anisotropy for low coercive fields (and hence low drive levels) and a high anisotropy for high remanence ratios.

Porosity will always reduce remanence, and therefore should be minimized. Porosity also acts to trap domain walls and thus influences coercive field values. Coercive fields vary as the product of anisotropy field times the porosity (Smit and Wijn, 1963).

Microstructure also influences the shape of the hysteresis loop. Domain wall energy is minimized when the wall intersects a nonmagnetic region. The minimum experienced is approximately proportional to the volume of the pore. Thus, the distribution of pores will influence wall motion and hence hysteresis loop shape. If the pores are randomly distributed the wall, once freed of its initial well (see Fig. 30), will jump to the next larger obstacle and not pause at each successive well. If the wall traverses a statistically large number of fairly uniform obstacles, a square loop will be formed. A rounded loop will result if the wall in its progress encounters increasingly strong obstacles to its motion.

The switching time for reversal of magnetization in memory cores is governed by an equation of the form

$$t_s = \frac{S_w}{H_m - H_c} \tag{33}$$

where t_s is switching time measured from the time when 10% of the switching current is reached to the time when the voltage induced by flux reversal has fallen to 10% of its maximum value. S_w is the switching constant of the core, and is generally in the range of 20–120 A/μsec m.

In coincident current switching H_m cannot be more than about $1\frac{2}{3}H_c$. Thus this equation could be written as

$$t_s \geq \frac{3S_w}{2H_c}, \tag{34}$$

and, therefore, short switching times require high coercive fields. However, practical coercive fields are limited by available drive fields and heating resulting from hysteresis losses. Thus, an inherent switching time limitation seems to emerge.

For switching applications not employing coincident currents, faster switching may be achieved by driving fields substantially above the coercive field. Small core size (cores down to 0.05 cm are used in practice) and small grain sizes (to minimize wall travel) also decrease switching times.

The switching constant S_w is found to vary directly with the anisotropy field, and thus increasing the coercive field by increasing anisotropy in the material will not yield the decrease in time predicted by cursory inspection of Eq. (34).

D. Summary

Desirable characteristics of square loop ferrites might be summarized as follows.

1. Moderate anisotropy to promote squareness while not raising coercive fields prohibitively.
2. Small (zero!) magnetostriction to enhance squareness.
3. Small magnetization to minimize the effects of demagnetizing fields arising at pores.
4. Minimum porosity to promote squareness.
5. Uniform distribution of pores to promote squareness.

Square loop ferrites are probably exemplified by the magnesium–manganese–zinc spinel ferrites. Compositions of the form

$$0.68MgFe_2O_4 \cdot 0.02MnFe_2O_4 \cdot 0.15ZnFe_2O_4 \cdot 0.15Mn_3O_4$$

produce desirable properties such as those listed below:

Coercive field H_c	150 A/m
Switching constant S_w	32 A/μsec m
Switching time t_s	0.6 μsec (for a 0.075-cm toroid)
M_R/M_m	0.93
H_T/H_M	0.7

In addition, such materials have a vanishingly small magnetostriction that makes them insensitive to mechanical stress (which is often inadvertently applied by switching wires in such small toroids). Lithium ferrites, nickel–zinc ferrites, and mixtures of all of the above are also found to be useful square loop compositions. In addition, yttrium–iron garnet has desirable square loop properties, but its cost is unjustifiably high for low-frequency devices.

IV. Conclusion

Much progress has been made in the past decade in the control of ferrite material properties both through chemical composition and preparation techniques. It is to be expected that such progress will continue, and probable areas of improvement are described briefly below.

a. Compositions. Present knowledge permits a high degree of control of the saturation magnetization of ferrites and a fair degree of control of its temperature dependence. Further exploration of the high Curie temperature materials (e.g., lithium ferrite) will no doubt occur. The degree of available control of magnetocrystalline anisotropy and magnetostriction is not so substantial. Both properties affect the performance of microwave and lower frequency devices. Cobalt doping of ferrites is known to control anisotropy, and manganese doping can control magnetostriction, but the exact mechanism for the latter effect is not known. Temperature dependence of both these effects is not presently controllable. Further work to more accurately describe these effects and to develop controls over their properties is needed. Magnetostrictive effects are especially deleterious to remanent state devices in all frequency ranges.

b. Preparation techniques. The initial permeability and hysteresis loop properties of ferrites are strongly influenced by the microstructure of the material, and hence by preparation methods used. Microwave properties, in particular dielectric and magnetic loss, are also sensitive to the method of fabrication. While great strides have been made in improving ferrite homogeneity and reproducibility, considerable room for improvement remains.

The newly developed hot-pressing techniques offer real hope of leading to 100% dense homogeneous ferrites. The use of grain-oriented cubic ferrites also remains largely unexplored. Some work has demonstrated the effect of grain orientation on microwave resonance properties in garnets (Rodrigue and Crouch, 1966), but the degree of orientation achieved to date has been small. Incorporation of hot pressing may allow better use of the thermodynamic process to improve the achievable orientation.

Improvements should be made in reducing porosity, in particular, eliminating pores within grains, and also in controlling grain size. Such property control would lead to better materials for devices using saturated materials because of the resultant ability to control spinwave coupling. For the unsaturated materials it would promote higher permeability at low frequency on the one hand and lower loss at high frequencies on the other.

Finally, from a theoretical standpoint considerable uncertainty remains concerning a quantitative description of domain wall dynamics and permeability. Often the relation between intrinsic material parameters and damping or restoring forces is not known. Improvements in material homogeniety and reproducibility should lead to an improved ability to determine such parameters and to place that theory on a more solid footing.

References

Abrahams, S. C., and Calhoun, B. A. (1953). *Acta Crystallogr.* **6**, 105.

Anderson, P. W. (1963a). *In* "Magnetism," (G. T. Rado and H. Suhl, eds.), Vol I, p. 25. Academic Press, New York.

Anderson, P. W. (1963b). *Solid State Phys.* **14**, 99.

Bertaut, F., and Forrat, F. (1956). *C. R. Acad. Sci. Paris* **242**, 382.

Bickford, L. R. Jr. (1950). *Phys. Rev.* **78**, 449.

Bickford, L. R. Jr., Pappis, J., and Stull, J. L. (1955). *Phys. Rev.* **99**, 1210.

Blankenship, A. C., and Hunt, R. L. (1966). *J. Appl. Phys.* **37**, 1066.

Bloembergen, N., and Wang, S. (1954). *Phys. Rev.* **93**, 72.

Bomar, S. H., (1971). Final Rep. on Contract No. DAAH03-69-C-0334, Georgia Inst. of Technol.

Bozorth, R. M., Tilden, E. F., and Williams, A. J. (1955). *Phys. Rev.* **99**, 1788.

Braun, P. B. (1952). *Nature (London)* **170**, 1123.

Clarricoats, P. J. B. (1961). "Microwave Ferrites," p. 44ff. Chapman and Hall, London.

Damon, R. W. (1953). *Rev. Mod. Phys.* **25**, 239.

Damon, R. W. (1963). *In* "Magnetism," (G. T. Rado and H. Suhl, eds.), Vol. I, p. 552. Academic Press, New York.

DeLau, J. G. M. (1968). *Proc. Brit. Ceram. Soc.* **10**, 275.

Dillon, J. F. Jr. (1958). *J. Appl. Phys.* **29**, 539.

Dillon, J. F. Jr. (1957). *Phys. Rev.* **105**, 757.

Folen, V. J., and Rado, G. T. (1956). *Bull. Am. Phys. Soc. Ser. 2* **1**, 132.

Geller, S., and Gilleo, M. A. (1957a). *Acta Crystallogr.* **10**, 239.

Geller, S., and Gilleo, M. A. (1957b). *J. Phys. Chem. Solids* **3**, 30.

Gorter, E. W. (1954). *Philips Res. Rep.* **9**, 295.

Gruintjes, G. S., and Oudemans, G. J., (1965). "Special Ceramics" (P. Popper, ed.), p. 289. Academic Press, New York.

Hardtl, K. H. (1975). *Bull. Am. Ceram. Soc.* **54**, 201.

Hastings, J. M., and Corliss, L. M. (1953). *Rev. Mod. Phys.* **25**, 114.

Healy, D. W. Jr. (1952). *Phys. Rev.* **86**, 1009.

Iida, S. (1960). *J. Appl. Phys.* **31**, 3515.

Iida, S. (1966). Private communication, Univ. of Tokyo, Tokyo, Japan.

Kanamori, J. (1963). *In* "Magnetism," Vol. 1. (G. T. Rado and H. Suhl, eds.), Academic Press, New York.

Kramers, H. A. (1934). *Physica* **1**, 182.

Miyata, N., and Funatogawa, Z. (1962). *J. Phys. Soc. Jpn. Suppl.* **17** (BI), 279.

Ohta, K. (1960). *Bull. Kobayashi Inst. Phys. Res.* **10**, 149.

Osborn, J. A. (1945). *Phys. Rev.* **67**, 351.

Pauthenet, R. (1956). *C. R. Acad. Sci.* Paris **243**, 1499.

Peruijn, D. J., and Peloschek, H. P. (1968). *Proc. Brit. Ceram. Soc.* **10**, 263.

Pippin, J. E. (1957). Thesis, Harvard Univ.

Rodrigue, G. P. (1969). *J. Appl. Phys.* **40**, 929.

Rodrigue, G. P., and Crouch, L. A., (1966). *J. Appl. Phys.* **37**, 923.

Rodrigue, G. P., Meyer, H., and Jones, R. V. (1960). *J. Appl. Phys.* **31**, 376S.

Schlomann, E., Green, J. J., and Milano, U., (1960). *J. Appl. Phys.* **31**, 386S.

Seavey, M. H., and Tannelwald, P. E. (1956). MIT Lincoln Lab. Q. Rep. on Solid State Res., p. 61.

Shenker, H., (1957). *Phys. Rev.* **107**, 1246.

Spencer, E. G., and LeCraw, R. C. (1959). *J. Appl. Phys.* **30**, 149S.

Smit, J., and Wjin, H. P. J. (1959). "Ferrites." Wiley, New York.

Suhl, H., (1957). *J. Phys. Chem. Solids* **1**, 209.

Tannenwald, P. E. (1955). *Phys. Rev.* **100**, 1713.

Verweel, J. (1971). *In* "Magnetic Properties of Materials" (J. Smitt, ed.), p. 98ff. McGraw-Hill, New York.

von Aulock, W. H. (1965). "Handbook of Microwave Ferrite Materials." Academic Press, New York.

Vrehen, Q. H. F., Van Groenou, A. B., and deLau, J. G. M. (1969). *Solid State Commun.* **7**, 117.

Walker, L. R. (1957). *Phys. Rev.* **105**, 390.

White, R. L., and Solt, I. M. (1956). *Phys. Rev.* **104**, 65.

Winkler, G. (1957). *In* "Magnetic Properties of Materials" (J. Smit, ed.), p. 26. McGraw-Hill, New York.

Wolf, W. P., and Rodrigue, G. P. (1958). *J. Appl. Phys.* **29**, 105.

Yoder, H. S., and Keith, M. L. (1951). *Am. Minerol.* **36**, 519.

A
B   7
C   8
D   9
E   0
F   1
G   2
H   3
I   4
J   5